CONCRETE MATERIALS

BUILDING MATERIALS SCIENCE SERIES

CONCRETE ADMIXTURES HANDBOOK; Properties, Science and Technology: edited by V.S. Ramachandran

HANDBOOK OF CONCRETE AGGREGATES; A Petrographic and Technological Evaluation: by Ludmila Dolar-Mantuani

CONCRETE ALKALI-AGGREGATE REACTIONS: edited by P.E. Grattan-Bellew

HANDBOOK OF FIBER-REINFORCED CONCRETE; Principles, Properties, Developments and Applications: by James J. Beaudoin

CONCRETE MATERIALS; Properties, Specifications and Testing; Second Edition: by Sandor Popovics

Related Titles

CORROSION AND CHEMICAL RESISTANT MASONRY MATERIALS HANDBOOK: by W.L. Sheppard, Jr.

CONCRETE MATERIALS

Properties, Specifications and Testing

Second Edition

by

Sandor Popovics

Baxter Professor of Civil Engineering
Drexel University
Philadelphia, Pennsylvania

np NOYES PUBLICATIONS
Park Ridge, New Jersey, U.S.A.

Library of Congress Catalog Card Number: 92-8953
ISBN: 0-8155-1308-9
Printed in the United States

Published in the United States of America by
Noyes Publications
Mill Road, Park Ridge, New Jersey 07656

10 9 8 7 6 5 4 3 2 1

Library of Congress Cataloging-in-Publication Data

Popovics, Sandor, 1921-
        Concrete materials: properties, specifications, and testing / by
    Sandor Popovics. -- 2nd ed.
            p.   cm.
        Includes bibliographical references and index.
        ISBN 0-8155-1308-9
        1. Concrete. I. Title.
    TA439.P768   1992
    666'.893--dc20   92-8953
                                                      CIP

Preface

The material called *concrete* in this book is made of three or four basic ingredients: (a) hydraulic cement, usually portland cement; (b) mineral aggregate(s); (c) mixing water; and (d) admixture(s), an optional ingredient.

These materials are the topic of the new edition of this book. Advancements in cement and concrete technology, including new admixtures, have made it necessary to add a significant amount of new information.

The audience for the book remains unchanged—college professors, researchers, consultants, design offices, testing laboratories, graduate students, and the large number of engineers and architects interested in concrete.

The general approach is *practice oriented,* in that the book aims to assist improved applications of concrete through explanations of technically important properties of concrete components and their relationship to concrete properties, by presenting recommendations for the proper selection and utilization of these materials. Certain theoretical aspects are also discussed but only to the extent that they are helpful for the understanding of the nature of concrete. For example, a simplified discussion of the hydration process of portland cement may be helpful for construction engineers when the work in question is unusual; or, the understanding of the basic mechanism of the frost damage of aggregates in concrete can make the difference between proper and improper selection of aggregates.

Generous use of *references* has been made for readers interested in further inquiry. Foreign references and standards have also been included to facilitate the effective use of the book outside the North American continent. Details have been omitted that can be found easily in the literature, such as the step-by-step descriptions of ASTM test methods. The *grading* of aggregates is discussed in considerable length, partly because the rapid depletion of aggregates will force us to use more and more aggregates of unusual gradings, and partly because the theoretical foundation of gradings, grading problems, and their possible solutions are barely mentioned in the literature beyond the elementary aspects.

It is the author's hope the book will contribute to the technically and/or economically improved use of this important construction material.

Philadelphia, Pennsylvania   Sandor Popovics
April, 1992

NOTICE

To the best of our knowledge the information in this publication is accurate; however, the Publisher does not assume any responsibility or liability for the accuracy or completeness of, or consequences arising from, such information. Mention of trade names or commercial products does not constitute endorsement or recommendation for use by the Publisher.

Final determination of the suitability of any information or product for use contemplated by any user, and the manner of that use, is the sole responsibility of the user. We recommend that anyone intending to rely on any recommendation of materials or procedures for concrete materials mentioned in this publication should satisfy himself as to such suitability, and that he can meet all applicable safety and health standards. Concrete materials could be potentially hazardous; therefore, we strongly recommend that users seek and adhere to the manufacturer's or supplier's current instructions for handling each material they use.

Contents

1. INTRODUCTION 1

2. PORTLAND CEMENT—TYPES, PROPERTIES AND
   SPECIFICATIONS 3
       Summary 3
   2.1   Introduction 4
   2.2   History of Portland Cement 4
   2.3   Manufacture 6
   2.4   Composition of Portland Cement Clinker 14
       2.4.1 Oxide Composition 14
       2.4.2 Major Constituents of Portland Cement 18
       2.4.3 Minor Constituents 23
   2.5   Types of Portland Cement 24
   2.6   Blended Portland Cements 29
   2.7   Technically Important Properties of Portland
       Cement—Testing and Specifications 31
       2.7.1 Fineness of Cement 36
       2.7.2 Normal Consistency 44
       2.7.3 Time of Setting 47
       2.7.4 Soundness 61
       2.7.5 Strength 63
           2.7.5.1 Test Methods 63
           2.7.5.2 Effects on Strength Development 70
       2.7.6 Heat of Hydration 75
       2.7.7 Volume Instability (Shrinkage) 83
       2.7.8 Other Properties 90
           2.7.8.1 Specific Gravity 90

2.7.8.2 Alkali Reactivity 92
2.7.8.3 Sulfate Resistance 92
2.7.8.4 Air Entrainment 92
2.7.8.5 Bleeding 94
2.7.8.6 Efflorescence 95
2.8 Uniformity of Cements 95
2.9 Sampling 98
2.10 Storage of Cements 100

3. HYDRATION OF PORTLAND CEMENT 103
Summary 103
3.1 Introduction 104
3.2 Reactions in Early Hydration and Setting 108
3.3 Reactions in the Hardening Process 112
3.4 Mechanism of Hydration 116
3.5 Structure of the Cement Paste 122
3.5.1 Fresh Paste 122
3.5.2 Hardened Paste 122
3.5.3 Porosity 123
3.5.4 Specific Surface 129
3.5.5 Models for the Structure of Cement Gel 133
3.6 Effect of Cement Composition on the
Strength Development—Mathematical Models 135
3.6.1 The Linear Model 136
3.6.2 The Exponential Model for Relative
Strength 138
3.6.3 Exponential Model for Strengths in
Stress Units 146
3.6.4 Generalization of the Exponential Model
for Curing Temperature 146
3.6.4.1 The Generalization 148
3.6.4.2 Comparison to Experimentally
Obtained Strengths 151
3.6.4.3 Characteristics of the Kinetics
of Hydration 153
3.6.4.4 Interpretation of the Formulas 153

4. HYDRAULIC CEMENTS OTHER THAN STANDARD
PORTLAND 158
Summary 158
4.1 Introduction 159
4.2 High–Alumina Cement 160
4.2.1 History, Manufacture, Hydration 160
4.2.2 Properties and Applications 161

4.3	**Expansive Cement**	166
	4.3.1 Composition	166
	4.3.2 Hydration and Properties	168
4.4	**Special Portland Cements**	171
	4.4.1 White Portland Cement	171
	4.4.2 Colored Cements	172
	4.4.3 Oil–Well Cements	173
	4.4.4 Rapid–Setting Portland Cements	173
	4.4.5 Regulated–Set Cement	174
	4.4.6 Waterproofed Cement	175
	4.4.7 Hydrophobic Cement	175
	4.4.8 Antibacterial Cement	176
	4.4.9 Barium Cements and Strontium Cements	176
4.5	**Other Hydraulic Cementing Materials**	177
	4.5.1 Hydraulic Lime	177
	4.5.2 Natural Cement	177
	4.5.3 Masonry Cement	178
	4.5.4 Supersulfated Cement	179
	4.5.5 Slag Cement	180
	4.5.6 MgO–Based Cements	181
	4.5.7 Phosphate Cements	182
4.6	**Latent Hydraulic Materials**	183
	4.6.1 General	183
	4.6.2 Granulated Blast–Furnace Slags	184
	4.6.3 Pozzolans	187
4.7	**Fly Ash and Silica Fume**	191
	4.7.1 General	191
	4.7.2 Fly Ash Properties	191
	4.7.3 Test Methods	193
	4.7.4 Strength Development and Hydration	197
	4.7.5 Effects of Fly Ash on Fresh Concrete	200
	4.7.6 Effects of Fly Ash on Hardened Concrete	201
	4.7.6.1 Strength	201
	4.7.6.2 Other Properties	202
	4.7.7 Proportioning	206
	4.7.8 Applications	207
	4.7.9 Silica Fume	208
4.8	**Selection of Cements**	211
4.9	**Future of Cements**	212
5.	**WATER**	214
	Summary	214
5.1	**Introduction**	215
5.2	**Mixing Water**	215

5.3   **Water for Curing and Washing** 219

6.   **ADMIXTURES** 221
       Summary 221
6.1   **Introduction** 222
6.2   **Classification** 223
6.3   **Air-Entraining Admixtures** 224
       6.3.1   Air Entrainment 224
       6.3.2   The Admixtures 225
       6.3.3   Applications 227
       6.3.4   The Entrained Air 228
       6.3.5   Improvement of Frost Resistance 234
       6.3.6   Test Methods 236
6.4   **Accelerating Admixtures** 236
       6.4.1   Acceleration 236
       6.4.2   The Admixtures 238
       6.4.3   Effects of Accelerators 239
       6.4.4   Mechanism of Acceleration by $CaCl_2$ ... 242
       6.4.5   Chloride-Free Accelerators 242
6.5   **Water-Reducing Admixtures and Set-
       Controlling Admixtures** 246
       6.5.1   Classification 246
       6.5.2   Conventional Water-Reducing Admixtures .. 249
       6.5.3   High-Range Water-Reducing Admixtures 252
       6.5.4   Set-Retarding Admixtures 254
6.6   **Polymers** 258
       6.6.1   General 258
       6.6.2   Latexes 259
       6.6.3   Epoxies 261
6.7   **Other Chemical Admixtures** 266
       6.7.1   Admixtures for Extreme Consistencies 266
       6.7.2   Miscellaneous Admixtures 267
6.8   **Finely Divided Mineral Admixtures** 269
6.9   **Storage, Sampling, and Testing** 270
6.10   **Future of Admixtures** 271

7.   **MINERAL AGGREGATES—GENERAL** 274
       Summary 274
7.1   **Introduction** 275
7.2   **Classification of Aggregates** 275
7.3   **Sampling of Aggregates** 280

8.   **MINERAL AGGREGATES—PHYSICAL PROPERTIES** 287
       Summary 287

8.1   Introduction 289
8.2   Specific Gravity and Solid Volume 289
8.3   Absorption, Moisture Content, and
      Permeability 297
8.4   Unit Weight, Voids Content, and Bulking 303
8.5   Strength, Toughness, Hardness, and Deformability 306
8.6   Thermal Properties 319
8.7   Durability and Soundness 321
      8.7.1   Mechanism of Frost Action in Aggregate 321
      8.7.2   Aggregate and the Frost Resistance of
              Concrete 322
      8.7.3   Aggregate and the Soundness of Concrete 325
      8.7.4   Methods of Testing Frost Resistance 327
      8.7.5   Methods of Testing Soundness 331
8.8   Porosity in Aggregates 333
      8.8.1   Pore Size 333
      8.8.2   Significance of Porosity 334
      8.8.3   Methods of Testing Porosity 335
8.9   Wear and Skid Resistance 339
      8.9.1   Wear Resistance 340
      8.9.2   Skid Resistance 343

9. CHEMICAL PROPERTIES OF AGGREGATES 349
      Summary 349
9.1   Introduction 350
9.2   Deleterious Materials 351
      9.2.1   Particles Finer Than a No. 200 (75–μm)
              Sieve 351
      9.2.2   Lightweight and Soft Particles 354
      9.2.3   Organic Impurities 355
      9.2.4   Other Impurities 356
      9.2.5   Specifications for Deleterious Materials 356
9.3   Reactivity of Concrete Aggregates 361
      9.3.1   Alkali–Silica Reaction 361
      9.3.2   Other Aggregate Reactions 367
      9.3.3   Test Methods and Evaluation of the Results 369
9.4   Deterioration of Aggregates by Chemical Attacks
      from Outside 372

10. GEOMETRIC PROPERTIES OF AGGREGATES 375
      Summary 375
10.1 Shape and Surface Texture of Particles 377
10.2 Particle Size 383
10.3 Sieves and Screens 386

10.4   Grading 389
10.5   Sieve Test 390
10.6   Grading Curves 391
10.7   Grading Representation in Triangular Diagram 396
10.8   Numerical Characterization of Grading 399
10.9   Average Particle Size of the Complete Grading 404

11. FINENESS MODULUS AND SPECIFIC SURFACE 411
        Summary 411
11.1   Fineness Modulus 412
11.2   Experimental Justification of the Fineness
        Modulus 420
11.3   Optimum Fineness Moduli 423
11.4   Specific Surface 430
11.5   Critique of the Specific Surface and Other
        Numerical Characteristics 436
11.6   Attempts to Improve Numerical Grading
        Characterization 438

12. GRADING EVALUATION AND SPECIFICATION 442
        Summary 442
12.1   Grading Evaluation 443
12.2   Grading Specification in General 444
12.3   Specification of Maximum Particle Size 447
12.4   Specification of the Sand and Fine Sand Contents 447
12.5   Specification of the Grading Curve 449
12.6   Grading Specification with Limit Curves 458
12.7   Grading Specification on Percentage Passing–
        Retained Basis 462
12.8   Grading Specifications with Particular Conditions 464
        12.8.1 Percentage Gradings 464
        12.8.2 Particle Interference 467
        12.8.3 Principle of Maximum Density 468
12.9   Specifications with Fineness Modulus 469
12.10   Critical Comparison of Various Methods for
        Grading Evaluation 472

13. INTERNAL STRUCTURE OF CONCRETE AND ITS
    OPTIMIZATION 475
        Summary 475
13.1   Aggregate Grading and the Internal Structure
        of Concrete 476
13.2   Need for Blending Aggregates 483
13.3   General Theory of Blending 485

13.4   Improvement of Grading 488
13.5   Blending of Two Aggregates 490
13.6   Graphical Methods for Blending Proportions 494
13.7   Graphical Method by Rothfuchs 494
13.8   British Method 497
13.9   The Triangular Method 504
13.10   Critical Comparison of the Methods for
        Blending Proportions 511

14. LIGHTWEIGHT AND HEAVYWEIGHT AGGREGATES 512
        Summary 512
14.1   Introduction 513
14.2   Lightweight Aggregates 514
14.3   Types of Lightweight Aggregates 517
        14.3.1 Pumice 517
        14.3.2 Foamed Blast–Furnace Slag 517
        14.3.3 Expanded Perlite 518
        14.3.4 Vermiculite 519
        14.3.5 Clays, Shales, and Slates 520
        14.3.6 Fly Ash 521
        14.3.7 Organic Materials 522
14.4   Possible Problems Related to Lightweight
        Aggregates 522
14.5   Requirements and Test Methods for Lightweight
        Aggregates 524
14.6   Heavyweight Aggregates 527
14.7   Possible Problems Related to Heavyweight
        Aggregates 530
14.8   Requirements and Test Methods for Heavyweight
        Aggregates 532

15. HANDLING AND SELECTION OF AGGREGATES 533
        Summary 533
15.1   Aggregate Handling 534
15.2   Selection of Aggregates 538
15.3   Future of Aggregates 542

BIBLIOGRAPHY 543

INDEX 641

1

Introduction

The material called *concrete* in this book is made of
three or four basic ingredients, usually called
concrete making materials. These *materials for*
concrete are the topic for this book and include:

Hydraulic cement, usually portland cement
Mineral aggregate(s)
Water
Admixture(s) (optional)

Concrete has been the construction material used
in the largest quantity for several decades. The
reason for its popularity can be found in the
excellent technical properties of concrete as well as
in the economy of this material. It is also
characteristic that the properties of concrete
ingredients have a major influence on the fresh as
well as hardened concrete. Therefore, the selection of
concrete-making materials for a given purpose is quite
important.

In order to make this selection intelligently, the
selecting person should be able to assess
concrete-making materials, and should know what to
select, how to select it, and why to select it in a
particular way. In other words, he or she should be
familiar with the available types of each of the
concrete-making materials; the significance and

application of this type in practice; its
concrete-making qualities and the effect of certain
factors on them; recommended sampling and handling
procedures; the underlying principles of the more
important test methods; and the principal points of
the pertinent specifications along with the usual
values of the technically important properties.

The content of the following chapters is an
organized summary of our present knowledge concerning
these topics on materials for concrete with a view to
making the best possible use of the available
materials.

2

Portland Cement—Types, Properties and Specifications

SUMMARY

Portland cement has become one of the most important construction materials during the last 150 years or so, primarily because concretes can be used advantageously for so many different purposes. Portland cement is produced from an appropriate combination of a lime-containing material, such as limestone, and clayey materials by burning this mixture, then grinding the resulting clinker along with a small amount of gypsum. A typical tiny portland cement grain consists of numerous microscopic crystals called *clinker minerals*, also called *cement compunds*. About three-fourths of these minerals are calcium silicates (alite and belite), the rest being calcium aluminate, iron compounds, and minor constituents. The amounts of these minerals can be measured directly, or calculated from oxide analysis data.

The various applications of concretes require portland cements of differing properties. These properties can be changed by altering the cement fineness and proportions of the cement compounds. On this basis, five main *types of portland cement* are recognized, *Types I through V*. In addition, there are *air-entraining portland cements* and *blended portland cements*. Cements with high alite and tricalcium aluminate contents develop high early strengths along with high early heat. Such cement is Type III. Cements with low alite and tricalcium aluminate contents, such as Types IV and V, have slow strength as well as heat developments and increased chemical resistance. Of the other compounds in portland cement, the so-called minor constituents, crystalline magnesia and the alkalies, can be harmful if present in excessive quantities.

In addition to chemical composition, a portland cement, to be acceptable, should comply with the requirements of fineness, time of setting, soundness, air content, and strengths, to which other requirements, heat of hydration for instance, may be added.

2.1   INTRODUCTION

The term *hydraulic cement* refers to a powdery material that reacts with water and, as a result, produces a strong as well as water-insoluble solid. *Portland cement (pc)* and *high-alumina cement (hac)*, among others, are hydraulic cements. Gypsum is not a member of this family of materials, because it is soluble in water; neither is lime, because its hardening is produced by the reaction with carbon dioxide.

By far the most significant class of hydraulic cements is portland cement, because its price and properties and, consequently, because of the quantity consumed. The term portland cement refers to the class of hydraulic cements the essential constituents of which are two calcium silicates.

2.2 HISTORY OF PORTLAND CEMENT

The name *cement* goes back to Roman times when a concrete-like masonry made of crushed stone pieces with burned lime and water as binding material was called *Opus Caementitium*. Later the combination of brick powder and volcanic tuff with burned lime used as hydraulic binder was called *Cementum* and *Cement*.

The importance of clay content concerning the hydraulic properties of a naturally occurring mixture of limestone and clay was discovered by J. Smeaton of England as he was looking for a water resistant mortar for the construction of the Eddystone lighttower by Plymouth. This material would be called *hydraulic lime* today but was named as *Roman cement* at that time. The French L. J. Vicat and the German J. F. John found independently that 25 to 30% of clay addition to limestone produces the best hydraulic lime.

The name *portland cement* was coined in 1824 by Joseph Aspdin, a bricklayer of Leeds, England, who took out a patent for "a cement of superior quality resembling Portland stone," a natural limestone quarried on the peninsula of Portland in England [Haegerman 1964]. Despite the name, however, the material was probably hydraulic lime because of the apparently low burning temperature used. The production of portland cement in the modern sense began about 20 years later by Isaac C. Johnson.

Further development was influenced significantly by W. Michaelis. He was the first to discuss in 1868 the most favorable composition of the raw materials for portland cement.

The first *high-early strength portland cement* was manufactured in the factory Loruns in Austria in 1912-13; the clinker was burned at higher temperature and the fineness of the cement was also increased. The first portland cement with improved *sulfate resistance* was patented by F. Ferrari in Italy in 1919. *White portland cement* was manufactured in small quantities already in the 1880's in Heidelberg Germany. The manufacturing of *oil well cement* began around 1930 when the increased well depths required cements with adequately long setting times under high temperature and pressure. The *expanding cement* appeared around 1920, and was developed by H. Lossier.

The first *high alumina cement* was manufactured on the basis of the patent of the French J. Bied. It was used by the French during the first World War for foundations of heavy artilleries [Locher 1984a].

The first portland cement to be made in the United States was produced by David Saylor at Coplay, Pennsylvania, in 1871 in vertical kilns somewhat similar to those used for burning lime. The increasing demand for both quality and quantity lead to the introduction in 1899 of the rotary kiln. From the turn of the century the production of portland cement in the United States has assumed gigantic proportions rising from less than 10 million barrels, 376 lb (170 kg) each, to almost 400 million barrels by the end of the 1960's. [Bogue 1955a] (Fig. 2.1) Thomas Edison contributed significantly to the development by the introduction of the large size rotary kilns. [Edison 1926] World production of portland cement at present is about six times the production of the United States. Further details concerning the development of portland cement are comprehensively described in several places in technical literature.[Bogue 1955b, Lea 1956, Steinour 1960b, Pollitt 1964]

Along with the quantity produced, the quality of the portland cement has also been improving. This is reflected in the continuous changes of the specifications for portland cement [Mehta 1978].

2.3   MANUFACTURE

The severity of the chemical restrictions for portland cements is such that calcium oxide, silica, alumina, and ferric oxide must be present in the raw material with narrowly defined limits, and other constituents, such as magnesia and alkalies, must not exceed specific limits.

The raw materials of portland cement consist principally of *limestone* or *chalk* (the calcareous component) as the principal source of CaO, and *shale* or *clay* (the argillaceous component) as the principal source of SiO_2, Al_2O_3 and Fe_2O_3. Marls and some other materials containing significant proportions of the four oxides are also frequently employed. Inevitably the use of naturally occurring materials introduces also a number of minor components. The chemical restrictions sometimes necessitate the introduction of other types of rock besides those immediately available, such as high-calcium limestone, sandstone, or iron. Benefaction of the available materials may also be necessary. It requires around 3200 lb (1450 kg) of raw material to provide a short ton (907 kg) of cement.

The rock portions of the raw material are obtained in quarries by blasting. Each blast can

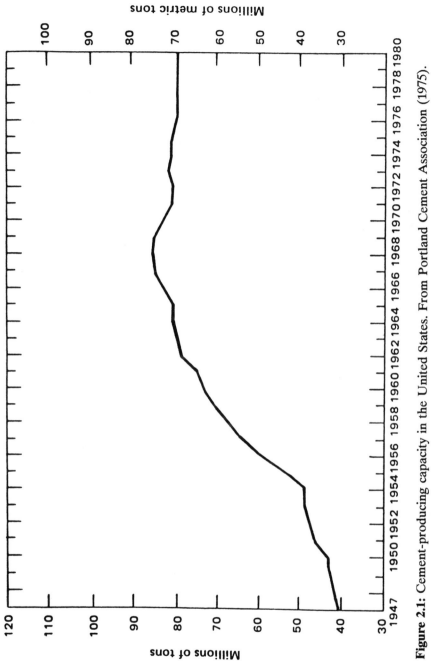

Figure 2.1: Cement-producing capacity in the United States. From Portland Cement Association (1975).

produce stone quantities up to 100,000 lb (45,000 kg) or even more. The stone pieces are quite large, therefore they are shipped to stone breakers by heavy trucks where the sizes are reduced to approximately 1 - 1.5 in (25 - 38 mm) sizes. The loose components of the raw material are obtained without blasting by appropriate machinery. The composition of each component must remain within specific narrow limits otherwise the quality and uniformity of the portland cement cannot be assured. When any of the compositions fluxuate within wider limits, the material must be homogenized. This takes place usually in large containers. The compositions of the homogenized components are continuously or frequently checked and, if necessary, corrections are made by additions of supplementary materials for the maintenance of the required compositions.

All raw materials must be ground to an impalpable powder and intimately mixed before burning. Blending of the rock may begin in the quarry and continue as the raw materials flow into each crusher or mill. Water is partially removed from the ground raw material by various processes. Often, however, the raw material is fed directly into the kiln, and the water is removed by evaporation aided by chains or baffle slates inserted in the back ends of the kiln. Then the material is burned into a clinker in a kiln.

The burning can be done either by *dry process* or *wet process* or *semi-dry process*. In the dry process each component of the raw material is batched precisely in the specific quantity and placed into a mill where they are ground to a fine powder. All grinding and blending operations are done without any addition of water. The raw material in the mill is further dried with heat coming from the kiln. Thus the moisture content will be reduced to 8-12 percent. If corrective materials are needed, these are added to the mill during the grinding period, that is, the final mixing is accomplished chiefly in the grinding mills. The traditional mill type is the rotational mill but presently cylinder mill is becoming popular because of their lower energy need, and less noise production. Large mills presently can produce 400 tons of raw powder hourly.

When the raw material contains more than 20% water to begin with, the final grinding and blending are brought about in a water slurry, and mixing is accomplished both in the grinding mills and by stirring in large vats. Some materials, like chalk, or clay, are made into slurry separately and mixed to the raw slurry in the vats.

In the semi-dry process, the blended raw mixture is fed into the kiln in the form of pellets with medium water content.

Although the wet process requires approximately 15% more energy than the dry process, in many cases it is mandatory to use the wet process for technological and/or ecological reasons. In all processes, rigid control of the composition of the final kiln feed is attained through continuous analysis made on raw materials, for instance by X-ray diffraction, at various stages of the operation and subsequent measured blending of mixtures to attain the required composition.

The rotary kiln has mostly replaced the vertical or shaft kiln in the U.S. Today's kilns vary in size from 60 - 500 ft (18-150m) in length and from 6-15 ft (1.8 to 4.5 m) in diameter; capacity ranges from 200 (34) to 4,000 barrels (680 metric tons) a day. The rotary kiln is set at an inclination of about 0.5 in/ft (40 mm/m) and rotated at a speed of between 30 and 90 revolutions per hour, causing the load to work its way downward toward the discharge end. The heat is usually introduced by a blast of ignited powdered coal and air, or less commonly with fuel oil or gas. The maximum temperature in the kiln is between 2280 and 3458°F (1250 and 1900°C). During the passage of the mixture down the length of the kiln, several reactions take place at various temperature levels. These are summarized in Table 2.1. This applied temperature is such that only a minor proportion of the reacting mixture, and only for 10 to 20 minutes, is in the molten condition. The liquid that forms during the burning process causes the charge to agglomerate into nodules of various sizes, usually 1/4 to 1 in (5 to 25 mm) in diameter and characteristically black, glistening hard. This material is called *portland cement clinker*.

Table 2.1: Reactions in the Kiln[a]

Temperature (°C)	Process	Thermal change
100 and below	Evaporation of free water	Endothermic
500 and above	Dehydroxylation of clay minerals	Endothermic
900 and above	Crystallization of products of clay mineral dehydroxylation	Exothermic
	Decomposition of $CaCO_3$	Endothermic
900–1200	Reaction between $CaCO_3$ or CaO and aluminosilicates	Exothermic
1250–1280	Beginning of liquid formation	Endothermic
Above 1280	Further liquid formation and completion of formation of cement compounds	Probaly endothermic on balance

[a]From Lea and Desch, *The Chemistry of Cement and Concrete*, 2nd ed., Edward Arnold (1956).

The clinker drops from the end of the kiln into some form of cooler which usually uses an air stream for cooling. The temperature of this air goes up to 800 to 900°C after which it is lead from the cooler back to the kiln for helping the burning. If there is any additional hot air remaining, this maybe used for the preheating of the raw mixture.

The cooled clinker is stored for a suitable period after which it is ground usually with 3 to 5% of gypsum ($CaSO_4.2H_2O$) to a fine powder. The mill consists of a body and grinding media which are cast iron or steel pieces inside the body. These are responsible for the grinding. In recent years combinations of compound or tube mills with centrifugal separators and bucket elevators, arranged in closed circuits with air separators, have come into use. The purpose of the air separators is to remove the cement particles of 200-mesh fineness from the mill because these retard the grinding of the remainder. After grinding, the cement is ready for packing in paper bags or in drums or for transporting in bulk [Bogue 1955a, Locher 1984b]. In the American practice a bag contains 94 lb (42.5 kg) of cement while in Europe a bag contains 50 kg (112 lb) of cement.

Details concerning the raw material and processes for portland cement manufacture can be found in the literature [Bogue 1955b, Lea 1956, Peray 1979, Witt 1966, Blanks 1955, Barton 1965, Eitel 1966, Kuhl 1961a, Labahn 1971, Kerton 1987]. A typical process flow in cement manufacturing is shown in Figs. 2.2. and 2.3.

Recent developments in cement manufacturing have consisted primarily of reducing energy requirements, shifting from petroleum products and natural gas to coal [Halstead 1981], reducing production costs, improving environmental protection around the cement factory, and raising the uniformly good quality of cement through automatization. For instance, when rotary kilns were first used, over 2 million Btu (0.5 million kcal) were required per barrel of cement produced. Today's average is about 1 million Btu (0.25 million kcal). This is the result of the use of pre-heaters, precalciners, roller mills, and other

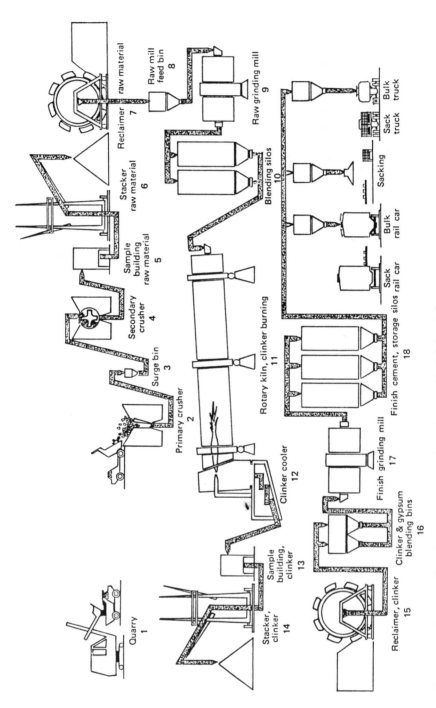

Figure 2.2: Process flow in cement manufacturing at Clarkdale, Arizona.

Figure 2.3: Another view of the process cement manufacturing at Clarkdale, Arizona.

modern equipment. Labor costs have been reduced by use of higher capacity machinery. In recent years research has been carried out in order to permit the use of raw materials containing higher proportions of certain components than have usually been considered allowable. Examples of this are the development of high-magnesia portland cements [Gaze 1978a,1978b], and the use of raw materials high in P_2O_5 [Taylor 1964b, Gutt 1969,1978a]. Control and instrumentation of kilns are also making great strides. Computer controls and complete closed-loop operations are in practical use at quite a few plants and are being considered or experimented with at many more. Sonic, gamma-ray and electronic indicators are in general use; and X-ray spectrography, which provides analysis in a few minutes that used to take 6 or 8 hr, permits immediate and accurate plant adjustments for quality control [Barton 1965].

2.4   COMPOSITION OF PORTLAND CEMENT CLINKER

The composition of a portland cement depends on the compositions and proportions of the raw materials as well as on details of the manufacturing process, such as burning temperature, cooling rate, etc. [Midgley 1964b]

2.4.1 Oxide Composition

About 95% of portland cement clinker is made of combinations of four oxides. These are: lime (CaO), silica(SiO_2), alumina (Al_2O_3), and iron oxide (Fe_2O_3). Other, so-called minor constituents or impurities include, among others, magnesia; sodium, and potassium oxides (the alkalies); titania; phosphorous and manganese oxides. The compositions of these oxides are presented in Table 2.2. Methods for chemical analysis of portland cement are described in ASTM C 114-85 and elsewhere in the literature. [Yamaguchi 1969, Chalmers 1964]

One way to characterize the chemical composition of portland cements in terms of oxides is shown in Figure

Table 2.2: Names, Compositions, and Abbreviations of Common Oxides as Well as Major Compounds in Portland Cement

Name	Composition	Molecular weight	Abbreviation	Mineral name
Lime	CaO	56	C	
Silica	SiO_2	60	S	
Alumina	Al_2O_3	102	A	
Iron	Fe_2O_3	160	F	
Water	H_2O	18	H	
Sulfuric anhydride	SO_3	80	$\bar{S}$	
Magnesia	MgO	40	M	
Soda	Na_2O	62	N	
Potassa	K_2O	94	K	
Tricalcium silicate	$3CaO \cdot SiO_2$	228	C_3S	Alite
beta-Dicalcium silicate	$2CaO \cdot SiO_2$	172	C_2S	Belite
Tricalcium aluminate	$3CaO \cdot Al_2O_3$	270	C_3A	
Tetracalcium aluminoferrite[a]	$4CaO \cdot Al_2O_3 \cdot Fe_2O_3$	486	C_4AF	Celite.

[a] Actually, the iron-containing phase is a solid solution of variable composition. C_4AF seems to be a fair average composition.

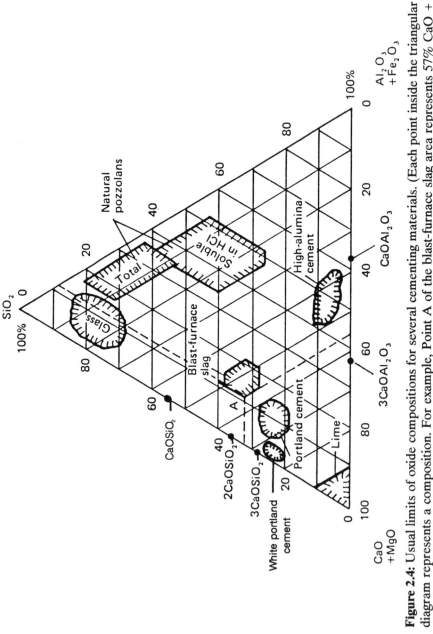

Figure 2.4: Usual limits of oxide compositions for several cementing materials. (Each point inside the triangular diagram represents a composition. For example, Point A of the blast-furnace slag area represents 57% CaO + MgO, 32% SiO_2, and 11% Al_2O_3 + Fe_2O_3. For further explanation of the triangular system, see Chapter 10.)

2.4. Another method is to use ratios. [Hayden 1954] For instance, the so-called Silica Ratio is

$$SR = \frac{SiO_2}{Al_2O_3 + Fe_2O_3} \qquad (2.1)$$

where the oxides represent the amounts determined by chemical analysis of the clinker expressed in percent by weight. Values of the silica ratio may vary approximately within 1.7 to 2.5, the usual value being within 2.0 and 2.5. A high value of SR suggests that the clinker may be difficult to homogenize adequately by heat treatment.

A second ratio is the Alumina Ratio:

$$AR = \frac{Al_2O_3}{Fe_2O_3} \qquad (2.2)$$

In the practically ferric oxide-free white portland cements this ratio is high. Its usual value in standard portland cements is between 1 and 3.

A third useful ratio is the Lime Saturation Factor:

$$LSF = \frac{CaO - 0.7SO_3}{(CaO)_{max}} = \frac{CaO - 0.7SO_3}{2.8SiO_2 + 1.2Al_2O_3 + 0.65Fe_2O_3} \qquad (2.3)$$

This factor is the ratio of the quantity of lime actually available in the cement for chemical binding to the maximum quantity of lime that can combine chemically with the other oxides in the cement at the clinkering temperature. In other words, this is the theoretically required lime quantity to produce a clinker in which all the other oxides are saturated with lime but without any free lime in it. The calculation of this theoretical maximum is shown in Example 2.2.

The lime saturation factor of commercial portland cements is around 0.85 - 0.90 [Czernin 1962, Locher

1976b, Spohn 1969].

Eqs. 2.1 through 2.3, and several other ratios, have been used successfully in raw-mix proportioning and have served to assure control of the quality of portland cement in manufacture. Where such control measures are not applied, problems may occur, for instance, overliming may result, with the consequent presence of uncombined (free) lime in the finished product. Free lime may cause a cement to be *unsound;* that is, it may cause excessive expansion and cracking of the hardened cement paste in the concrete (Section 2.7.4). The reason for the unsoundness is that the hydration of CaO, overburned at the high temperature of the cement kiln, is slow and expansive in nature which can cause excessive volume increase and stresses. Incidentally, underliming is also undesirable because such portland cements have slow and low strength-productions [Barton 1965].

2.4.2 Major Constituents of Portland Cement

Although most of the substances of which portland cements are composed contain three of more elements in a state of combination, one may introduce a considerable simplification into their study by regarding them as being built up of the previously mentioned oxides.

The publication of Le Chatelier's research [1905] led to recognition that the four compounds can be considered as the four major constituents in portland cement clinker. These are:

Name:	Composition:	Symbol:
Tricalcium silicate	$3CaO \cdot SiO_2$	C_3S
Dicalcium silicate	$2CaO \cdot SiO_2$	C_2S
Tricalcium aluminate	$3CaO \cdot Al_2O_3$	C_3A
Tetracalcium aluminoferrite	$4CaO \cdot Al_2O_3 \cdot Fe_2O_3$	C_4AF

These compounds are shown at the bottom of Table 2.2. Note that the calcium silicates give the cement its hydraulic character, that is, the property of hardening by reaction with water.

Most of these compounds, the so-called clinker minerals, form crystals which are so small that the particles in the ground clinker contain more than one crystal [Regourd 1974]. The porosity inside of a particle is usually negligible [Powers 1968]. Clinker minerals can be identified under a microscope if a piece of clinker is cut, polished, and suitably etched [Midgley 1964a], or by X-rays [Dent-Glasser 1964, Peter 1971], or other methods [Eitel 1966, Midgley 1964b, Farmer 1964]. Some of the compounds can form more than one type of crystal form which can influence the hydration characteristics of the compound. [Jeffery 1964, Guinier 1969, Highway Res. B. 1972a, Lafuma 1965, Nurse 1966, Taylor 1964d] For instance, in the case of dicalcium silicate, only the beta form has cementing value under normal hardening conditions. The gamma-dicalcium silicate develops sizable strength only under autoclave curing. The tricalcium aluminate can also crystallize both in cubic and in orthorhombic forms the former being less vulnerable to sulfate attack [Mehta 1980a, Mortureux 1980] It should be noted that a micro-crystalline, glassy material can also be identified in the clinker which is aluminate for the most part.

The amount of clinker minerals, that is, the *compound composition*, can also be calculated approximately from the oxide analysis data. Bogue was the first to offer formulas for such computations. [Bogue 1929]. The formulas recommended by ASTM C 150-85 for portland cements having an Alumina Ratio of 0.64 or more are similar to the original Bogue formulas, and are as follows:

Tricalcium silicate:

$$C_3S = 4.071CaO - (7.600SiO_2 + 6.718Al_2O_3 + 1.430Fe_2O_3$$

$$+ 2.852SO_3) \tag{2.4}$$

where all the quantities are expressed in percent by weight.

Dicalcium silicate:

$$C_2S = 2.867 \; SiO_2 - 0.7544 \; C_3S \qquad (2.5)$$

Tricalcium aluminate:

$$C_3A = 2.650 \; Al_2O_3 - 1.692 \; Fe_2O_3 \qquad (2.6)$$

Tetracalcium aluminoferrite:

$$C_4AF = 3.043 \; Fe_2O_3 \qquad (2.7)$$

Equations 2.4 - 2.7 reveal that despite the linear relationships, the compound composition is greatly affected by relatively small changes in the oxide composition.

Example 2.1

If the CaO content in a cement clinker is decreased from 66 to 63% with an increase in SiO_2 from 20 to 22%, and corresponding increases in the other oxides, the calculated C_3S content will decrease from 65 to 33%, while the C_2S content will increase from 8 to 33%.
In view of these large changes in the calculated values, it seems a better practice to characterize the portland cement by the compound rather than by the oxide composition.

The Bogue formulas are based on two major assumptions, namely, that (a) the four major constituents have the exact composition of C_3S, C_2S, C_3A, and C_4AF; and (b) equilibrium is reached at the clinkering temperature and maintained during cooling. Other formulas have also been offered that were derived from other assumptions [Lea 1956, Hansen 1968, Dahl 1939]. Since, however, there are always deviations in the actual portland cement from the assumed conditions, a calculation of this kind provides only an approximate answer, called the *potential* compound composition of portland cement. Comparative investigations with microscopic tests and

X-ray diffraction analysis [Nurse 1969, Yamaguchi 1974, Brown 1948, Copeland 1959, Soroka 1979] show, for instance, that the Bogue method of calculation regularly overestimates tricalcium aluminate contents [Midgley 1964b, Brunauer 1959, Moore 1977, Berger 1966], and, at least in one series, also overestimated the amounts of dicalcium silicate [Spohn 1969]. Nevertheless such calculations came to be universally adopted because they are simple and yet meaningful. For instance, they proved to be very useful for many practical purposes, including the development of the various types of portland cement.

Note also that the calculated compound composition cannot establish whether the constituents, for instance the C_2S, are present in the preferred crystal forms.

Example 2.2

An application of the major constituents is illustrated below by the derivation of the lime saturation factor presented earlier as Eq. 2.3. The portland cement constituent that contains the most CaO is the C_3S. The CaO content of this is 73.7% and the silica content is 26.3%. Since 73.7/26.3 = 2.80, the maximum amount of lime that a silica molecule can bond is $(CaO_{max})' = 2.80$ SiO_2. Thus, if the portland cement consisted only of lime and silica, the lime saturation factor would be

$$LSF = \frac{CaO}{2.8SiO_2}$$

In reality, however, the portland cement also contains Al_2O_3, Fe_2O_3, and sulfate which are all capable to bond lime. Since C_4AF contains the most lime, let us assume for the sake of the calculation of the theoretical upper limit of the lime content in a clinker that (a) all the alumina and ferrite are in this form; and (b) $C_4AF = C_2A$ + C_2F. In the first term of the right hand side

CaO bonds 1.2 Al_2O_3, whereas in the second term CaO bonds 0.65 Fe_2O_3. Therefore, the maximum amount of lime that the alumina and ferrite can bond is

$$(CaO_{max})" = 1.2\ Al_2O_3 + 0.65\ Fe_2O_3$$

That is, the maximum amount of CaO that SiO_2, Al_2O_3 and Fe_2O_3 together can bond chemically is

$$CaO_{max} = (CaO_{max})' + (CaO_{max})" =$$

$$= 2.8SiO_2 + 1.2Al_2O_3 + 0.65Fe_2O_3$$

This is the denominator of the lime saturation factor in Eq. 2.3. The presence of $CaSO_4$ is taken into consideration in the numerator.

The two calcium silicates form about 70 - 80% of a portland cement. The internal structure of these silicates can be described as SiO_4 tetrahedrons separated and connected by calcium ions.

In reality, the clincker minerals in a portland cement are not in the form of pure compounds. The calcium silicates, for instance, contain small amounts of alumina, magnesia, and possibly some other oxides [Berger 1966]. Since these impurities can influence the crystal forms and other properties of the compound, it is reasonable to call these compounds, as they occur in portland cement, by their mineral names (alite, etc.) originated by Tornebohm [1897]. These names are also shown in Table 2.2.

Each of the clinker minerals has important individual characteristics, as was shown first by Bates and Klein [1917]. *Alite* paste attains the greater part of its strength in a week and little increase occurs at longer ages. The heat development produced by reactions between alite and water ("heat of hydration") is also quite intensive. *Belite* produces little strength until after several weeks, but gains steadily in strength at later ages until it approaches equality with alite. Also, the development of its heat of hydration is much slower. *Tricalcium aluminate* alone attains very little strength, but in mixes with calcium silicates shows a favorable effect on early strength development. It is not clear yet how tricalcium aluminate contributes to the hardening and strength. A possible mechanism is that interacts with the reactions of the calcium silicates as a sort of *catalyst* [Popovics 1974a,1976a]. In any case, the reaction of tricalcium aluminate with water is very rapid, causing a flash setting, that is, quick setting of the cement paste accompanied by a vigorous heat evolution. Gypsum in the amount of 3-6% should be added to the clinker to slow down this reaction and control the time of setting as well as the strength development. Tricalcium aluminate of the clinker in large quantities also reduces the resistance of the hardened paste against sulfates and other chemical attacks. The *celite* hydrates rapidly, accompanied by marked heat evolution, but the reactions are less intensive than those of tricalcium aluminate. The ferrite phase may contribute to the strength development of portland cement at later ages, although this mechanism is not clear. It is due to the presence of the ferrite phase that portland cement derives its characteristic gray color. Portland cement without iron is white.

2.4.3   Minor Constituents

The term "minor" refers to the quantity of these constituents in portland cement rather than to their importance. They may have significant effects on the quality of cement. This is particularly true for *magnesia* and the *alkalies*. Magnesia usually occurs in an uncombined state in the portland cement. [Mohan 1988] If it is present in an excessive quantity (about 5% or so), especially in crystalline form as periclase, it may cause the cement to be unsound similarly to the effect of free lime. The alkalies are usually combined with the major compounds. However, they may react with active silicate of certain aggregates, causing extensive expansion and cracking.(Section 9.3) This aggregate-alkali reaction is not harmful when the total amount of alkali is less than 0.6% by weight of the cement. Such cements are called *low-alkali* cements. An increase in the alkali content may also be associated with an increase in the drying shrinkage of the hardening paste [Blaine 1969]. Alkalies also increase the compressive strength up to the age of 7 days but reduce it at later ages [Blaine 1968, Strunge 1985]. This effect is similar to the effect of C_3A on the strength development.

Calcium sulfate compounds were first used in portland cement to regulate setting time. However, they are now also used to control the rate at which the cement paste develops strength at early ages and to control its drying shrinkage. For each clinker there is an optimum amount of SO_3 [Hansen 1988] which can be checked by the method given in ASTM C 563-84.

A usual item in the chemical analysis of portland cement is the *ignition loss*. This represents the percentage weight loss suffered by a cement sample after heating to $1000^\circ C$ ($1832^\circ F$). The substances expelled from the cement by heating are mainly water and carbon dioxide. Ignition losses higher than approximately 5% may indicate faulty quality, especially low early strength development.

An *insoluble residue* is also determined in chemical analyses. This is dissolved in concentrated hydrochloric acid. This residue is inert, therefore its quantity in portland cement is limited by specifications. For instance, the permissible upper limit, somewhat arbitrarily, in ASTM C 150 is 0.75%, and in B.S. 12 is 1.5%. Insoluble residues are derived mostly from the added gypsum, although a part may be unreacted silica.

2.5 TYPES OF PORTLAND CEMENT

The wide range of application of concrete has made it necessary to produce portland cements of different properties. This is done primarily by altering the proportions of the cement compounds through the composition of the raw material and certain details of the manufacturing process (burning temperature, etc.) [Kravchenko 1974]. In other words, all portland cements have the same constituents; it is only the differences in the proportions of these constituents that determine the individual type of cement. For instance, to obtain high early strengths, relatively high alite and C_3A (up to 15% potential) contents are required. On the other hand, a portland cement of low heat of hydration, such as would be used in a massive dam construction, requires a relatively high belite content (at least 40%) at the expense of the alite (maximum 35%) and C_3A (maximum 7%) contents.

Currently five main types of portland cement are recognized in the United States. These standard cement types are described in Table 2.3.

Type I, or, according to the British terminology, *ordinary* portland cement, is the general purpose cement that is used when the special properties of the other four types are not required. Concrete blocks, floors, reinforced frames, beams, and slabs are typical examples. *Type II*, or *modified* cement is a modification of Type I to increase resistance to sulfate attack and decrease the rate of heat evolution. It is used, for instance, in concrete pipes, pavements, and foundations. It is also popular for the production of high-strength concrete. [Peterman 1986] *Type III*, or *high early strength* cement is used when rapid strength development in concrete is essential, as in precast plants, winter concreting, and repairs. *Type IV*, or *low heat*, and *Type V*, or *sulfate resistant*, cements can be considered extreme and special cases of the Type II cement. Type IV has been used for massive dams and other large concrete structures to reduce the cracking tendencies resulting from accumulated heat of hydration. Type V cement has been used for structures where the concrete may be in contact with soils and groundwater containing larger amounts of sulfates. Such uses include canal linings, culverts, and foundations. Because of their special nature, Types IV and V are not commonly carried in stock and usually are made on special order. Type IV cement has not been used much in recent years because improved construction techniques have made the application of this slowly hardening cement unnecessary in mass concretes. Informative data collected by the Bureau of Reclamation and quoted by Neville [Neville 1963] concerning the compositions of the various types of

Table 2.3: Standard Types of Portland Cements and Blended Portland Cements

Portland cements and air-entraining portland cements	
Type I	For use in general concrete construction when the special properties specified for types II, III, IV, and V are not required
Type IA	Air-entraining cement for the same uses as type I, where air entrainment is desired
Type II	For use in general concrete construction exposed to moderate sulfate action, or where moderate heat of hydration is required
Type IIA	Air-entraining cement for the same uses as type II, where air entrainment is desired
Type III	For use when high early strength is required
Type IIIA	Air-entraining cement for the same uses as type III, where air entrainment is desired
Type IV	For use when a low heat of hydration is required
Type V	For use when high sulfate resistance is required

Portland blast-furnace slag cements and portland pozzolan cements	
Type IS^a	Corresponds to type I portland cement
Type IS-A^a	Corresponds to type IA air-entraining portland cement
Type IP^a	Corresponds to type I portland cement
Type IP-A^a	Corresponds to type IA air-entraining portland cement
Type P^b	For use in concrete construction where high strengths at early ages are not required
Type PA^b	Air-entraining portland-pozzolan cement for use in concrete construction where high strengths at early ages are not required

[a] Moderate sulfate resistance or moderate heat of hydration, or both, may be specified by adding the suffixes (MS) or (MH), or both, to the selected type designation.

[b] Moderate sulfate resistance or low heat of hydration, or both, may be specified by adding the suffixes (MS) or (LH), or both, to the selected type designation.

portland cement are shown in Table 2.4. Further data on the composition of American portland cements can be found in the literature.[Clifton 1971, ACI 1985b]

There are also Types I, II, and III cements that are modified by the addition of air-entraing admixtures (Section 6.3) during the manufacturing period. These are called *air-entraining cements* and are designated as Types IA, IIA, and IIIA, respectively. The air-entraining cements were used mainly in structures where there was expectation that the concrete would be exposed to frost action. Note, however, that the development of automatic admixture dispensers greatly reduced the importance of air-entraining cements because it is not possible to control the air content in concrete at the desired level when they are used.

It can be seen from Table 2.4 that C_3S and C_2S form 70-80% of all portland cements. Yet there may often be wide differences between cements of the same type. Also, there may be overlapping among the compositions, and thus among the properties of portland cements of different types.

The difference in cement composition is one of the reasons that an admixture can produce different results when used with different portland cements, even when the cements belong to the same type.

All types of portland cement described in ASTM C 150 as well as blended portland cements described in ASTM C 595 (Section 2.6) can be used in concretes cured by *steam curing* at atmospheric pressure, or by most of the accelerating heat methods. The cement type to be used is determined in the same manner as normal curing. That is, Types I and III are used most frequently but where special conditions warrant, such as sulfate exposure, Types II and V or blended cements should be used. Air entraining cements can also be used interchangeably with comparable results. Note, however, that different cements of the same type may have even more different behavior when cured under accelerated conditions than under normal curing. [ACI 1980] Also, early strengths can be dramatically increased during heat curing by using cements with SO_3 contents somewhat higher than usual.

Western hemisphere countries follow the five

Table 2.4: Typical Values of Compound Composition of Portland Cements of Different Types[a]

Cement	Type of data	Compound composition (%)								Number of samples
		C_3S	C_2S	C_3A	C_4AF	$CaSO_4$	Free CaO	MgO	Ignition Loss	
Type I	Max.	67	31	14	12	3.4	1.5	3.8	2.3	21
	Min.	42	8	5	6	2.6	0.0	0.7	0.6	
	Mean	49	25	12	8	2.9	0.8	2.4	1.2	
Type II	Max.	55	39	8	16	3.4	1.8	4.4	2.0	28
	Min.	37	19	4	6	2.1	0.1	1.5	0.5	
	Mean	46	29	6	12	2.8	0.6	3.0	1.0	
Type III	Max.	70	38	17	10	4.6	4.2	4.8	2.7	5
	Min.	34	0	7	6	2.2	0.1	1.0	1.1	
	Mean	56	15	12	8	3.9	1.3	2.6	1.9	
Type IV	Max.	44	57	7	18	3.5	0.9	4.1	1.9	16
	Min.	21	34	3	6	2.6	0.0	1.0	0.6	
	Mean	30	46	5	13	2.9	0.3	2.7	1.0	
Type V	Max.	54	49	5	15	3.9	0.6	2.3	1.2	22
	Min.	35	24	1	6	2.4	0.1	0.7	0.8	
	Mean	43	36	4	12	2.7	0.4	1.6	1.0	

[a]From [Neville 1963].

cement types of Table 2.3, whereas European countries usually do not. It is a general practice everywhere, however, to produce at least three types of portland cement: one for general use as *ordinary portland cement*, one for high-early strength as *rapid-hardening portland cement*, and another for improved sulfate resistance as *sulfate-resistant portland cement*.

2.6 BLENDED PORTLAND CEMENTS

The term *blended portland cement* may be defined as a hydraulic cement consisting of an intimate and uniform mixture of a portland cement and replacement material(s). It is produced either by intergriding portland cement clinker and the replacement material(s), by blending portland cement and finely divided replacement material(s), or a combination of intergriding and blending. (The properties of the replacement materials are discussed in Section 4.6.) In certain classes the replacement material(s), such as natural pozzolan, fly ash, slag, etc., are the minor component, whereas in other classes they are present in larger quantity than the portland cement. In any case, the general properties of blended portland cements are similar to the properties of comparable (plain) portland cement [Mather 1957, Klieger 1967, Kokobu 1969, 1974, Frohnsdorff 1986]. The advantage of adding a pozzolanic material to the clinker is that it can combine chemically with the lime and alkalies liberated from the portland cement paste during the hardening period. Thus the resistance of concrete against certain chemical effects is improved and the expansion caused by excessive aggregate-alkali reaction is reduced. In addition, blended portland cements generally provide a lower heat of hydration, and an improvement in the properties of fresh concrete, as compared to portland cement. On the other hand, a blended protland cement may give lower strengths at early ages although the ultimate strength is not reduced when extended moisture curing is provided.

ASTM C 595-85 recognizes five classes of blended

portland cement:

1.*Portland blast-furnace slag cement*. It is a blend of portland cement and fine granulated blast furnace slag in which the slag constituent is between 25 and 70 % of the weight of portland blast-furnace slag cement. It may come in air-entraining and non-air-entraining form.

2. *Slag-modified portland cement*. It is a blend of portland cement and fine granulated blast furnace slag in which the slag constituent is less than 25 % of the weight of the slag-modified portland cement. It may come in air-entraining and non-air-entraining form.

3. *Portland-pozzolan cement*. It is a blend of portland or portland blast-furnace slag cement and fine pozzolan in which the pozzolan constituent is between 15 and 40 % of the weight of the portland-pozzolan cement. It may come in air-entraining and non-air-entraining form.

4. *Pozzolan-modified portland cement*. It is a blend of portland or portland blast-furnace slag cement and fine pozzolan in which the pozzolan constituent is less than 15 % of the weight of the pozzolan-modified portland cement. It may come in air-entraining and non-air-entraining form.

5. *Slag cement*. It is a blend of granulated blast-furnace slag and portland cement, or hydrated lime, or both, in which the slag constituent is at least 70 % of the weight of the slag cement. It may come in air-entraining and non-air-entraining form.

Some of the standard types of these cements are shown in Table 2.3. The slag as well as the pozzolanic material usable in blended cements should comply with specifications given in ASTM C 595-85.

Blended portland cements are also manufactured in Europe. Portland blast-furnace slag cement, for instance, is used in Germany under the name *Eisenportlandzement*, and in France under the name *ciment Portland de fer*. Europeans also produce blended cements in which the major constituent is the slag (blast-furnace slag portland cement) [Schroder 1969, Satarin 1974]. These are *Hochofenzement* and *ciment de haut fourneau*, respectively. The usual composition limits for blast-furnace slags and for pozzolans are shown in Fig.2.4.

A modification of the usual process for making portland blast-furnace slag cement resulted in *Trief cement*. In this process the slag, after granulating, is ground wet and stored as a wet slurry. It is kept as a separate constituent until the concrete is being mixed, that is, when the portland cement, slag slurry, and aggregate are added together. The advantages claimed are a saving in fuel for drying the slag and a greater efficiency of grinding in the wet state. With finely ground slag, good strengths can still be obtained with low proportions of portland cement [Lea 1956, Orchard 1973a].

Pastes made with blended cements of high slag content, and especially those of supersulfated cements, produce lower pH values (from 11 to 12.5, and from 10.5 to 11, respectively) than portland cements. However, theses are still high enough to protect the embedded reinforcement with normal procedures provided that no corrosive ions, such as Cl, are present [Nurse 1964, Gouda 1975].

Properties and applications of pozzolanic materials as well as those of blended cements are discussed in the literature [Barton 1965, Popovics 1968e, Kramer 1960, Faber 1967, Turriziani 1960, Malquori 1960, Fifth Intrn. Symp. 1969].

2.7 TECHNICALLY IMPORTANT PROPERTIES OF PORTLAND CEMENT - TESTING AND SPECIFICATIONS

Portland cement must have a number of qualities to be acceptable. It must contribute to workability of the concrete or mortar; must set and get hard in a specific time; must have good finishing characteristics; must produce a concrete or mortar that will serve without undo deterioration; must adhere to the aggregate particles and to the reinforcement; and must be compatible with admixtures [Price 1974]. In order to fulfill all these requirements, pertinent standards usually contain specifications for chemical composition and for the physical properties of portland cements.

Most of the pertinent *chemical requirements* for portland cement have already been mentioned here in connection with the composition of portland cement clinker. These are important mainly to the manufacturer.

From the standpoint of the user, *physical properties* provide more direct information concerning the acceptability of a cement. The physical properties of portland cement covered by ASTM specifications are *fineness, time of setting, soundness, air content,* and *strength.* There are also tentative requirements concerning *heat of hydration* and *false set* (Tables 2.5 and 2.6). The required physical properties for blended cements are presented in Table 2.7. It may be interesting to note that the first recorded national specifications for portland cement, adopted by Germany in 1878, already contained requirements for fineness, soundness, and strength [Woods 1959]. The first specifications for cement in the United States were recommended by the American Society of Civil Engineers in 1885.

The measured value of any of the cement properties is greatly influenced by the testing conditions, which may include temperature, water content, time of mixing, type of testing equipment, and so forth. All these factors must be standardized in order to obtain comparable and meaningful test results. In the following, the ASTM test methods and requirements will mainly be discussed. Summaries concerning other methods can be found elsewhere [Barton 1965, Neville 1981, Orchard 1973a, Duriez 1961, Graf 1957, Troxell 1968, Cembureau 1958,1961, Gilliland 1968, Wischers 1976b]. The literature also contains considerable data concerning the various technically important properties of American portland cements [Clifton 1971, Lerch 1948, McMillan 1949, Verbeck 1950, Jackson 1951,1958, Klieger 1957, Tyler 1960, Mather 1967, U.S. Dept.of Inter. 1949] as well as test methods [ASTM 1974].

Table 2.5: Standard Physical Requirements for Portland Cements (ASTM C 150-89)

Cement Type [A]	I	IA	II	IIA	III	IIIA	IV	V
Air content of mortar, [B] volume %:								
max	12	22	12	22	12	22	12	12
min	...	16	...	16	...	16	...	...
Fineness, [C] specific surface, m²/kg (alternative methods):								
Turbidimeter test, min	160	160	160	160	...	...	160	160
Air permeability test, min	280	280	280	280	...	...	280	280
Autoclave expansion, max, %	0.80	0.80	0.80	0.80	0.80	0.80	0.80	0.80
Strength, not less than the values shown for the ages indicated below: [D]								
Compressive strength, psi (MPa):								
1 day	...	...	...	...	1800 (12.4)	1450 (10.0)	...	...
3 days	1800 (12.4)	1450 (10.0)	1500 (10.3) 1000 (6.9)[F]	1200 (8.3) 800 (5.5)[F]	3500 (24.1)	2800 (19.3)	...	1200 (8.3)
7 days	2800 (19.3)	2250 (15.5)	2500 (17.2) 1700 (11.7)[F]	2000 (13.8) 1350 (9.3)[F]	...	...	1000 (6.9)	2200 (15.2)
28 days	...	...	...	...	...	...	2500 (17.2)	3000 (20.7)
Time of setting (alternative methods): [E]								
Gillmore test:								
Initial set, min, not less than	60	60	60	60	60	60	60	60
Final set, min, not more than	600	600	600	600	600	600	600	600
Vicat test: [G]								
Time of setting, min, not less than	45	45	45	45	45	45	45	45
Time of setting, min, not more than	375	375	375	375	375	375	375	375

[A] See Note.

[B] Compliance with the requirements of this specification does not necessarily ensure that the desired air content will be obtained in concrete.

[C] Either of the two alternative fineness methods may be used at the option of the testing laboratory. However, when the sample fails to meet the requirements of the air-permeability test, the turbidimeter test shall be used, and the requirements in this table for the turbidimetric method shall govern.

[D] The strength at any specified test age shall be not less than that attained at any previous specified test age.

[E] The purchaser should specify the type of setting-time test required. In case he does not so specify, the requirements of the Vicat test only shall govern.

[F] When the optional heat of hydration or the chemical limit on the sum of the tricalcium silicate and tricalcium aluminate is specified.

[G] The time of setting is that described as initial setting time in Test Method C 191.

Table 2.6: Optional Physical Requirements^A (ASTM C 150-89)

Cement Type	I	IA	II	IIA	III	IIIA	IV	V
False set, final penetration, min, %	50	50	50	50	50	50	50	50
Heat of hydration:								
7 days, max, cal/g (kJ/kg)	...	...	70 (290)^B	70 (290)^B	...	...	60^C (250)	...
28 days, max, cal/g (kJ/kg)	...	...	...	...	...	...	70^C (290)	...
Strength, not less than the values shown:								
Compressive strength, psi (MPa)								
28 days	4000 (27.6)	3200 (22.1)	4000 (27.6) 3200^B (22.1)^B	3200 (22.1) 2560^B (17.7)^B	...	...	...	...
Sulfate expansion,^{D} 14 days, max, %	...	...	...	...	...	...	...	0.040

^{A} These optional requirements apply only if specifically requested. Availability should be verified.

^{B} The optional limit for the sum of the tricalcium silicate and tricalcium aluminate in Table 2 shall not be requested when this optional limit is requested. These strength requirements apply when either heat of hydration or the sum of tricalcium silicate and tricalcium aluminate requirements are requested.

^{C} When the heat of hydration limit is specified, it shall be instead of the limits of C3S, C2S, and C3A.

^{D} When the sulfate expansion is specified, it shall be instead of the limits of C_3A and $C_4AF + 2 C_3A$.

Table 2.7: Physical Requirements for Blended Cements (ASTM C 595-86)

Cement Type	I(SM), IS, I(PM), IP	I(SM)-A, IS-A, I(PM)-A, IP-A	IS(MS), IP(MS)	IS-A(MS), IP-A(MS)	S	SA	P	PA
Fineness	[A]	[A]	[A]	[A]	[A]	[A]	[A]	[A]
Autoclave expansion max, %	0.50	0.50	0.50	0.50	0.50	0.50	0.50	0.50
Autoclave contraction, max, %	0.20	0.20	0.20	0.20	0.20	0.20	0.20	0.20
Time of setting, Vicat test:[C]								
Set, minutes, not less than	45	45	45	45	45	45	45	45
Set, h, not more than	7	7	7	7	7	7	7	7
Air content of mortar (Method C 185), volume %	12 max	19 ± 3	12 max	19 ± 3	12 max	19 ± 3	12 max	19 ± 3
Compressive strength, min, psi (MPa):								
3 days	1800 (12.4)	1450 (9.9)	1500 (10.3)	1200 (8.3)	...	...	...	...
7 days	2800 (19.3)	2250 (15.5)	2500 (17.2)	2000 (13.8)	600 (4.1)	500 (3.4)	1500 (10.3)	1250 (8.6)
28 days	3500 (24.1)	2800 (19.3)	3500 (24.1)	~2800 (19.3)	1500 (10.3)	1250 (8.6)	3000 (20.7)	2500 (17.2)
Heat of hydration:[D]								
7 days, max, cal/g (kJ/kg)	70 (293)	70 (293)	70 (293)	70 (293)	...	...	60 (251)	60 (251)
28 days, max, cal/g (kJ/kg)	80 (335)	80 (335)	80 (335)	80 (335)	...	...	70 (293)	70 (293)
Water requirement, max weight % of cement	...	...	...	...	...	...	64	56
Drying shrinkage, max, %	...	...	...	...	...	...	0.15	0.15
Mortar expansion:[E]								
At age of 14 days, max, %	0.020	0.020	0.020	0.020	0.020	0.020	0.020	0.020
At age of 8 weeks, max, %	0.060	0.060	0.060	0.060	0.060	0.060	0.060	0.060

[A] Both amount retained when wet sieved on 45-μm (No. 325) sieve and specific surface by air permeability apparatus, cm²/g, shall be reported on all mill test reports requested under 14.4.

[B] The specimens shall remain firm and hard and show no signs of distortion, cracking, checking, pitting, or disintegration when subjected to the autoclave expansion test.

[C] Time of setting refers to initial setting time in Method C 191.

[D] Applicable only when moderate (MH) or low (LH) heat of hydration is specified, in which case the strength requirements shall be 80 % of the values shown in the table.

[E] The test for mortar expansion is an optional requirement to be applied only at the purchaser's request and should not be requested unless the cement will be used with alkali-reactive aggregate.

It should be recognized that existing methods of testing cement are, at best, crude tools, and it must not be expected that exact agreement will be secured in tests by different operators, or by the same operator at different times, even when the tests are made on the same sample and under the same conditions. Many steps are involved in making cement tests, and the results are sometimes greatly influenced by seemingly minor variations in procedure. Therefore, most ASTM test specifications contain limits for the acceptable fluctuations of test results. These are called *precision* and *bias statements*.

In order to improve the reliability of the test results, ASTM has produced a manual [ASTM 1988] the purpose of which is to emphasize those factors that may affect results of cement tests and call attention to less apparent influences that are important but sometimes overlooked.

2.7.1 Fineness of Cement

The term *fineness* or *fineness of grinding* refers to the average size of the cement particles. A higher fineness means a more finely ground cement, smaller particles. The significance of fineness lies in the fact that it affects several technically important properties of cement and concrete. For instance, the higher the fineness, the higher compressive strengths are developed at early ages. [Gruenwald 1939, Higginson 1970, Kondo 1972, Reinsdorf 1962]. Also, finer cement bleeds less than a coarser cement [Powers 1968] (Section 2.7.8.5 and Fig. 2.5), contributes to better workability [Vivian 1966], and shows less autoclave expansion as well. For these reasons the fineness of cement was steadily increased until about the middle of this century. Since then this increase has leveled off, partly because higher fineness is more expensive, and partly because an overly fine cement (Blaine specific surface over 500 m^2/kg, that is, 5000 cm^2/g) has certain technical disadvantages, such as higher shrinkage tendencies, stronger reaction with alkali-reactive aggregates, higher water demand, and poor storability. Some

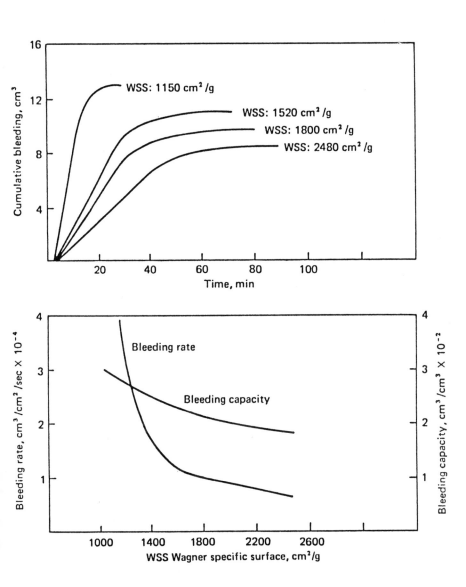

Figure 2.5: Bleeding of cement paste decreases as the fineness is increased [Blanks 1955].

current thoughts concerning fineness are presented in a recent ASTM publication [ASTM 1969a].

The most complete information about fineness is provided by the *particle-size distribution* of the cement. Particle sizes usually vary between 1 and 200 ∝m. The details of the particle-size distribution can be determined by air elutriation, by means of a hydrometer [Klein 1941], or a turbidimeter [Hime 1965], or micrometer and microscopic sizing or with laser technique in which the average dimensions or representative particles are established by direct measurements [Kantro 1964]. These methods are rarely used because they are quite time consuming [Blanks 1955]. Besides, when the Stoke's law is used for the calculation of the particle size distribution, the interpretation of the results require great care because results are influenced not only by the particle sizes but also the shapes and specific gravities of the particles. [Matouschek 1947]. Further details about methods are presented by Orchard [Orchard 1973b].

There are simpler methods for the estimation of the particle-size distribution from two points of the distribution curve [Herdan 1960]. The most frequently used such method is probably the formula of Rosin and Rammler [Rosin 1933]:

$$y = 100e^{-bd^n} \tag{2.8}$$

where   y = total percentage, by weight, of the particles that are greater than d
   d = size of the cement particle in microns
   b,n = experimental parameters.

Equation 2.8 provides a straight line in a log-log(100/y) versus log d system of coordinates with slope n, so it is easy to construct when two points are know. Experimental results support this formula from 5 ∝m up for the most cements. The degree of fit is illustrated in Fig. 2.6 for three cements. In these examples the values of *b* for Type I, II, and III cements are 0.045, 0.020, and 0.100, respectively, whereas the values of *n* are 0.925, 1.22, and

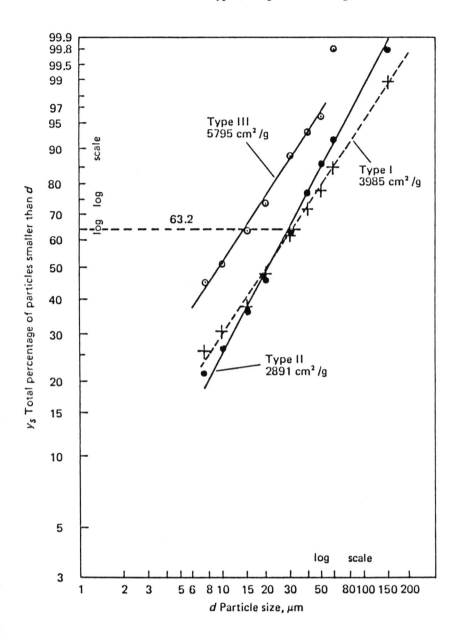

Figure 2.6: Particle-size distributions of three portland cements in the Rosin-Rammler system. The specific surfaces were determined by an air-permeability method. Data from Lerch and Ford [Lerch 1948].

0.887, respectively. Parameter b is characteristic
of the amount of the finest particles, whereas n is
characteristic of the range of the particle-size
distribution. The particle size corresponding to y
=36.8% (=100/e), that is, to y_s = 63.2%, can also be
used conveniently as an indicator of the degree of
pulverization [Herdan 1960].

There are several other methods available for
the measurement of fineness [Kester 1963]. The oldest
and simplest method is to *sieve* a cement sample
through fine sieves and record as a percentage the
amount withheld on each [ASTM C 184-83 and C 430-83].
Since, however, more than 95% of a typical cement
passes the No.200 sieve (one net opening equals 74
μm), this test provides little useful information
about the quantity of the most important cement
particles, namely those that are smaller than 10-15
μm. Therefore, the sieve residues have been dropped
from the ASTM requirements. Instead, it has become
customary to characterize the fineness of a cement by
the *specific surface* in m²/kg, which is the surface
area of the particles per kilogram of cement in
square meters. The finer the cement, the higher the
specific surface. If, for instance, a volume V
consists of a single cube of size a, that is , V =
a^3, then its surface area is $6a^2$. If the same volume
is made up of eight cubes of size $a/2$, then the
surface area, as well as the specific surface, will
be doubled. Since the hydration starts at the surface
of the cement particles, the higher the specific
surface, the more intensive become the early portions
of the hydration. Incidentally, the specific surface
is inversely proportional to the harmonic average
size of the particle size distribution [Popovics
1962a], as demonstrated in Chap.11.

The specific surface of a portland cement is
usually determined either by an optical method or by
an air-permeability method, although it can also be
measured by gas (nitrogen) absorption. Optical methods include, for instance, the Klein turbidimeter and
the Wagner turbidimeter (Fig. 2.7) methods. The
Lea-Nurse apparatus as well as the Blaine apparatus
(Fig. 2.8) utilize the air-permeability principle.

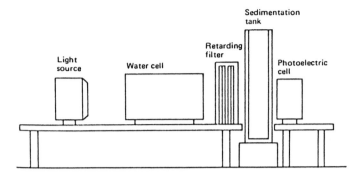

Figure 2.7: Schematic representation of the Wagner turbidimeter apparatus for determining the specific surface of cement. Reproduced from Orchard, *Concrete Technology*. Volume 2, 3rd edition by permission of the author and Applied Science Publishers [Orchard 1973b].

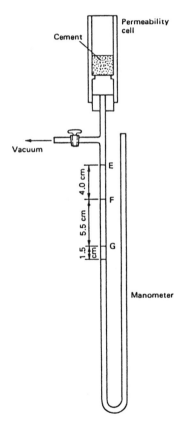

Figure 2.8: Schematic representation of the Blaine apparatus for determining the specific surface of cement by the air-permeability method [Orchard 1973b].

The accepted standard methods in the United States are the *Wagner turbidimeter method* [ASTM C 115-86], and the *Blaine air-permeability method* [ASTM C 204-84].

The Wagner turbidimeter determines the rate of settlement of the cement particles suspended in a fluid by measuring photoelectrically the change in turbidity. A sample of portland cement is dispersed in kerosene in a tall glass container; then a beam of light is passed through the suspension and the percentage of the light transmitted is measured by a photocell at a given elevation on a stated time schedule. A value for the specific surface is calculated from the data obtained, which calculation is based on certain simplifying assumptions.

In the Blaine method a cement sample of about 2.8 g is compacted into a cell to a porosity of 50%. Then a known volume of air is passed though the compacted sample at a prescribed average pressure, starting from the E mark of the manometer (Fig. 2.8), with the rate of flow diminishing steadily. The time for the flow to take place between marks F and G is measured in seconds, the square root of which is proportional with the specific surface.

Comparative tests have demonstrated that the specific surface as determined by the Blaine method is almost double that determined by the Wagner method. This is due mainly to the oversimplifying assumption applied in the Wagner method, namely, that the average size of the cement fraction smaller than 7.5 μm is 3.75 μm. Nevertheless, either method provides an acceptable picture of the relative variation in the fineness of cement, and for practical comparative purpose this is sufficient.

The specific surface of a cement as measured by *gas adsorption* is considerably higher than that by other methods. This is due mainly to the usual presence of a small amount of hydration product adhering to the surface of dry cement particles. The immense specific surface of the hydration product, as will be discussed later, is measured by gas adsorption but it is not measured by the other methods [Powers 1968]. Thus the high specific surface obtained by gas adsorption is unrealistic as far as cement fineness is concerned.

Strength results have shown good correlation with the specific surface of the cement used [Kondo 1972, Alexander 1972]. Nevertheless, it appears that the particle-size distribution of a cement provides more reliable information about the effects of cement fineness [Hraste 1974]. This is so because experimental data show that a portland cement with one size grading develops lower strengths at later ages than the same cement with a grading of identical specific surface but that contains finer and coarser particles [Vivian 1966, Kuhl 1961b]. Although the reason for this is not clear, one can speculate that the coarser cement particles also play a role in the hardening of cement pastes, for instance, through their slower but longer hydration, or because their presence reduces the percent of voids between the cement particles, or because proper combinations of finer and coarser cement particles may optimize the rheological properties of cement paste [Skvara 1976] similarly to the optimization of concrete workability by the grading of aggregate. In any case, the hydraulic activities of cement particles larger than say, 25 μm are small, thus it is possible to substitute for these by less expensive, suitable inert particles, such as limestone particles, without hurting the strength development [Vivian 1966].

Measurements also indicate that the C_3S content is greater in the finer-size fractions of a cement than in the coarser fractions, at the expense of the C_2S content [Matouschek 1947]. This should be considered when the effects of fineness on strength development and some other properties of the cement are investigated.

Minimum acceptable values for the specific surface are shown in Table 2.5. These values are valid both for portland cements and for blended portland cements. The beneficial effect of cement fineness on the reduction of the bleeding rate is mainly responsible for these limits on the minimum permissible specific surface. The specific surface of modern commercial portland cements ranges approximately from 280 to 360 m^2/kg by the air-permeability method. For Type III cements the range is approximately 360 - 450 m^2/kg.

2.7.2   Normal Consistency

In several standard tests concerning the quality of portland cement, such as in the soundness, time of setting, and tensile strength tests, the amount of mixing water specified is related to the water content needed to bring the paste to a standard condition of wetness, called *normal consistency* [Dise 1968]. Normal consistency is determined by ASTM C 187 by a trial-and-error method as follows: A number of cement pastes are made with different amounts of mixing water. Each paste is placed in turn in the standard mold and centered under the 300-g rod B of the apparatus developed by Vicat [Vicat 1818] shown in Fig. 2.9.   Then the plunger end C (10 mm in diameter) is brought in contact with the surface of the paste and quickly released. The paste is of normal consistency when the rod settles to a point 10±1 mm below the original surface of the paste in 30 sec. The amount of water required for normal consistency is given as the percent of the weight of the dry cement.   This water requirement increases with increasing fineness, but is also influenced by the cement composition [Blaine 1965]. The number of paste samples of this trial-and-error method can be reduced by using the nomogram of Fig. 2.10 [Popovics 1963,1966a,1968a], or pertinent tabulated values [Popovics 1968c].

There are no requirements specified for the water content at normal consistency. The usual values range from about 22 to 28% for commercial portland cements, somewhat higher for portland cements with natural pozzolans, and lower for high alumina cements.

A comparison of experimental results [Blaine 1965] reveals that a cement with a higher water requirement for normal consistency does not necessarily need a higher water content to produce a specified slump.   In other words, the result of the normal consistency test cannot be used for the

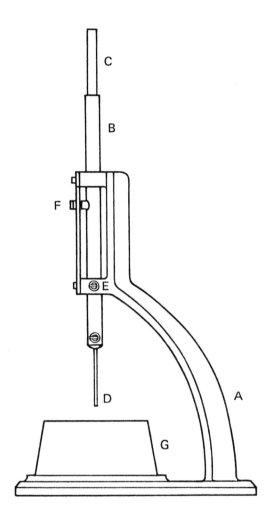

Figure 2.9: Vicat apparatus for testing normal consistency and time of setting of cement pastes. From ASTM C 191-82.

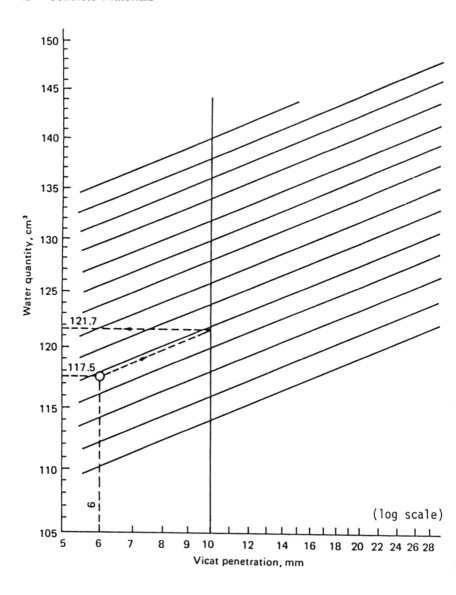

Figure 2.10: Chart for the estimation of water content of cement paste (with 500 g of cement) needed for a specified penetration [Popovics 1963, 1966a, 1968a]. (Vicat test according to ASTM C 187-85, or B.S. 12, or DIN 1165.) *Example:* If a water content of 117.5 cm^3 results in a penetration of 6 mm, then the water content needed for the penetration of 10 mm is 121.7 cm^3. (Note that C 187-85 specifies 650 g of cement for this test.)

prediction of water need of the same cement in a concrete. The reason for this paradox is that the two tests measure two different properties of the cement paste: The Vicat test measures primarily the viscosity [Cusens 1973], whereas the result of the slump test is influenced mostly by the lubricating capability of the paste. These two paste properties are not correlated, or at least the relationship is not a simple linear one [Popovics 1982a].

2.7.3   Time of Setting

When cement is mixed with water of around 20-35%, the result is a paste. This paste displays considerable plasticity and maintains it for a period of time called the *dormant period*. After a while, however, the paste starts stiffening, less and less plasticity can be observed, and finally all the plasticity is gone; that is the paste becomes brittle, although it is still without any sizable strength. This stiffening process is called *setting* and is the result of a series of reactions between the cement and water that will be discussed in Chapter 3. One should recognize that the stiffening is not a drying process; it takes place even if the fresh cement paste is kept under water. The gain of strength, that is, the *hardening* process, takes place subsequent to the setting. One can say, therefore, with a certain justification that the setting period is the zero stage of strength development in the cement-water mixture during which the reactions are accelerating, whereas the hardening is the first and second stages during which the reactions are decelerating. A schematic presentation of this inter- pretation is shown in Fig. 2.11.

It is customary to talk about *initial* setting, which is basically the beginning of the stiffening, and *final setting* which is marked by the disappeareance of plasticity. The setting process should not start too early, because the freshly mixed concrete should remain in a plastic condition for a sufficient period to permit satisfactory compaction and finishing after transporting and placing. On the other hand, too long a setting process is also

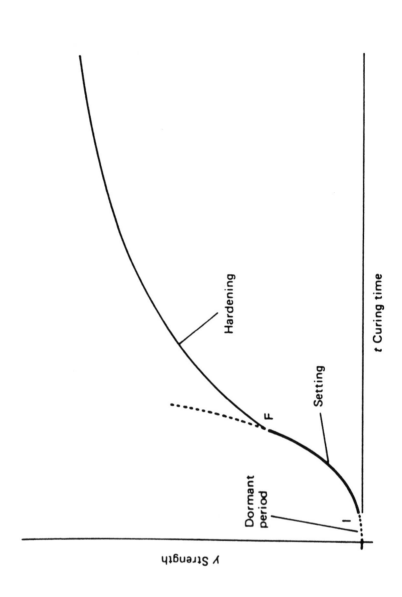

Figure 2.11: Illustration of the terms *setting* and *hardening*. I, initial setting; F, final setting. (Not to scale.)

undesirable, because this will cause excessive bleeding and a useless delay in strength development after
finishing. Of the two concepts, the initial setting
has higher significance.

There is no strict dividing line between setting
and hardening. Any distinction is more or less
arbitrary and poorly defined. So are the terms
initial setting and final setting. Nevertheless, for
practical purposes, it is convenient to have test
methods for the approximate determination of the time
when stiffening starts, and when plasticity is gone.

The time of set of a paste or concrete is usually
determined by measuring repeatedly the changes in its
resistance to penetration by specified small rods or
needles, although slump, heat development [Kleinlogel
1933], ultrasonic pulse velocity [Whitehurst 1951],
shearing test [Bombled 1967], and electrical
resistivity measurements [Calleja 1952, Szuk 1962]
are also applicable for this purpose. There are two
penetration methods standardized for portland cement
pastes in the United States. One applies the *Vicat
apparatus* [ASTM C 191-82], the other, developed by
General Gillmore [Gillmore 1874], the *Gillmore
needles* [ASTM C 266-87]. The Vicat apparatus is
primarily for laboratory use, whereas the Gillmore
needles are for field tests. For blended portland
cements only the Vicat method is specified.

For determination of the initial set, the needle
end D (1 mm in diameter) of the rod of the Vicat
apparatus (Fig. 2.9) is used in a manner similar to
the test procedure followed when determining the
normal consistency. When the paste of normal consistency stiffens sufficiently for the needle to
penetrate only to a depth of 25 mm in 30 sec, initial
set is considered to have taken place. The time of
initial set is expressed as the time elapsed since
the mixing water was added to the cement. Final set
is usually considered to have taken place when paste
will support the needle without appreciable indentation.

The standard forms of the Vicat apparatus used in
European countries differ slightly from the form
accepted in the United States.

The Gillmore method applies to needles of pattern

similar to that shown in fig. 2.12. The *N*   needle,
1/12   in (2.12 mm) in diameter and weighing 0.5 lb
(113.4 g), is used for the determination of the
initial set, and the needle 1/24 in (1.06 mm) in
diameter and weighing 1 lb (452.6 g) is used for the
final set. From the cement paste at normal consistency a pat about 3 in (75 mm) in diameter and 0.5
in (12.5 mm) in thickness is prepared. The initial
set is considered to have taken place when the light
needle, gently applied to the surface of the pat,
fails to make an appreciable indentation. The final
set is considered to have taken place when the heavy
needle fails to make an appreciable indentation.

The times of set determined by the Gillmore
needles are longer than those determined by the Vicat
apparatus.

ASTM C 403-88 provides a procedure for determining the *time of setting of concrete* with slump
greater than zero by measuring the penetration
resistance of the mortar that is sieved from the
concrete mixture, by specified rods. The penetration
resistance is calculated as the force required to
cause a 1-in (25.4-mm) depth of penetration of the
needle divided by the area of the bearing face of the
needle. Time of initial setting is defined as the
elapsed time, after initial contact of cement and
water, required for the mortar to reach a penetration
resistance of 500 psi (3.45 MPa). Time of final
setting is the elapsed time required for the mortar
to reach a penetration resistance of 4000 psi (27.6
MPa)

The shape of the convex portion of the stiffenig
- versus - curing time curve in Fig. 2.11 suggests
that it can be approximated either by an exponential
curve or by a power function. The exponential
approximation has been applied by Polivka and Klein
for the Proctor penetration resistance versus curing
time relation [Polivka 1960]. The power function for
the approximation of the development of various
processes accompanying the setting is [Popovics
1971b, 1982a]:

$$y = at^i \qquad\qquad (2.9)$$

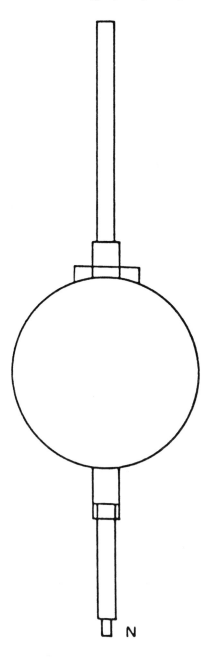

Figure 2.12: A Gillmore needle for testing the time of setting of cement pastes (ASTM C 266-77).

where y = a physical setting characteristic of cement
         paste or concrete, such as the Vicat
         penetration depth
      t = time elapsed since mixing, that is, curing
         time
   a, i = empirical parameters.

The a parameter is a function of the cement
composition, the composition of the paste or concrete
as well as the curing and testing conditions, and is
characteristic of the initial set. The i parameter
is a function of the test method employed, the amount
of certain admixtures present, and perhaps the composition of the cement but, significantly, is independent to a great extent of the type and amount of
cement, the water-cement ratio (that is, the ratio of
the amount of water to the the amount of cement), in
certain cases of the curing temperature, and so
forth. In a physical sense i represents the approximate value of the measured rate of stiffening
[Popovics 1968d].

As long as i is constant, Eq. 2.9 provides a
family of parallel straight lines in a log-log system
of coordinates. Figure 2.13 shows that experimental
results by the standard Vicat test with various
pastes of Type I portland cement approximate the
parallelism well within the limits of 0.04 and 0.8
in (1 and 20 mm) penetrations. The common slope of
these lines provides the value of $i = -10$, which
value was confirmed for pastes of a Type III cement.
Figure 2.14 shows again a good agreement between
repeated slump measurements and Eq. 2.9. Proctor
penetrations, pulse velocity, and electrical
resistance measurements (Fig. 2.15) have also shown
good fits with Eq. 2.9. The i values obtained for
these test methods with commercial cements are
presented in Table 2.8 [Popovics 1971b]. The validity
of Eq. 2.9 is restricted in most cases to early ages.
 A higher absolute value of i for a given test
method indicates a faster stiffening, which also
means that the time of set is lengthened relatively
more, or shortened relatively less, than that of the
final set. For instance, low temperature seems to

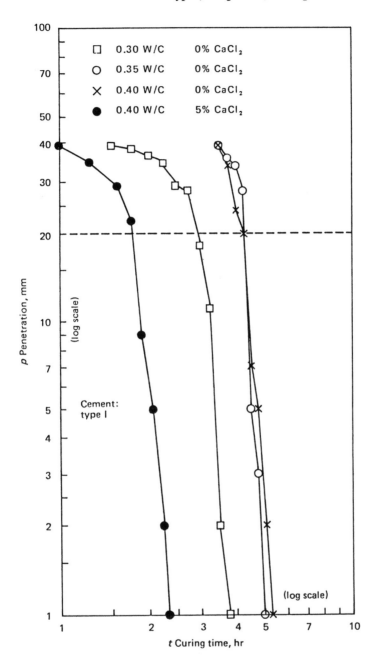

Figure 2.13: Typical curves of Vicat penetration for cement pastes setting at room temperature [Popovics 1971b].

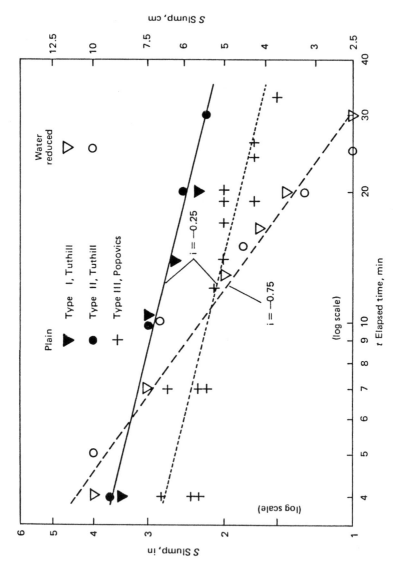

Figure 2.14: Slump loss of concretes with and without a given water-reducing admixture as a function of the elapsed time. Curing temperature: 90° to 95°F (32° to 35°C). Cement content: 520 lb/yd^3 (310 kg/m^3). [Popovics 1971b].

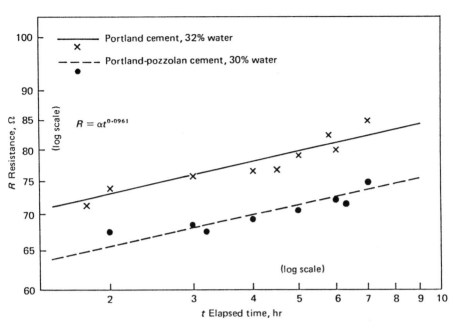

Figure 2.15: Electrical resistance of cement pastes as a function of cement type and elapsed time. Data from Szuk and Naray-Szabo [Szuk 1958].

Table 2.8: Values of i for Eq. (2.9)–Rates of Stiffening of Fresh CementPastes and Concretes as Measured by Various Methods[a]

Test method	i
Standard Vicat needle for pastes (ASTM C 191-74)	−10
Standard slump test for concrete (ASTM C 143-71)	−0.25
Standard Proctor penetration resistance for concrete (ASTM C 403-70)	7
Pulse velocity for concrete	1
Electrical resistance	0.1

[a]From [Popovics 1971b].

reduce significantly the absolute   value of i , as shown in Fig, 2.16 [Voellmy 1956] for the Vicat test. For different mixtures where i is identical, changes in the time of initial setting and final setting are proportional. Note that this proportionality is not expected strictly from the results obtained with the standard Vicat test, since Eq. 2.9 is not valid for penetrations over 0.8 in (20 mm); but even so, if i = -10 were valid for every cement tested by Vicat method, the related setting values would provide a straight line with a slope of approximately 0.75 in the system of time of initial set versus time of final set. Scattering around such a straight line, as shown in Fig. 2.17 for German cements [Graf 1960], demonstrates the effect of the composition of the cement on the value of i, that is, on the rate of setting. Present-day American cements are not expected to show such a large scattering about i=-10, but it is more important that Eq. 2.9 is applicable for the setting of cements even when the value of i differs from the pertinent value given in Table 2.8. The practical consequence of this is that Eq. 2.9 can be used with the appropriate i value as a simple tool for the interpretation of the effects of various factors, including admixtures, on the setting.

The primary factors that influence the times of setting of a given cement are curing temperature (Figs. 2.16, 2.18, and 2.19); water-cement ratio, (Fig. 2.20); [Reinsdorf 1962] admixtures (Fig. 2.13); and fineness of the cement (Fig. 2.21), [Popovics 1982a, Fattuhi 1985, Reinsdorf 1962]

Standard requirements concerning the time of Setting are presented in Table 2.5. The standard initial set of commercial portland cements ranges approximately form 2 to 4 hr, and final set from 5 to 8 hr. In concrete mixtures the set usually occurs later because of the higher water-cement ratios in concretes. Under special conditions, however, such as construction at high temperature, a concrete may set early. In the latter case a suitable admixture or other protective methods should be used to delay the setting period appropriately.

Portland cements that meet the standard requirement for initial set are called *normal*

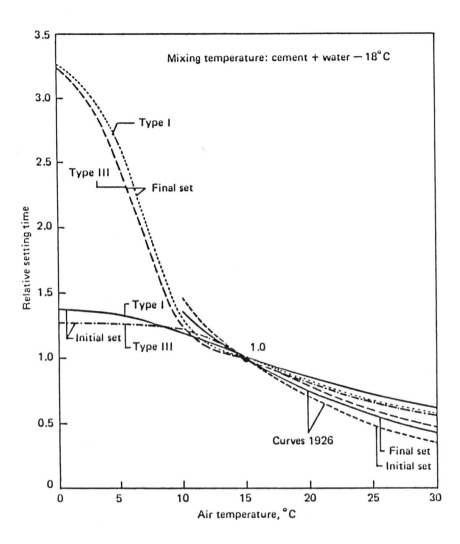

Figure 2.16: Influence of the curing temperature on the settings of type I and type II portland cement pastes [Voellmy 1956].

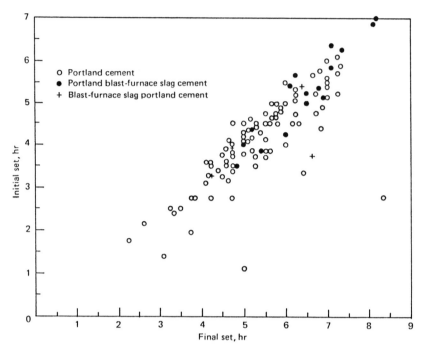

Figure 2.17: Relationship between the initial set and final set of various portland, portland blast-furnace slag, and blast-furnace slag cements tested according to the German standard DIN 1164 [Graf 1960].

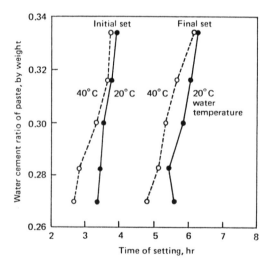

Figure 2.18: Influence of water temperature and the water-cement ratio on the setting of a type I portland cement paste [Yokomichi 1956].

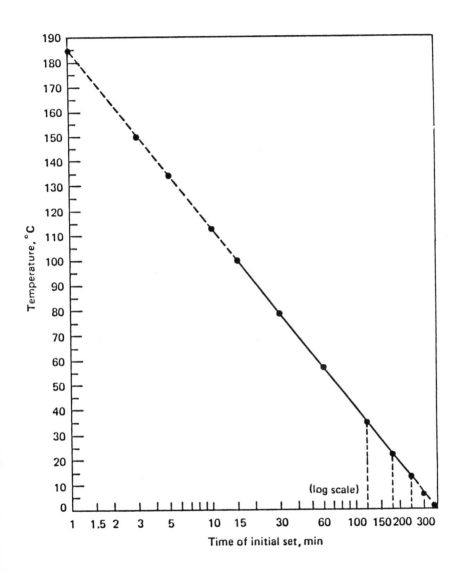

Figure 2.19: Time of initial set of a portland cement as a function of the temperature of the paste and curing. From Duriez and Arrambide [Duriez 1961].

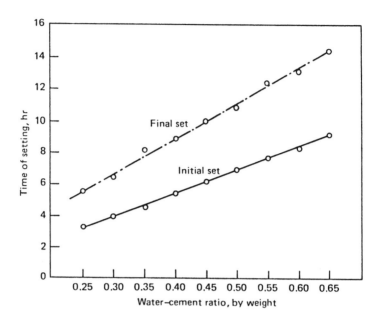

Figure 2.20: Influence of water-cement ratio on the setting of a portland cement paste [Wesche 1974].

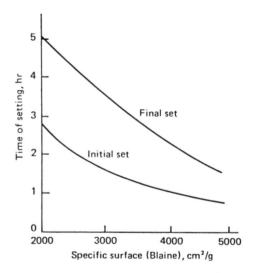

Figure 2.21: Influence of cement fineness on the setting of a type III cement paste at 20°C [Ujhelyi 1973].

setting, whereas cements that fail to pass the initial set requirement are called *quick setting*. Quick-setting cements are used only for special jobs, such as repair of leakage in a tunnel. The speed of setting does not indicate a corresponding rapidity of gain in strength. On the contrary, quick setting results, as a rule, in strength reduction later on.

False set (grab set, rubber set, etc.) is a special kind of quick set where a remixing of the paste or concrete, without addition of water, restores plasticity until the mixture sets in the normal manner and without loss of strength. ASTM C 451-83 and C 359-83 prescribe procedures for the determination of false set in portland cement paste and in Ottawa-sand mortar, respectively. Both methods are based on repeated penetration measurements: one immediately after mixing, another after a waiting time of 5 min or more, and then immediately after remixing, another penetration measurement.

Special cements may require different criteria for the evaluation of the setting process. For instance, setting of oil-well cements is usually measured in terms of change in viscosity at elevated temperature.

2.7.4 Soundness

Unsoundness is the harmful property that occurs when the hardened cement paste develops an undue expansion that is manifested by cracking of the mass. Such expansion can be produced by several factors, for instance by misadjusted gypsum content in the cement [Kuhl 1961a], but the usual cause of unsoundness is the presence of free or uncombined lime. Excess quantities of crystalline magnesia, as periclase, in cement also causes unsoundness. Hydrated lime and hydrated magnesia occupy larger volumes than the original oxides, therefore expansion takes place during their hydration. When such hydration occurs before the final set of the cement paste, no detrimental effect results. If, however, the hydration is slow, hence the expansion continues after the final set, which is the usual case with overburned free lime and crystalline magnesia in portland cement, this may cause disintegration of the hardened paste or concrete.

Since unsoundness of cement is not apparent until after a period of months, it is essential to test the soundness in an accelerated manner. This was done formerly by means of a pat test in which a pat of cement paste was subjected to steam of boiling water in order to accelerate the hydration of lime and magnesia. It was found, however, that this method was not sensitive enough to expansion caused by crystalline magnesia [Gonnerman 1953]. It was to overcome this difficulty that the present standard *autoclave test* (ASTM C 151-84) was introduced. (An autoclave is essentially a high-pressure steam boiler). In this test, the specimen is a bar of cement paste of normal consistency, 1 in square (6.45 cm^2) in cross section and 10 in (254 mm) long, which is cured in humid air for 24 hr. Then it is placed in an autoclave where the steam temperature is raised to 420° F (216°C) in 1 hr, which is equivalent to a steam pressure of 295 psi (2.0 MPa). The temperature and pressure are maintained for 3 hr. The high steam temperature and pressure accelerate the hydration of lime and magnesia, therefore the development of the expansion. After cooling, the length of the bar is measured. Expansion due to autoclaving must not excede 0.8% for portland cement, 0.5% for portland-pozzolan cement, and 0.2% for portland blast-furnace slag cement.

Under the British Standard specifications (B.S. 12), a cement that fails the soundness test may be aerated and retested.

Experiments have shown that, besides CaO and MgO, C$_3$A contributes the most significantly to the autoclave expansion of portland cement pastes [Blaine 1966c]. The results of the autoclave test are also influenced by the early strength development of the cement. For instance, fine-ground cement that develops strength rapidly may develop sufficient strength during the 1-day curing prior to making the autoclave test to resist expansion. On the other hand, a cement that passes the autoclave test may fail when it is blended with a pozzolan because the early strength of the blend is less than that of the plain portland cement [Price 1974]. Unsound portland cements are very rare nowadays.

2.7.5   Strength

2.7.5.1 Test Methods The *strength-developing ability* of a cement is its most sought-after property because, after all, concrete is a construction material. The strength development, that is, hardening of the cement paste is the result of the cement hydration, the subsequent development of bonds in the hydration products, and gradual reduction of the internal porosity. (Chapter 3) It starts essentially after the completion of setting (Fig. 2.11), and may be divided into two stages, the first and second stages of strength development.

The best engineering information about strength-developing ability can be obtained by making concrete of appropriate composition with the cement in question and testing the *concrete strength* under strictly controlled conditions. There are certain standards, the British B.S. 12:1958 for instance, that specify both mortar and concrete as alternatives for the strength test of cement. However, expediency usually dictates the less time-consuming *testing of mortar* for quality-control purposes in laboratories. Such mortar specimens are prescribed by ASTM both for compression- and tensile-strength determinations. Strength tests on cement pastes do not necessarily evaluate adequately the concrete-making properties of the cement under test, therefore they are rarely used.

According to ASTM C 109-84, standard compressive strength tests are made on 2-in or 50-mm mortar cubes composed of 1 part cement and 2.75 parts, by weight, of graded (fine) Ottawa sand sized between the No. 30 (600 μm) and No. 40 (450 μm) sieves, and complying with other requirements of the standard. The needed amount of mixing water is determined by trial and error as follows: A number of standard mortars are made with different amounts of mixing water; a molded specimen of each of the fresh mortars is jigged in turn 25 times on a special plate, called

standard flow table [ASTM C 230-83]; when the
increase in diameter of the mortar specimen is
between 100 and 115% of the original diameter, the
water content is considered to be proper for the
preparation of the mortar cubes. As a guide in
preparing trial batches, a water content of about 47%
of the weight of the dry cement is suggested for
air-entraining portland cements, and 49% for non-
air-entraining portland cements. The number of trial
mixes can be reduced by using the nomogram of Fig.
2.22.

After mixing and compacting the mortar in the
cube molds, the specimens are placed for *curing* in a
moist room having 73.4°F (23°C) temperature, and
stored there in the molds on the base plates from 20
to 24 hours. Then the specimens, except those for the
24-hour test, are immersed in saturated lime water
until break. The specimens are removed from the
storage just a short time before *testing*, and wiped
to a surface-dry condition. The load is applied to
specimen faces that were in contact with the true
plane surfaces of the mold. The specimen is placed in
the testing machine below the center of the upper
bearing block without any cushioning or bedding
materials. After an initial loading, the rate of load
application is adjusted so that the remainder of the
load reach the maximum value, without interruption,
in not less than 20 or no more than 80 s. The total
maximum load indicated by the testing machine is
recorded, and the individual compressive strength is
calculated as {the maximum load}/{cross-sectional
area}. The average compressive strength is *calculated*
as the arithmetic average of all *acceptable* test
specimens made from the same sample and tested at the
same period, and reported to the nearest 10 psi (70
kPa). Specimens are *not acceptable* that are
manifestly faulty, or that give strengths differing
by more than 10 % from the average value of all test
specimens made from the same sample and tested at the
same period.

As will be discussed later in this chapter, the
strength differences between various portland cements
may be quite large due to the differences in the
compositions as well as in the manufacturing

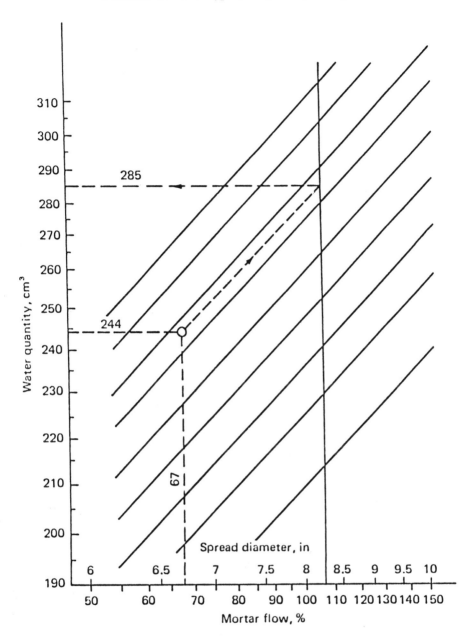

Figure 2.22: Chart for the estimation of water content of standard mortar (with 500 g of cement) needed for a specified flow [Popovics 1963, 1966, 1968a]. (Flow test mortar according to ASTM 230-83,, or DIN 1165.) *Example:* If a water content of 244 cm³ results in a flow of 67%, then the water content needed for a flow of 108% is 285 cm³.

processes. These differences include fluctuations due to the test method used as well as the inherent non-uniformity of the cement samples.

As far as the *uncertainty of the test methods* is concerned, most standardized mortar methods have a coefficient of variation of less than 3% within laboratories for compressive strength. More specific are the *precision statements* concerning the test method specified in ASTM C 109-84. They are applicable to mortars made with Type I, II, or III cement tested at 3 or 7 days. The two precision statements are applicable when a test result is the average of compressive strength tests of three cubes molded from a single batch of mortar and tested at the same age.

1. The multilaboratory coefficient of variation has been found to be 7.3%. Therefore, results of properly conducted tests of single batches by two different laboratories should not differ by more than 20.6 % of their average.

2. The single-laboratory coefficient of variation has been found to be 3.8 %. Therefore, results of two properly conducted tests of single batches of mortar made with the same materials either on the same day or within the same week should not differ from each other by more than 10.7 % of their average.

The appropriate limit is likely somewhat larger for 1-day tests and slightly smaller at ages greater than 7 days.

When information is desired on the *inherent strength uniformity* of a cement, the statistical method specified in ASTM C 917-88 can be used.

According to ASTM C 190-85, *tensile-strength* tests are made on 3-in (76.2-mm) long mortar briquetes having a minimum net cross-section area of 1 in^2 (6.45 cm^2). The mortar is composed of 1 part cement to 3 parts, by weight, standard (coarse) Ottawa sand sized between the No 20. (850 μm) and No. 30 (600 μm) sieves. The quantity of water to be used with this and similar mortars can be computed from the water requirement of the cement as well as that of the aggregate, and from the mix proportion of the mortar. The formula offered in ASTM C 190-85 is

$$w \% = \frac{2P}{3(n + 1)} + K \qquad (2.10)$$

where w% = quantity of water required for the
            mortar, percent of the *dry constituents*
      P = quantity of water needed for standard
            normal consistency of the paste of the
            cement, percent of the *weight of the*
            cement
      n = aggregate-cement ratio, by weight
      K = parameter the magnitude of which depends
            on the characteristics of the sand.

For C 190 Ottawa-sand mortars, n = 3 and K = 6.5.
     The prescribed curing of the briquettes is
similar to the curing of standard 2-in (50mm) cubes.
     Another mortar method for checking the strength-
developing ability of a cement utilizes *prisms* as
specimens. First the *flexural strength* of the
specimen is determined, then the compressive strength
by using as test specimens the portions of the prism
made for and broken in flexure. This method has been
accepted tentatively as an alternative by the ASTM
(C 348-80 and C 349-82), prescribing 1.575 by 1.575
by 6.30 in (40 by 40 by 160 mm) prisms, and the
plastic mortar specified for the 2-in (50mm) cube
standard compression specimen. The use of similar
prism methods in Europe is quite widespread. The ISO
R 679 also specifies this prism, but the grading of
the sand ranges from 0.08 to 2.0 mm (0.003 to 0.08
in), the *water-cement ratio*, that is, the ratio of
the weight of water to the weight of cement in the
mixture, is 0.5 by weight, and prisms are compacted
by jolting.
     Each of these strength methods has certain
advantages and weaknesses. From the standpoint of
testing, the determination of tensile strength as
well as that of the flexural strength are the
simplest and least expensive. They require far
simpler equipment than the compression test.
Therefore, the tensile and flexural tests are
suitable for field laboratories. Also, the use of

prism specimens obviates the separate preparation of two different sets of specimens, namely, one for tension and the other for compression. In addition, it has been reported that the test methods using mortars of more plastic consistency are likely to have less testing variance than the less plastic or stiff mortars [Popovics 1955a]. ("Testing variance" is the portion of the total variance of the strength results that is caused by random testing errors and uncertainties.)

A more fundamental question is which one of the mortar methods indicates best the concrete-making quality of the cement under test. Most concrete technologists would agree that the results of the briquette test are the least informative. Compressive-strength data obtained from standard mortar show better correlation with concrete strengths than tensile data, particularly when when the water-cement ratio is the same in the mortar and concrete [Popovics 1953]. When the water-cement ratios differ in the two kinds of mixture, the concrete strength-mortar strength ratio depends on the testing age [Popovics 1967c]. For compressive strength:

$$f_{cc,t} = 0.673 \ f_{cm,t} + 1080 \ \log(0.8t) \qquad (2.11)$$

and for flexural strength:

$$f_{fc,t} = 0.58 \ f_{fm,t} + 110 \ \log(10t) \qquad (2.12)$$

where

$f_{,cc,t}$ and $f_{fc,t}$ = compressive strength and
flexural strength of concrete,
respectively, at the age of t days, psi
$f_{cm,t}$ and $f_{fm,t}$ = compressive strength and
flexural strength of Ottawa-sand mortar,
respectively, at the age of, t days, psi
t = age of the specimen at testing, days.

Eq. 2.11 is illustrated in Figure 2.23.
Results of comparative strength tests performed according to the cement standards of various

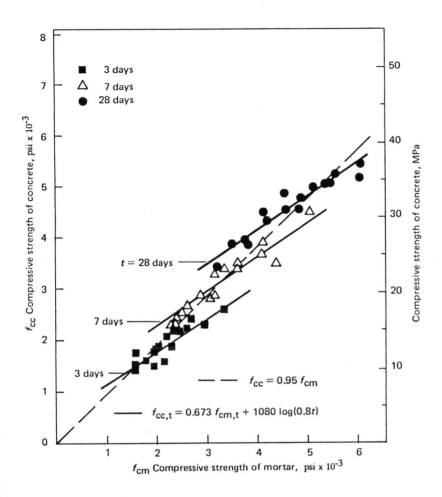

Figure 2.23: Observed and calculated values for the relationship between the compressive strengths of constant-slump concrete and corresponding ASTM C 109 mortar at various ages [Popovics 1967c]. The dimension of the values of f in the formulas is psi. The coefficients of the formulas are a function of the water-cement ratios.

countries are available in the literature [Meyer 1966, Walz 1963, Foster 1968].

Recent experiments also seem to show the advantage of the ISO mortar method over the ASTM C 109-87 method [Wischers 1972, Dutron 1974, Markestad 1965], presumably because the water-cement ratio as well as the porosity of the ISO mortar are similar to those in concretes generally used.

2.7.5.2   Effects   on   Strength   Development Typical strength developments of the *five standard types* of portland cement are shown in Fig. 2.24. Higher C_3S and C_3A contents, as well as higher fineness improve the strength development at early ages. Among these, the fineness is the most effective as far as the compressive strength is concerned, so that often a "high-early strength" cement is merely one which is ground finer than a Type I cement from the same clinker. [Popovics 1981a] Figure 2.25 illustrates the effect of fineness and supports the observation [Alexander 1972] that the relationship between the specific surface and the developed strength of a cement is practically linear within wide limits. Note that (a) high early strength may be detrimental to strength increases at later ages;   (b) in contradistinction to the compressive strength, the *flexural* and *tensile strengths* are hardly affected by the cement fineness. [Graf 1960]. Further information about the effect of the fineness on strength can be found in the literature. [Gruenwald 1939, Higginson 1970].   Other factors that influence the strength development are *water-cement ratio* [Locher 1977]; *curing temperature* [Popovics 1987a] (Figs. 2.26 and 2.27);   other *curing conditions;* and certain *admixtures* (Fig. 2.28).

It may be worhtwhile to mention that one unit of weight, such as one pound or kg, of Type I portland cement develops the maximum compressive strength under normal conditions when the water-cement ratio is approximately 0.5 by weight. In other words, 0.5 by weight is the most economical water-cement ratio for such cements. [Popovics 1990b]

Strengths of mortars *cured in autoclave* do not show good correlation with comparable strengths obtained under normal curing conditions (Fig. 2.29).

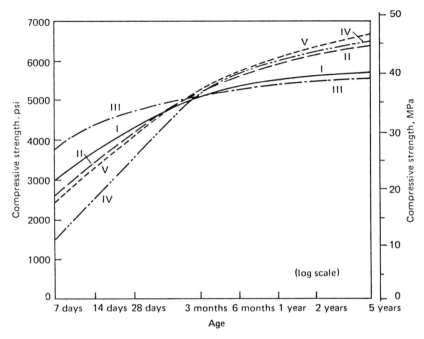

Figure 2.24: Effect of type of cement on compressive strength of concrete. Aggregate 0 to 1.5 in (37.5 mm); 6 sacks (564 lb) cement per yd³ (335 kg/m³); 6 by 12 in cylinders (15 by 30 cm); standard curing [U.S. Bureau of Reclamation 1966].

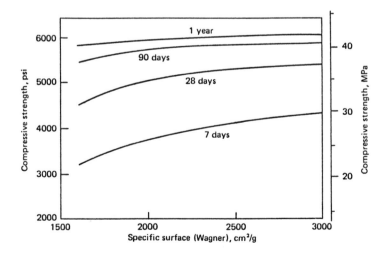

Figure 2.25: Illustration of the effect of cement fineness on the compressive strength of concrete at various ages, W/C = 0.56 by weight [Price 1951].

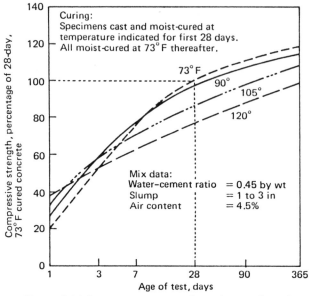

Figure 2.26: Effect of high temperatures of casting and moist curing on compressive strength of concrete at ages up to 1 yr [Portland Cement Association 1965].

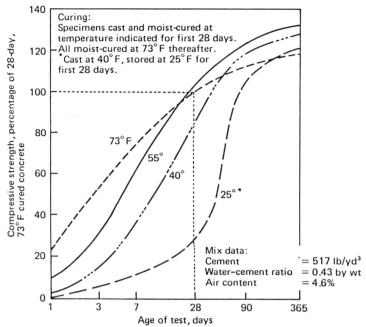

Figure 2.27: Effect of low temperatures of casting and moist curing on compressive strength of concrete at ages up to 1 yr [Portland Cement Association 1965].

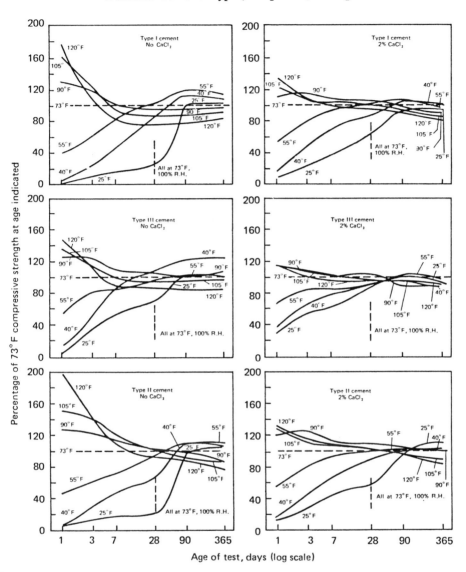

Figure 2.28: Effect of temperature on compressive strength of concretes made with different types of cement with and without an accelerator. Air content of all concretes, 4.5 ± 0.5% (neutralized Vinsol resin solution added at mixer); cement content of all concretes, 5.5 sacks per yd^3. Within the individual boxes, the net water-cement ratios of the concretes are approximately equal. From [Klieger 1958].

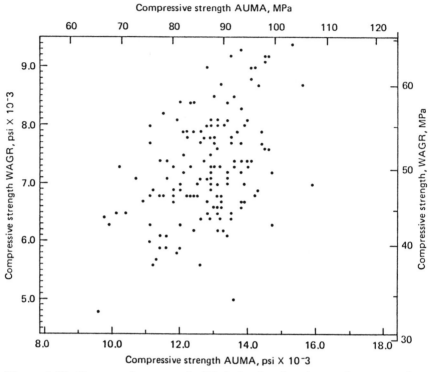

Figure 2.29: Compressive strength (28-day) of moist air-cured mortar cubes (WAGR) vs compressive strength of cubes of identical mortars, moist-cured for 24 hr, then autoclaved at 10 atm for 4 hr, then oven-dried (AUMA) [Blaine 1968].

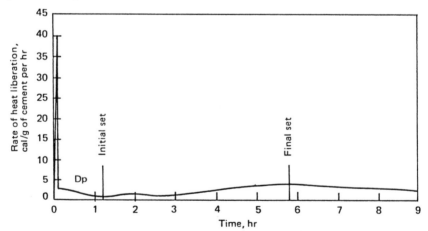

Figure 2.30: Rate of early heat liberation of portland cement paste as a function of time. Dp = dormant period [Powers 1968].

[Blaine 1968] This is also true, although to a lesser degree, for most low-pressure steam curings. Another practical consequence of this is that a cement that provides high strengths under normal curing conditions will not necessarily provide relatively high strengths when cured in low-pressure steam or in autoclave. A rule of thumb is that portland cements containing about 10% C_3A and about 45% C_3S are usually the most suitable for steam curing [Budnikov 1966]. A higher than average gypsum content is also advantageous for early strengths of steam-cured concretes. A better procedure is to test such cements in specimens that are submitted to steam or autoclave curing, respectively [Ujhelyi 1973]. Similarly, trial mixes are recommended whenever a cement is planned for some special purpose. Otherwise, typical examples for the utilization of the five standard cement types are given in Table 2.3.

Minimum acceptable values by ASTM for the standard compressive and tensile strengths of portland cements and air-entrainig portland cements are shown in Table 2.5.

2.7.6 Heat of Hydration

Most of the reactions occurring during the hydration of cement are exothermic; that is, they develop heat. This heat is called *heat of hydration*. Heat of hydration is usually unimportant from a practical standpoint. In certain cases, however, it is advantageous, for instance during winter concreting. More important, in other cases it can be very harmful, such as in mass concretes. In these cases it is desirable to know the heat-producing capacity of the available cements in order to choose the most suitable cement for the given purpose.

Portland cements develop a certain amount of heat during the short period (Fig. 2.30) beginning when cement and water are first brought into contact. Soon, however, the heat development subsides to a low level and becomes intensive again after a dormant period that usually lasts 40 to 140 min.

The most common method for determining the heat

of hydration is by measuring the heat of solution of the dry cement powder and the heat of solution of a separate portion of the cement that has been partially hydrated for a known period of time, the difference between these values being the heat of hydration for the respective hydration period. ASTM C 186-86 specifies hydrofluoric acid (HF) and nitric acid (HNO_3) as solvents. Results of approximately 1-cal/g accuracy can be obtained with this method on portland cements. There is higher uncertainty in the application of the heat-of-solution methods to cements containing pozzolan, or to supersulfated cements, since a portion of the pozzolan (or the calcium sulfate and slag, respectively) usually remains insoluble in the acid and it is not quite the same for the cement powder and the hydrated cement [Lea 1956].

A simpler but less accurate method for the determination of heat evolution is based on placing a sample, usually concrete, in an adiabatic calorimeter, that is, a calorimeter from which no heat loss can occur [Berman 1963]. This situation provides conditions similar to the interior of a large concrete mass. The heat generated by the hydrating sample raises the temperature in the calorimeter, which can be recorded as a temperature - versus - time curve. The temperature rise can be converted into calories per gram of cement. Such a curve is shown as curve A in Fig. 2.31. The method is applicable up to 7 days or even 28 days, although the effect of cumulative errors becomes more and more appreciable with increasing age.

The *rate of heat development* can be characterized by the magnitude of heat development during early setting, but this is just a rough estimate. A much more accurate method is one that utilizes an isothermic calorimeter [Kuhl 1961a]. This method yields a value for heat development per unit time. Such a rate-versus-time curve is shown as curve I in Fig. 2.31. Curve I is the differential curve of curve A, from which follows that curve A has a point of inflection at $t = S$ where curve I has a maximum; and the development of heat of hydration is an accelerating process up to period S, after which it

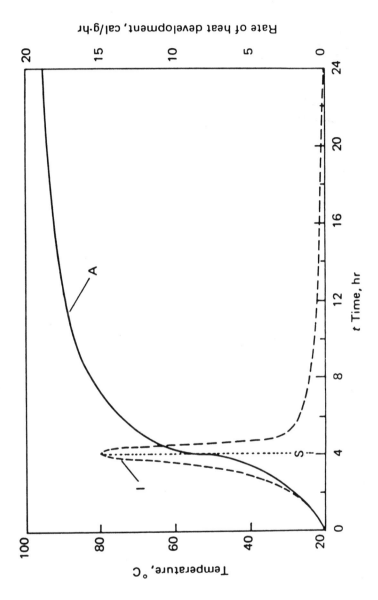

Figure 2.31: Two kinds of presentation of the development of heat of hydration. Curve A is for the heat of hydration vs time, and curve I is for the rate-vs-time relations [Kleinlogel 1933].

is decelerating. Comparative measurements indicate that the S time is practically identical with the time of final setting of the paste determined by the Vicat method [Kleinlogel 1933].

For most practical purposes the rate of heat development is more important than the total heat of hydration because the same amount of heat produced over a longer period is less effective. The application of a low-heat cement or a pozzolan, both reduce and delay the development of heat of hydration.

Heat development shows certain correlations with other consequences of hydration. This is demonstrated for nonevaporable water content and water sorption in Fig. 2.32 and for compressive strength in Fig. 2.33 [Verbeck 1950]. This means that factors that increase the early strength (compound composition, fineness, temperature (Fig. 2.34), water-cement ratio, etc.) tend to increase the rate of development of heat of hydration. This is demonstrated for the compound composition in Fig.2.35 [Blaine 1966b]; or for later ages with the following equations:

$$H_{70} = 0.714 \ C_3S + 2.40 \ C_3A + 56.0 \qquad (2.13)$$

and

$$H_{MC} = 0.714 \ C_3S + 2.00 \ C_3A + 57.0 \qquad (2.14)$$

where H_{70} and H_{MC} = heat of hydration of cement pastes cured at 70 °F (21 °C) and under simulated mass storage, respectively, for 6.5 years, cal/g of cement
C_3S and C_3A = potential quantity of C_3S and C_3A, respectively, percent.

Eqs. 2.13 and 2.14 are illustrated in Figs. 2.36 and 2.37.

Note that identical results are not to be expected from alternative methods in which the testing conditions differ. Incidentally, the ratio of compressive strength to heat of hydration is an index to the adaptability of the cement to mass concrete

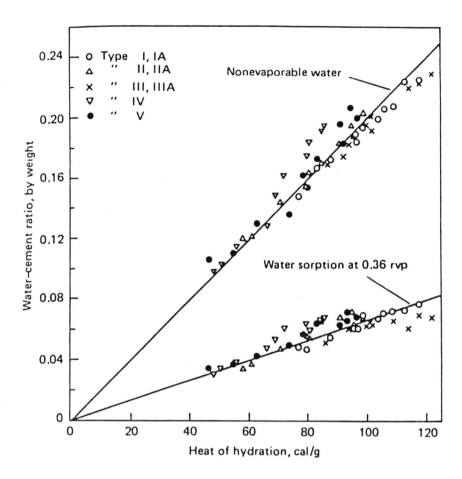

Figure 2.32: Relationship between heat of hydration and either nonevaporable water or water sorption at 0.36 relative vapor pressure. Average results for each of five types of portland cement, pastes cured at 70°F and mass-cured, ages from 3 days to 6.5 years. From Verbeck and Foster, *Proceedings, ASTM,* vol. 50, pp. 1235-1257 [Verbeck 1950]. Reprinted by permission of American Society for Testing and Materials, Copyright.

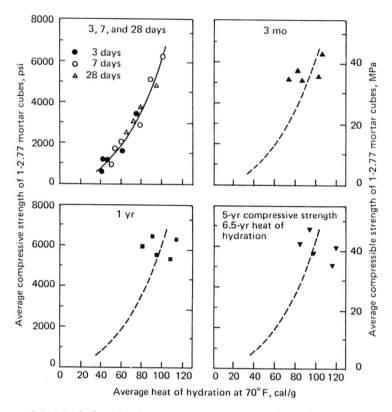

Figure 2.33: Relationship between average strength and average heat of hydration of the five types of non-air-entraining cement at various ages. From Verbeck and Foster, *Proceedings, ASTM,* vol. 50, pp. 1235-1257, 1950 [Verbeck 1950]. Reprinted by permission of American Society for Testing and Materials, Copyright.

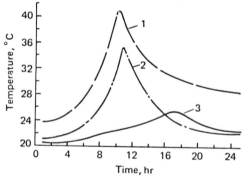

Figure 2.34: Heat evolution curves for a type IV portland cement after Mchedlov-Petrosyan et al. Curve 1, paste of a fresh portland cement mixed at 25°C. Curve 2, the same as curve 1, but mixed at 18°C. Curve 3, after storing (aeration) the cement for 3 months. From [Eitel 1966].

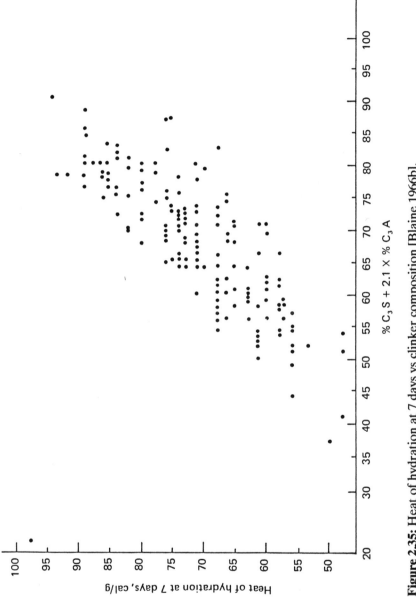

Figure 2.35: Heat of hydration at 7 days vs clinker composition [Blaine 1966b].

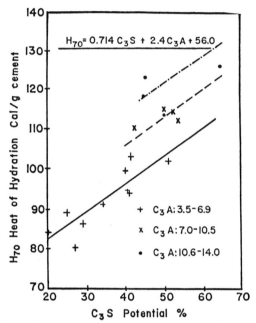

Figure 2.36: Heat of hydration as a function of clinker composition-I. The paste specimens were cured at 70°F (21°C) for 6.5 years. The experimental data were taken from [Verbeck 1950].

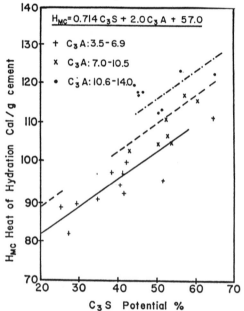

Figure 2.37: Heat of hydration as a function of clinker composition-II. The paste specimens were cured under simulated mass storage for 6.5 years. The experimental data were taken from [Verbeck 1950].

construction: The higher the ratio, the more suitable the cement.

The heat of hydration is influenced by fineness at early ages more than later on. For an increase of 100 m^2/kg (Wagner method), the heat evolution increases by some 4-5 cal/g at 1 day, 1-3 cal at 7 and 28 days, and less than 1 cal at 1 yr. Aeration or prehydration of the cement decreases the heat evolution (Fig. 2.34) by some 6 cal/g at 7 and 28 days, and 4 cal at 1 yr for each 1% increase in ignition loss. These, however, are average figures, and for individual cements the values may be halved or doubled [Lea 1956, U.S.Dept.of Int. 1949].

The rates and total amounts of heat liberated are also increased by pressure at early ages, indicating an intensified hydration (Fig. 2.38). This can be important when the cement is used in deep oil wells.

Comparison of ranges of heats of hydration of the different ASTM types of portland cement is presented in Fig. 2.39.

2.7.7 Volume Instability (Shrinkage)

An important characteristic of cement paste is that an unloaded specimen undergoes several kinds of volume change from the presetting period through a very mature stage. This can appear as an increase in volume, called *expansion* , or a reduction in volume, called *shrinkage*. Volume instability, especially shrinkage, is perhaps the major drawback of cement paste, and consequently of mortars and concretes. The reason for this is that these volume changes produce undesirable stresses in the mass that, in many cases, can be so high as to cause intensive cracking. Some volume changes are discussed below; others, such as volume changes caused by loading, temperature, chemical and physical attacks, and so forth, are discussed in other chapters.

1. When the rate of evaporation exceeds the rate of bleeding (Section 2.7.8.5), the surface of the compacted fresh paste, mortar, or concrete may be driven downward. Evaporation beyond this stage produces hydrostatic stresses in the mass which

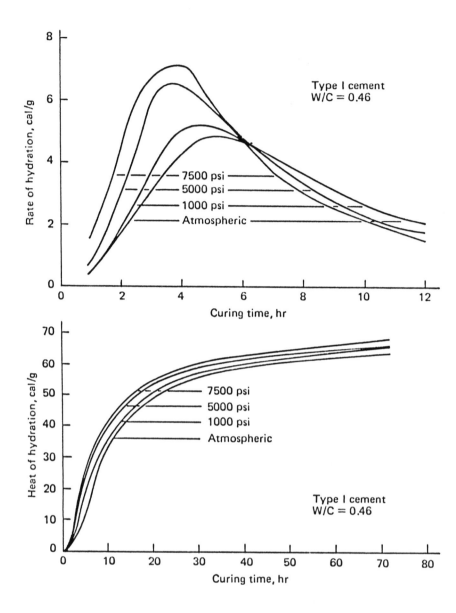

Figure 2.38: Effect of pressure on heat of hydration at 100°F (37.8°C) [Ludwig 1956].

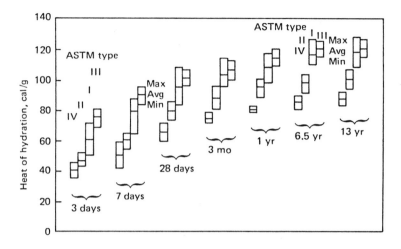

Figure 2.39: Ranges in heats of hydration of the different ASTM types of cement. (0.40 water-cement ratio; cured at 21°C) [Copeland 1960].

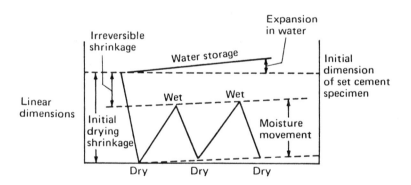

Figure 2.40: Diagrammatic representation of changes in dimensions of cement pastes, mortars, and concretes with moisture conditions [Lea 1956].

cause lateral shrinkage, and in certain cases the development of a pattern of cracks. Factors that increase the rate of evaporation, such as wind velocity, also increase the magnitude of volume reduction. Since this phenomenon occurs while the mixture is plastic, it is called *plastic shrinkage*, and *plastic shrinkage cracking* [Lerch 1957, L'Hermite 1962]. The magnitude of this volume usually is of the order of 1% of the absolute volume of dry cement in the mixture [Swayze 1942].

2. The period of bleeding and plastic shrinkage is followed by an expansion of the paste. The rate of expansion is high at first and then gradually diminishes during a period of 20 hr or so. Powers thinks that this expansion may be the result of disruption of the early gel coating on the cement grains during setting. [Powers 1961]. The magnitude of paste expansion that may occur during the first day is at least 0.09%, which is considerably greater than the magnitude that occurs during a prolonged period of wet curing *after* the first day [Powers 1968].

3. A third, and perhaps the most important kind of volume change takes place after the first day, that is, during the hardening of the cement paste. This is called *drying shrinkage* because it is the result of loss or gain of water in the hardened paste. As the water content increases, the hardened paste expands, and drying produces shrinkage. A portion of this volume, or length, change is irreversible, as illustrated in Fig. 2.40. The magnitude of the drying shrinkage can be considerably greater then the elastic deformation a concrete suffers when subjected to normal constructional stresses of around 1000 psi (7 MPa), thus they can produce intensive cracking in the unprotected hardening paste.

Although this type of shrinkage is related to drying, the loss of free water, which takes place first (see Fig. 3.1), causes little or no shrinkage. Only as the drying continues, and adsorbed water is removed, does volume reduction begin. The magnitude of the shrinkage in an unrestrained

paste is approximately equal to the loss of a
water layer one molecule thick from the surface
of all gel particles [Neville 1981].
Therefore, the larger the specific surface of the
hydration products, the greater the drying
shrinkage. For instance, a cement paste cured in
autoclave has a low internal specific surface
and shrinks much less than a similar paste cured
normally. Thus, the drying shrinkage is the
result of changes in the physical structure of
the gel of hydration products rather than in its
chemical and mineralogical character.
The total drying shrinkage of cement pastes is
rather variable, as shown in Fig. 2.41. Cements
with higher fineness shrink more probably because
of their higher rate of hydration. Specimens
stored in water may expand up to 0.1% in 3 months
The rate at which movements occur also depends on
the minimum dimension of the specimen. Influences of several factors on the drying shrinkage
of portland cement pastes are illustrated in
Figs. 2.42 and 2.43.
4.*Carbonation shrinkage* occurs again in hardened
portland cement pastes. This shrinkage is caused
by the reactions between the constituents of
hydrated portland cement and CO_2. The reactions
take place even at small concentrations of CO_2,
such as are present in the air. According to
Powers, carbonation shrinkage is probably caused
by the increased compressibility of the hardened
paste that results from dissolving of $Ca(OH)_2$
crystals in places where they are under
compression by the drying shrinkage, and
depositing of $CaCO_3$ in spaces free from stresses
[Powers 1962]. Humidity during exposure to CO_2 is
a major factor influencing carbonation shrinkage.
As Fig. 2.44 shows, the maximum carbonation
shrinkage of a mortar with Type I cement occurs
at about 50% relative humidity, whereas it is
negligible at 100% or relative humidities below
25% [Verbeck 1958]. Figure 2.44 also demonstrates
that the sequence of drying and carbonation
affects the magnitude of total shrinkage. Concen-

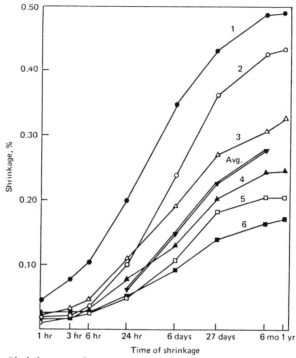

Figure 2.41: Shrinkages of pastes of normal consistency of six portland cements and average shrinkage as a function of age. The pastes were cured for 24 hr in moist air and then exposed to laboratory air [Blaine 1969].

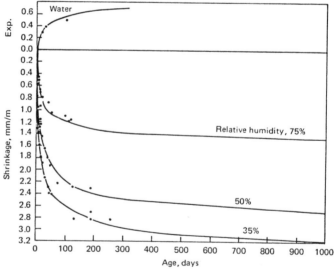

Figure 2.42: Expansion of cement paste in water and shrinkage in air at various relative humidities. The storing in air started at the age of 1 day [Wesche 1974].

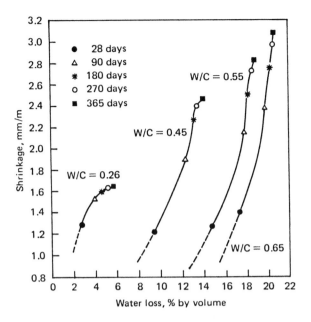

Figure 2.43: Shrinkage of portland cement paste as a function of the water-cement ratio and water loss [Wesche 1974].

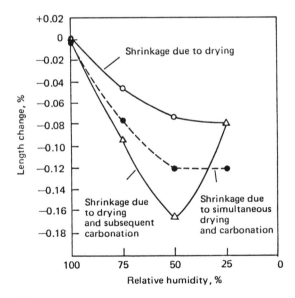

Figure 2.44: Effect of sequence of drying and carbonation on shrinkage [Verbeck 1958].

tration of CO_2 as well as size of the specimen also have important effects. Carbonation shrinkage of specimens cured in autoclave is again very small. Carbonation of cement paste also results in increased strength and reduced permeability.

5. Although the tendency of a hardening cement paste, mortar, or concrete to crack is influenced greatly by the magnitude of its shrinkage, nevertheless there are other influencing factors, two of which are the tensile strength and the modulus of elasticity of the paste. Due to the complexity of the problem, direct measurement of the cracking tendency may be justified, for instance by the method specified in France (AFNOR-P 15-402 and 434). Here a ring of 18.5 mm (0.73 in) thickness and 40 mm(1.57 in) height cross-section is formed from cement paste around a steel core, and the time of the development of the first crack is recorded as the characteristic of the cracking tendency of that paste. It was shown by this method that, for instance, the use of calcium chloride as well as higher cement fineness decrease the cracking time.[Balazs 1979]

2.7.8 Other Properties

2.7.8.1 **Specific gravity** The *specific gravity* is usually determined from the displacement principle by using kerosene in a Le Chaterlier flask as described in ASTM C 188-84 (Fig. 2.45). It should be noted, however, that the specific gravity of a cement determined in water, by using either the Ford method [Ford 1958] or the Steinour method [Steinour 1945], is greater by about 0.06 than the specific gravity determined in kerosene. This difference is the consequence of the reactions taking place between cement and water. Therefore, the calculation of the solid volume of cement in a fresh paste or concrete can be based on a specific gravity of 3.21 when the actual ASTM specific gravity of the cement is 3.15 [Powers 1968]. Although the specific gravity is useful in proportioning and

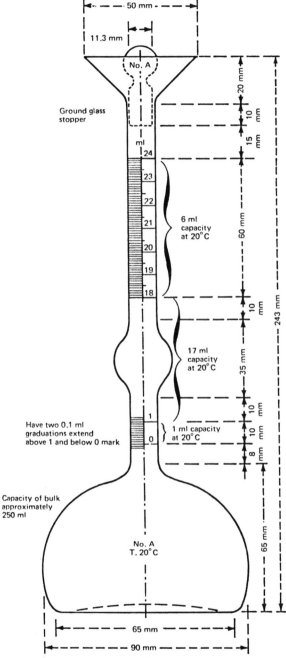

Figure 2.45: Le Chatelier flask for specific gravity test. From ASTM C 188-84. Reprinted by permission of the American Society for Testing and Materials, Copyright.

controlling concrete mixtures, most recent cement specifications no **longer** contain required values for it. The specific gravity of portland cements ranges from 3.10 to 3.23. When the value for a given cement is not known, it is usual to assume 3.15. The specific gravity of blended cements, particularly that of portland-pozzolan cements, may be less.

   2.7.8.2 Alkali reactivity The suscep-tibility of cement-aggregate combinations to expansive reactions involving the alkalies, that is the potential *alkali reactivity,* can be determined by several methods, including measuring the linear expansion of mortar bars under conditions prescribed in ASTM C 227-87. This phenomenon is discussed later in Section 9.3.

   2.7.8.3 Sulfate resistance The potential *sulfate resistance* of a portland cement can be determined by measuring the expansion of mortar bars made from a mixture of portland cement and gypsum in such proportions that the mixture has a SO_3 content of 7.0% by weight (ASTM C 452-85). There is considerable experimental evidence correlating sulfate resistance with the C_3A content of the cement, as shown, for instance, in Fig. 2.46 [Blaine 1966a, Swenson 1968]. A portion of the large spread of the data may be attributed to the observation that not only the quantity of C_3A but also its prevailing crystal form have influence on the sulfate resistance. C_3A in cubic form is less vulnerable to sulfate attack than in orthorombic form. [Mehta 1980a, Mortureux 1980]

   Effects of other reagents on mortars, that is, *chemical resistance* of mortars, can be investigated by determining changes in the weight, appearance, or compressive strength of the mortar specimens after exposure of the specimens to the reagent (ASTM C 267-65, discontinued). Sonic methods can also be used for this purpose.

   2.7.8.4 Air entrainment The *air-entrainment* test for cements described in ASTM C 185-85 is a gravimetric method. The

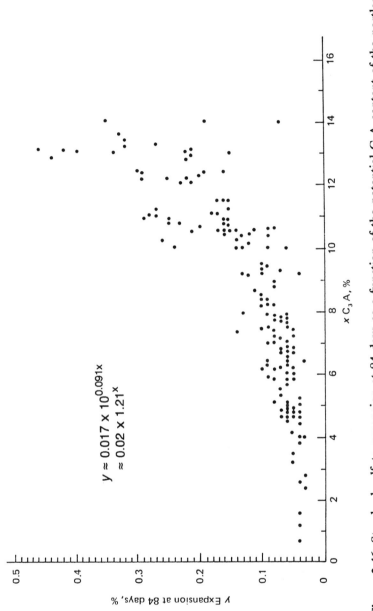

Figure 2.46: Standard sulfate expansion at 84 days as a function of the potential C₃A content of the portland cements [Blaine 1966a]. As an average, any increase of 3.5 percentage points in the C₃A content, such as from 4 to 7.5%, would approximately double the sulfate expansion at 84 days.

purpose of this test in the case of air-entraining cement is to determine if the air-entraining admixture is present in such quantity that a reasonable air content will be obtained in a concrete under the usual conditions. In the case of non-air-entraining cement, the purpose is to ensure that the cement will not entrain undesirable air and that the concrete will contain only that small amount of entrapped air that may result from incomplete consolidation.

In this method a mortar is made with the cement and coarse standard Ottawa sand in the proportion of 1:4 by weight. A water content is selected that gives a flow from 80 to 90% on 10 drops of the standard flow table. After mixing, the mortar is placed in a metal cup, compacted lightly, then the air content is calculated from the measured actual density and from the theoretical air-free density of the mixture.

ASTM C 150-85 requires all five types of non-air-entraining cement to have air contents less than 12% in the C 185 test; they usually range from 5 to 10%. It also calls for standard mortar air content of 19 ± 3% for air-entraining cements. Presumably a concrete made with an air-entraining cement meeting this requirement will have enough air under usual conditions to be reasonably resistant to freezing and thawing action but not so much that its strength will be seriously reduced [Dolch 1968]. Persistent difficulties with this air-entrainment test have been experienced by some user

2.7.8.5 **Bleeding** When a cement paste, mortar or concrete, after consolidation but still plastic, is allowed to stand unagitated, water usually appears at the surface. This autogenous flow of mixing water within, or its emergence from freshly placed mixtures is called *bleeding* or water gain [ACI 1985d]. Bleeding takes place during the period preceding setting. If the paste is protected against evaporation or other water loss, the level of the gained water is the same as the original surface of the paste. This indicates that bleeding is cause either by settlement of the solid particles within the plastic mixture in which case it is called *normal bleeding*, or by drainage of mixing water, in which case it is called *channeled bleeding* [Powers 1939, 1968]. Excessive bleeding is undesirable. Bleeding is affected by the quality and quantity of the materials in the mixture, such as the water-cement ratio, fineness of the cement (Fig. 2.5) and air entrainment, as well as by outside factors, such as high-speed mixing [Popovics 1973a]. Long setting period can also cause excessive bleeding.

A standard method for testing the bleeding of cement pastes and mortars is specified in ASTM C 243-85. This is based on the collection and measurement of the water gain through liquid displacement (Fig. 2.47). Fig. 2.48 presents a typical result of a bleeding test on a cement paste.

2.7.8.6 Efflorescence *Efflorescence* results from the deposition on the surface of concrete, masonry, or brickwork of salts dissolved out of the cement. For lack of a standard test for the tendency of cement to cause this trouble, the following test is recommended [Lea 1956, Walker 1958]:

A 2-in (50-mm) cube of mortar of 1:5 mix is made with coarse sand and dried at the age of 24 hr. It is then stood in a small depth of water and a piece of wet blotting paper is place on top of it. A 2-in (50-mm) cube of limestone (porous type) or brick (free of efflorescing salts) is then placed on top of the blotting paper. Any salts liable to cause efflorescene will be dissolved out of the cement and deposited on the surface of the limestone or brick. Inspection will show the intensity of this action, on which the cement can be judged.

2.8   UNIFORMITY   OF   CEMENTS

Lack of uniformity in cement is an important cause of nonuniform concrete strengths. Average strengths of portland cements of the same type but from different sources can vary substantially. In one investigation, for instance, the range of the compressive strength

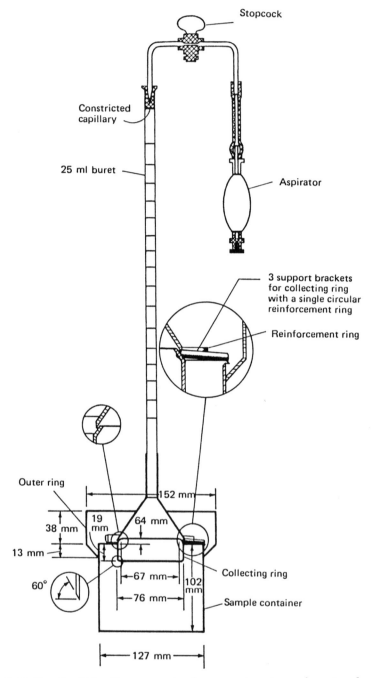

Figure 2.47: Standard bleeding apparatus for cement pastes and mortars based on the principle of liquid displacement. From ASTM C 243-85.

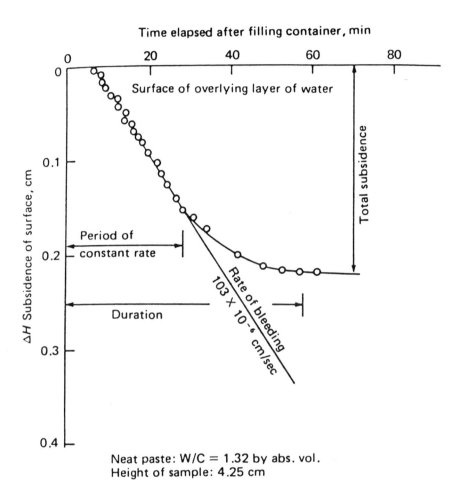

Figure 2.48: Typical bleeding curve after Powers [Powers 1939]. (The points start to the right of the origin on the time scale because of the time elapsed before the first reading could be obtained.)

for 10 Type I cements at 28 days was 36% of the
median strength [Walker 1961]. Even within one brand
there may be a considerable strength variation
[Campbell 1968, Walker 1958]. This means that the
variations in strength, and probably in the overall
quality of cement, from shipment to shipment from one
plant may require as much as 10% additional cement to
maintain a given minimum concrete strength level as
compared to a uniform cement of the same average
strength from another plant.

A portion of the total strength variation
originates from testing errors, as reflected by the
pertinent precision statements discussed earlier.
(Section 2.7.5.1) When information is desired on the
inherent strength uniformity of a cement, the
statistical method specified in ASTM C 917-88 can be
used. Note that according to Figure 2.49 the
standard mortar strengths follow, within one brand, a
truncated normal distribution with a good
approximation, provided that there are no regular
errors in the manufacturing process. [Popovics 1953]

When a clinker is produced that is somewhat off
the desired composition, it can be blended with a
good clinker in a proportion that will not seriously
affect the target strength set by the mill. Cement
clinker that has been wet by rain can be ground finer
to maintain the strength. Therefore, it seems
reasonable to expect that the strengths of cements
from a given plant should have a coefficient of
variation from shipment to shipment of less than 6%
[Price 1974, Walker 1958].

2.9 SAMPLING

Methods covering procedures for sampling cement
are described in ASTM C 183-83a. This specification
recognizes several kinds of cement sample as follows:

1. A cement sample secured in one operation is
   termed a *grab sample*.
2. A sample obtained by means of an automatic
   sampling device that continuously samples a
   cement stream is termed *continuous sample*.

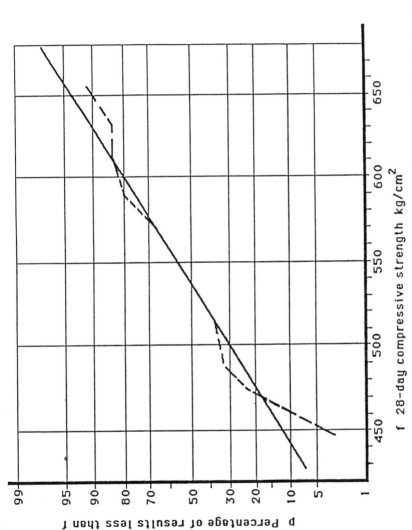

Figure 2.49: Comparison of the distribution of strengths of standard mortar cubes to normal distribution. Continuous line: theoretical normal distribution. Broken line: actual distribution. Number of tests: 153. 1 kg/cm^2 = 142 psi = 0.098 MPa. [Popovics 1953].

3. Grab samples taken at prescribed intervals, or individual continuous samples may be combined to form a *composite sample* that is that period of time.
4. Combined samples in which physical and chemical tests are to be made are termed *test samples*.

Individual samples to be composited should weigh at least 2.5 kg (5 lb). Test samples should weigh at least 5 kg (10 lb).

The cement may be sampled by any of the applicable methods described as follows [ACI 1985b]:

1. From the conveyor delivering to bulk storage from each 200 barrels or less from passing over the conveyor.
2. From bulk storage at point of discharge from each 500 barrels in the bin or silo.
3. From bulk storage and bulk shipment by means of a slotted tube sampler or sampling pipe from well distributed points and various depths of the cement.
4. From packaged cement by means of a tube sampler from a bag in each 25 barrels.
5. From bulk shipment of a car or truck from a minimum of three well-distributed points.

2.10 STORAGE OF CEMENTS

Cement may be stored for an indefinite period of time as long as it is protected from moisture, including the moisture in air. To reduce the possibility of damage to cement from unfavorable storage conditions, all cement should be stored in weathertight, properly ventilated structures [ACI 1985a, 1985b]. Damage is indicated by lumpiness, increase in loss of ignition, reduction in specific surface, or falling off in strengths, particularly in early strengths. The effect of the length of storage on compressive strength is illustrated in Fig. 2.50, and that on setting in Fig. 2.51 [Kleinlogel 1941]. At the time of use, cement should contain no lumps that cannot be broken by light pressure between the

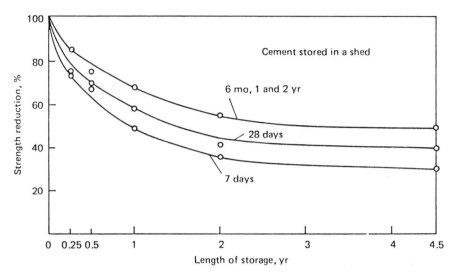

Figure 2.50: Effect of storage length of a cement on the compressive strength of a 1:5 concrete at various ages [Kleinlogel 1941].

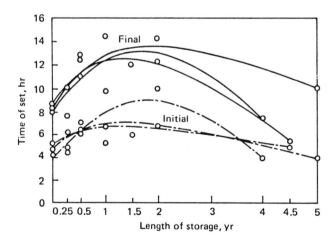

Figure 2.51: Effects of storage length on the setting of three portland cements [Kleinlogel 1941].

fingers. The removal of hard lumps by screening does not always restore the quality of the cement to its original level. If there is any doubt, the reclaimed cement should be tested.

Care should be exercised in transferring cement from the carrier to the storage space to protect it from becoming wet or contaminated with foreign material [Waddell 1962].

Storage facilities for *bulk cement* should include separate compartment for each type and brand of cement used. The interior of a cement silo should be smooth, with a minimum bottom slope of 50° from the horizontal for a circular silo and 55-60° for a rectangular silo. Silos not of circular construction should be equipped with nonclogging air-diffuser flow pads through which small quantities of low-pressure air 3 - 5 psi (approximately 0.2 - 0.4 kg/cm^2) may be introduced intermittently to loosen cement that has setteled tightly in the silos. Storage silos should be drawn down frequently, preferably once per month, to prevent cement caking. [Waddell 1974].

Each bin compartment from which cement is batched should include a separate gate, screw conveyor, air slide, rotary feeder, or other conveyance that effectively combines characteristics of constant flow with precise cutoff to obtain accurate automatic batching and weighing of cement.

Care must be used to prevent cement being transferred to the wrong silo, and effective methods must be used for eliminating dust nuisance during loading and transferring.

Sacked cement should be stacked on pallets or similar platforms to permit proper circulation of air. For the storage period of less than 60 days, it is recommended that the cement be stacked no higher than 14 sacks, and for long periods, no higher than 7 sacks. It is also recommended that the oldest cement be used first.

3

Hydration of Portland Cement

SUMMARY

Both the setting and the hardening of a portland cement paste are the result of a series of simultaneous and consecutive reactions between the water and the constituents of the cement. These reactions all together are covered by the term *hydration of portland cement*.

The two most important chemical reactions during the early part of the hydration are (1) the reaction between the C_3A as well as the gypsum of the cement and water; and (2) the hydration of alite in the cement and water. When the portland cement is insufficiently retarded, the reactions are too fast, causing an undesirable "quick set". The predominant reactions at later ages are the hydrations of the calcium silicates, which continue for many months at a diminishing rate.

In the course of hydration every portland cement grain breaks up into millions of particles, forming mostly a poorly crystallized, porous solid, so-called CSH (calcium silicate hydrate) gel. The mechanism of gel formation is as follows: On contact with the still unhydrated part of a cement grain, water dissolves a portion of it; this solution diffuses out from the grain surface toward larger spaces through the very small pores of the solid shell of previously created hydration products around the cement grains; then the new hydration products precipitate from the solution. The very fine texture and the resulting high specific surface are the most significant characteristics of the CSH gel. One cubic centimeter solid volume of cement can develop at least 2 cm^3 hydration products. Therefore, as a consequence of hydration, the volume of solids within the boundaries of paste specimen increases, producing interlocking laths and a reduction in the overall porosity of the paste. These are the primary sources of the stiffening and strength development of the cement paste.

The complex reactions between water and cement, thus the chemical composition as well as the structure of the hardened paste, vary to a certain extent with time, temperature, water-cement ratio, and several other factors. Regardless of the conditions, however, the hardened paste always contains a considerable amount of pores of different sizes, namely, gel pores, capillary pores, and air voids.

3.1   INTRODUCTION

When portland cement is mixed with a limited amount of water, the cement particles get dispersed in the water. The result is *cement paste,* which is a more or less plastic, mudlike material. The water-filled spaces between the cement particles in the fresh paste may be regarded as an interconnected capillary system. The amount of water used affects decisively not only the plasticity or consistency but also practically every important property of the fresh or hardened paste. The primary reason for this is that the less water there is in a unit of cement paste, that is, the smaller is the ratio of the amount of water to the amount of cement, the so-called *water-cement* ratio, the higher becomes the concentration of cement particles in a well-compacted fresh paste through a more closely packed internal structure. The packing, of course, is also affected by the fineness and particle-size distribution of the cement. Whatever the structure of a hardened cement paste, it is one that grew out of the structure of the fresh mixture of cement and water.

The aggregate particles in a mortar or concrete are embedded in and held together by the matrix of porous hardened paste. Therefore, most of the technically important properties of concrete, such as strength, shrinkage, and permeability, are determined primarily though, of course, not solely by the properties of the hardened portland cement paste matrix [Brunauer 1963].

The setting and hardening processes, as discussed in the preceding chapter, are the results of a series of simultaneous and consecutive reactions between water and the constituents of portland cement. These reactions together are covered by the term *hydration of portland cement*. It should be understood that in actuality the cement constituents react with the *aqueous phase*, which is the water solution of various salts, rather than water, and the term "water" is used here only for the sake of brevity.

As determined, for instance by X-rays, among the main constituents the fastest is the hydration of C_3A, the slowest is that of belite. The hydration of alite and that of celite occupy a place somewhere between these two extremes. It is not known yet what causes these differences in reactivities, or, in general, why certain compounds (C_3S, $ß-C_2S$, etc.) act as hydraulic cements, whereas others, of broadly similar composition (such as $\gamma-C_2S$), do not. One reason may be that in all the reactive compounds of portland cement the coordination of the calcium is irregular, therefore unstable, whereas the coordination of Ca^{2+} in the $\gamma-C_2S$ is symmetrical, thus stable. [Taylor 1964b] Another possible reason is that calcium could exist in structures in an *active* or *inactive* form. At higher temperatures, as required for the formation of cement compounds, the coordination of calcium may be lower than normal, therefore unstable at ordinary temperatures. Consequently, the structure of the compound tends either to form a new polymorfic arrangement, or to hydrate. In either case the new structure would contain calcium in its normal, inactive, coordination of six. [Lea 1956]

In the course of hydration the cement paste

becomes a stonelike material. One would think — as Brunauer put it so eloquently — that the millions of small cement grains would have coalesced into one piece; but the opposite is closer to the truth [Brunauer 1963]. Every portland cement particle breaks up into millions and millions of particles, because the constituents that collectively make up cement gradually disappear and are replaced by their hydration products, forming a very fine-textured solid, so-called *gel*. (The term gel designates a rigid aggregation of colloidal material.) The hydration of cement compounds is exothermic. The heat developed is called the *heat of hydration* discussed in section 2.7.6 [Mchedlow-Petrosyan 1974].

Water in hydrated pastes may be arbitrarily classified into *evaporable water*, and *nonevaporable water*, the latter being water not removed under standard drying conditions. The first term is intended to cover the *free water* and the major portion of the *physically adsorbed* water, whereas the second term is for the *chemically bound* water (Fig. 3.1).

A simplified description of the hydration process is presented below. The details of the chemistry of the reactions are dealt with mainly through references because (1) it is a very complicated topic for most engineers and concrete technologists; (2) there are excellent, comprehensive books on the subject [Eitel 1966, Bogue 1955b, Lea 1956, Kuhl 1961b, Taylor 1964a]; (3) the chemistry of hydration seems important mainly to the extent that it influences the amount of hydration products and the rates of their development; and (4) it is possible to identify most hydration products without any knowledge of the hydration characteristics of the individual constituents that comprise portland cement.

The course of hydration, its rate and deceleration, that is, the kinetics of hydration, can be followed directly by x-ray quantitative analysis [Gard 1964], infrared spectroscopy [Farmer 1964], differential thermal analysis [Mackenzie 1964], silylation through the measurement of the size and quantity of the formed oligomers [Tamas 1976], and

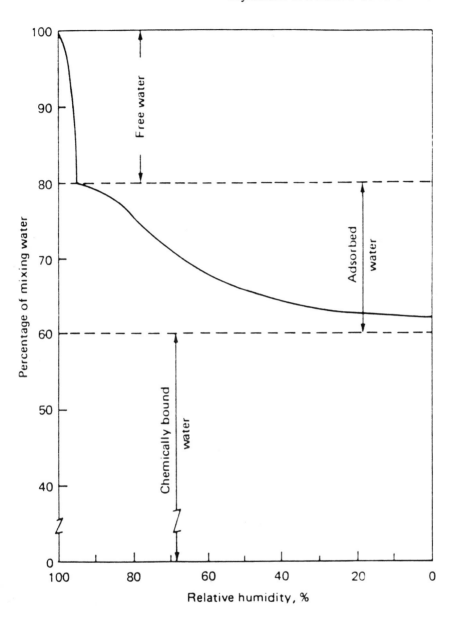

Figure 3.1: Approximate quantities of the various forms of water in a well-hydrated cement paste. The specimens were stored until equilibrium at various relative humidities [Duriez 1961].

indirectly by the determination of the nonevaporable water [Glasser 1964], thermal gravimetric analysis of the liberated calcium hydroxide [Tamas 1966], heat evolution, specific surface of the cement gel [Kantro 1964], strength measurements, etc. In addition to these traditional methods, more recent investigation procedures are also available [Wieker 1974].

3.2 REACTIONS IN EARLY HYDRATION AND SETTING

The measurement of heat evolution is particularly suitable for the investigation of the early stages of cement hydration [Adams 1976]. During a short period beginning when portland cement and water are first brought into contact at room temperature, and during the time of mixing, relatively rapid chemical reactions occur primarily between the C_3A of the cement and water (Fig.2.30). Simultaneously, the water rapidly becomes saturated with calcium hydroxide produced by hydrolysis of alite as well as of calcium aluminate, by $CaSO_4 \cdot 2H_2O$ from the gypsum, and by other compounds [Seligmann 1969]. The pH of this solution is in the neighborhood of 13. Eventually crystalline calcium hydroxide appears. Then the gypsum reacts with calcium aluminates to form solid *calcium aluminate sulfate hydrates*, for instance, *ettringite*, which act as a protective coating on the surfaces of the cement particles, thus slowing down the reactions. [Verbeck 1865, Hansen WC 1988, Skalny 1978] A large fraction of the SO_3 in cement paste reacts in the first few minutes, as has been established by Lerch in his pioneering investigation on the role of gypsum in portland cement hydration [Lerch 1946]. The presence of the ferrite phase leads to substitution of Fe and Al in all hydration products. The alkalies in cement clinker also dissolve rapidly, and persist as sulfates or hydroxides accelerating the hydration. Some of the other minor constituents of portland cement may also influence the early hydration process [Sakurai 1969]. The extent of hydration of the other main con-

stituents of portland cement is relatively very small
[Copeland 1964, Seligmann 1963, Kalousek 1974,
Steinour 1958].

Within 5 min, as the protective coating thickens
(Fig. 3.2), the rate of reaction subsides to a low
level; there is then a period during which the paste
normally remains plastic. This period, which has been
called the *dormant period*, normally lasts 40-120 min,
depending on the characteristics of cement (Fig.
2.11).

The fraction of cement used up in the initial
reactions is small, perhaps 1%, and the solid part of
the reaction products, which is most of it, adheres
to the surfaces of the cement grains [Powers 1964].
These surfaces, however, are cleaned by dissolution,
usually within 2 hr, causing the dormant period to
come to an end. Subsequently, a period of accelerating chemical reaction sets in again, which
usually lasts about 3 hr. During and after this time,
the paste gradually loses its plasticity and, if it
is a paste prepared according to standard test
methods, passes through arbitrarily defined degrees
of stiffness known as *initial set* and *final set* as
discussed in Chap. 2.

When the portland cement is insufficiently
retarded, the time of initial setting is considerably
less than 1 hr. This usually undesirable phenomenon
is called *quick set* or *flash set*. The chemical
reactions involved liberate a large amount of heat
and the set cannot be overcome by remixing the paste.
Quick set may be due to insufficient or faulty gypsum
in the portland cement, or to improper chemical composition of the clinker. Quick set can also be
produced intentionally by adding certain admixtures
to a normal-setting cement, or by raising the temperature of the paste. In contradistinction to quick
set, *false set* develops little heat and can be
overcome by remixing or by a longer initial mixing
time. There are several factors that may promote
false set, the most common probably being heating of
the cement during grinding. It is likely that a part
of the gypsum in the portland cement is dehydrated
during the grinding process to form the hemihydrate
which, having contacted with water, forms gypsum

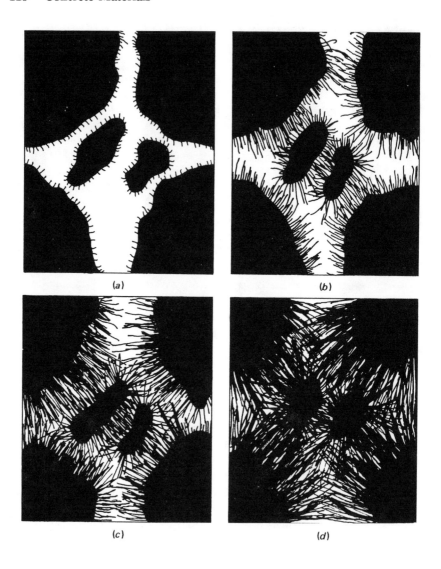

Figure 3.2: Schematic representation of the internal structure of a cement paste in various phases of hydration (not to scale). (a) Dormant period. (b) and (c) Setting. (d) Hardening. Black mass: unhydrated cement. Fibers: hydration products. White mass: pores.

crystals. Under certain conditions these crystals can form a network that causes the paste to stiffen, or sometimes to set. If the paste is mixed until the period of this stiffening is complete, there will be no false set. If the period of stiffening occurs after the paste is mixed, false set will be fully developed. If the mixing extends into the period of stiffening, the stiffening will be partially developed [Copeland 1964, Kalousek 1974, Hansen 1988].

It appears likely that the chemical reactions that are predominant during the regular setting period are (1) the reaction between C_3A and $CaSO_4$ to form a *calcium aluminate sulfate hydrate* at a slow rate as the liquid part in the paste is saturated with calcium hydroxide and gypsum; and (2) the hydration of calcium silicates, mainly of alite present in the cement, producing a poorly crystallized, porous but stable gel, called *CSH* (*calcium silicate hydrate*) *gel*, or sometimes *tobermorite gel* (Fig. 3.2b and 3.2c). In this gel are embedded several more or less well-crystallized hydrates, mainly *calcium hydroxide*, and unhydrated cement particles. Water is also present in the system in various states. The produced gel particles are so small that they are invisible under an ordinary microscope; they are visible, however, under an electron microscope.

The hydration of alite appears to be much more important in the setting process than that of the aluminates, since electron micrographic examinations by Copeland and Shulz [Copeland 1962] show that at about the time of initial set the alite has hydrated sufficiently that laths of CSH gel can begin to interlock, whereas final set is caused by sizable interlocking of CSH gel particles. Partly this interlocking, partly the reduction of the amount of free water by the hydration produce the accelerating stiffening of the paste during the setting [Locher 1974a]. Belite is far less reactive than alite, thus it has no significant role in this early hydration process.

3.3 REACTIONS IN THE HARDENING PROCESS

After final set, chemical reactions continue at a diminishing rate (Fig. 2.11) until one or more of the conditions necessary to reaction are lacking. This stage of hydration is called the *hardening* process, during which the predominant reaction is the continuing hydration of the calcium silicates. The decrease in rate is the result of two effects: (1) the surface area of unhydrated cement particles decreases as the smaller particles become completely hydrated and the larger particles become smaller; and (2) a layer of CSH gel forms on the surfaces of the cement particles, slowing down further reaction by forming a protective coating.

The chemical compounds found in the gel of hydrated cement are complex; most are impure in the sense that they contain elements not ordinarily given in their chemical formulas, and they do not have exactly the same composition when formed under differing conditions, especially with respect to temperature conditions and original water-cement ratio [Powers 1966]. Also, the various chemical reactions may influence one another or the various compounds of portland cement may themselves interact with one another during hydration. For instance, considerable experimental evidence indicates that the greater the amount of C_3A present in the cement, the faster will be the rate of hydration of the calcium silicates [Popovics 1967b, 1968b, 1969a, 1974a, Copeland 1969], as demonstrated in Section 3.6. This suggests one of two things: (1) the C_3A actually acts as a catalyst on the calcium silicates during hydration; or (2) the complex role of C_3A, for instance, as advanced by Hansen [Hansen 1970], can be approximated statistically as if the C_3A acted as such a catalyst [Popovics 1974a, 1976a, 1980]. Although the accelerating action of gypsum on alite hydration has also been shown [Copeland 1964], Alexander found no consistent evidence that the sensitivity of C_3S to C_3A varies with gypsum content.

It has been noticed that both the lime-silica (C/S), and the bound water-silica (H/S) ratios of the

CSH gel display a regular variation with time in the course of hydration. This can be explained by a change in reactivity of the silicate anions [Tamas 1973]. The lime-silica ratio of calcium silicate hydrate formed starts out with a high value at the beginning of hydration, then decreases sharply, which may be followed by an increase as the hydration proceeds, and finally levels off gradually as shown for C_3S in Fig. 3.3 [Kantro 1962]. Also, the value of the lime-silica ratio appears to increase with increasing temperature up to $100^{\circ}C$ [Idorn 1969] as well as with decreasing water-cement ratio. Results for the water-silica ratio are consistent with the lime-silica results [Copeland 1969]. These variations in composition are accompanied by changes in the morphology as well as in the x-ray diffraction picture of the gel. Thus, one can distinguish between CSH(I) gel with C/S ratio < 1.5 and with a foillike structure, and CSH(II) gel with C/S ratio > 1.5 and with a mostly fibrous structure [Taylor 1969, Ludwig 1974]. All this means is that the reaction between water and, say, alite varies to a certain extent with time, temperature, water-cement ratio, and several other factors. Nevertheless, it may be worthwhile to mention that examination of hardened cement pastes after complete hydration provides approximate information concerning the final compositions and the related reactions of the hydrated compounds of portland cement. For instance, using the symbols of Table 2.2 in Chap. 2 [Brunauer 1964a],

$$2C_3S + H_6 = C_3S_2H_3 + 3Ca(OH)_2 \qquad (3.1)$$

where $C_3S_2H_3$ represents a CSH gel; or

$$C_3A + H_{10} + CaSO_4.2H_2O = C_3A.CaSO_4.H_{12} \qquad (3.2)$$

where the right side of the equation is calcium aluminate monosulfate hydrate [Hansen 1988]. Note a further simplification applied in these equations, namely, that the reactions of the cement compounds

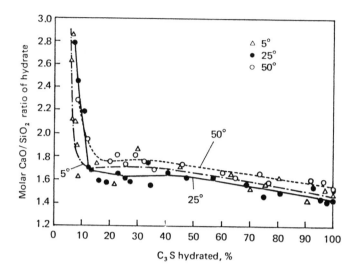

Figure 3.3: Molar lime-silica ratio of CSH gel as a function of the percentage hydration of C_3S and curing temperature [Kantro 1962]. Reprinted with permission from *Journal of Physical Chemistry*, vol. 66, no. 10, October 1962. Copyright by the American Chemical Society.

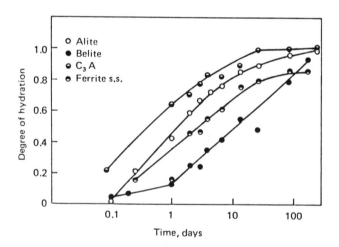

Figure 3.4: Degree of hydration of the constituents in a type I cement as a function of time [Copeland 1964].

take place with water. In reality, the liquid phase is the water solution of various salts.

The main difference between the hydration of tricalcium silicate and that of dicalcium silicate is that the former develops more $Ca(OH)_2$. Also, x-ray analysis indicates that the CSH gel developed from dicalcium silicate has lower CaO/SiO_2 ratio than that developed from tricalcium silicate during the first half of the hydration process, but this difference becomes negligible as the process approaches complete hydration [Kantro 1962]. In the early stages of hydration of portland cement, C_3A usually produces ettringite, that is, calcium aluminate trisulfate hydrates. The compositions of the other hydrated calcium aluminates and ferrites are quite complex [Schwiete 1969].

In contrast to some of the statements above, certain earlier experimental results seemed to suggest [Powers 1956] that the same products are formed at all stages of hydration of the portland cement; that is, the fractional rate of hydration of all compounds (alite, belite, etc.) in a given cement is the same. One can visualize this hydration mechanism by imagining that a cement particle is made up of layers of identical composition; upon contact with water, first all the compounds in the outside layer hydrate before the hydration in the next layer underneath can start; the hydration in the third layer can start only after the completion of the hydration of all the compounds in the second layer, and so on. More recent, refined measurements, however, have disproved the assumption of equal fractional rate [Copeland 1964]. As Fig. 3.4 shows, the C_3A crystals in a portland cement particle hydrate more intensively than the alite crystals, which, in turn, hydrate more intensively than the belite component.

Soroka [1979] combined the two views showing that during the stage of hydration when the rates of chemical reactions control the rate of hydration, the compounds hydrate at their own individual rates; later, however, when the diffusion takes over the control of the rate of hydration, the fractional rates of hydration of all compounds are the same in a given cement.

Measurements have also demonstrated that the rate, or fractional rate, of hydration of any component in cement is affected by the composition of the cement. For instance, the rates of hydration of C_3S and C_2S increase with increasing C_3A content, as has been discussed earlier. Thus, the hydration characteristics, including the strength and heat developments, of the pure individual cement compounds have limited value in the quantitative characterization of the hydration characteristics of a portland cement.

3.4   MECHANISM OF HYDRATION

The hydration process is essentially one of solution or perhaps solid-liquid reaction [Hansen 1962], followed by diffusion and crystallization. This presently accepted view is, in a sense, a combination of the two main theories that have been advanced for almost a century to explain the hardening of cements. The oldest is the crystalline theory put forward by Le Chatelier [Le Chatelier 1905], which ascribes the development of cementing action to the intergrowth of interlocking crystals, as occurs, for instance, in the setting of gypsum plasters. In the colloidal theory, which was first put forward by Michaelis [Michaelis 1907], cohesion is considered to result from the precipitation of a colloidal gelatinous mass as hydration product. This gel hardens as it loses water, either by external drying or by inner suction by hydration of the inner unhydrated cores of the cement grains [Lea 1956, Kuhl 1961a].

The modern view concerning the mechanism of the continuing hydration can be visualized in a simplified way. In the hardening paste the unhydrated parts of the cement grains are coated with a solid shell of previously created hydration products, the cement gel, containing very small, so-called *gel pores*. [Locher 1974a] On contact with this still unhydrated part, water reacts with and/or dissolves a portion of it forming a supersaturated solution. This solution diffuses out from the grain surface through the gel pores. Subsequently, new hydration products precipitate from the solution on the outer surface of the gel coating in the available air- or water-filled spaces between grains that are large enough to allow nucleation of new solid phase. In other words, the developing hydration products gradually replace the water and/or air between the cement grains and bind the grains together. This mechanism has been confirmed by microscopic observation [Williamson 1972, Skalny 1980]. The solid hydration products always contain a considerable amont of very small pores [Locher 1974a].

In the early stages of hydration there is plenty of water available for the reactions, thus the hydration rate is controlled by the cement; that is, the rate-determining step at this stage is the chemical reaction. This may be called the *first stage of strength development*. Later, however, the gel coating becomes so thick that the diffusion through it becomes slower than the slowest step in the chemical reaction proper. This means that not enough water can reach the unhydrated part of the cement for activation of all the available cement; therefore, the available water quantity, that is the diffusion, controls the hydration rate. This is the *second stage of strength development*. Since C_2S is far less reactive than C_3S, one may say that a thicker gel coating must build up on it before diffusion becomes rate-controlling [Brunauer 1964b]. The rate of chemical reactions strongly increases with the increase of paste temperature (Fig. 3.5) [Verbeck 1969]; the rate of a diffusion-controlled hydration process shows small temperature dependence. Therefore, the hydration of C_2S is expected to be more temperature-dependent, and for a longer period, than the hydration of C_3S.

During the process of hydration, a specimen of cement paste remains relatively volume-constant. More precisely, the external volume remains more or less constant but the volume of solids within the boundaries of the specimen increases, causing an overall reduction in the porosity of the paste. This reduction in porosity is one of the two primary sources

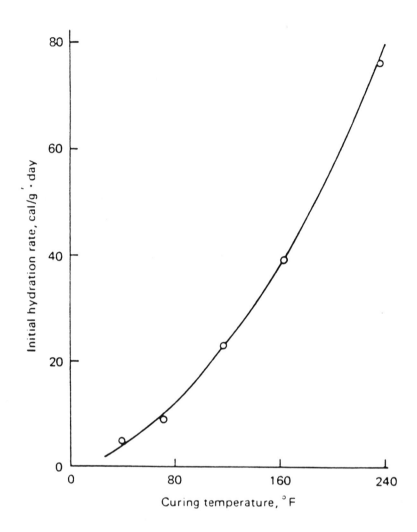

Figure 3.5: Initial rate of cement hydration as measured by the rate of heat development, as a function of curing temperature [Verbeck 1969].

of the increase in strength of the hardening cement paste. This is so because the developing hydration products gradually replace the liquid between cement grains, thus bind the grains together as well as reduce the overall porosity of the paste. [Double 1977] This binding and the reduction in porosity are the two primary sources of the increase in strength of the hardening cement paste.

The density of a fully hydrated portland cement is about 2.13 g/cm^3. This and other pertinent measurements show that 1 cm^3 of cement, solid volume, can develop at least 2 cm^3 of hydration products. Yet the volume of the hydration products formed is less than the sum of the volumes of cement and water that react to form it, because by chemical combination the volume of water is decreased by about one quarter [Mikhailov 1960]. So the hydration product does not fill completely the volume made available for it. This is sometimes called "contraction" [Czernin 1962]. On the other hand, if the paste is in contact with an external supply of water, more water will be drawn into the paste and hydration will proceed for a certain period of time, filling more and more available space with hydration products (Figs. 3.6 and 3.7). If an external source of water is not in contact with the paste, it will use up the free ("capillary") water, or self-desiccate, as the hydration proceeds [Copeland 1964] even when the evaporation of the free water is prevented. It is conceivable in such a case that the water content may fall to such a low value that hydration is stopped before the cement is completely hydrated, as shown in Fig. 3.6.

The mechanism of hydration presented implies that the hydration process stops when (1) no more water is available for reaction; (2) no more unhydrated cement is available; and (3) diffusion can no longer take place because of lack of spaces of sufficiently large dimension for nucleation of new solid hydration products, or because the coating is too impervious, or too thick. Powers has calculated from the double volume of hydration products that the water-cement ratio in a paste must be at least somewhere 0.35 and 0.40 by weight to provide enough space for all the

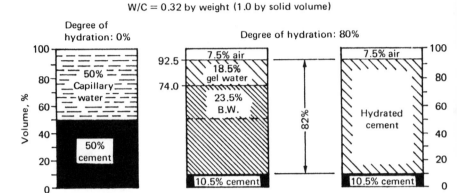

Figure 3.6: Schematic presentation of two stages of hydration of a fully compacted cement paste with a water-cement ratio of 0.32 by weight. When curing conditions are applied where neither access to water nor evaporation are possible, the reactions stop at about 80% of complete hydration because all the available capillary water has been used up. If cured under water, 7.5% of additional water would have been available by filling up the emptied pores, which could have hydrated another 6% of the cement. Further hydration, however, is impossible because there are no more spaces available for new products [Czernin 1962].

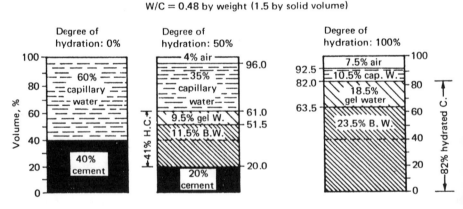

Figure 3.7: Schematic presentation of three stages of hydration of a fully compacted cement paste with a water-cement ratio of 0.48 by weight. Here the water and space conditions permit complete hydration even when curing is applied such that neither access to water nor evaporation is possible. In this case 18% capillary pores remain in the specimen, of which 10.5% are filled with water. If cured under water, all the 18% pores would have been filled up with water; the other volumes would remain unchanged [Czernin 1962].

hydration products that can be derived from the cement.

It follows, therefore, that the more closely the cement particles are packed in the fresh paste, that is, the lower is the water-cement ratio, the sooner the setting and hardening processes will start, resulting, for instance, in relatively high early strengths; but also, the more intensively the reactions will decelerate and come to an end. An additional pertinent effect is that the degree of supersaturation of the liquid phase of a cement paste increases with a decrease in the water-cement ratio, and this increase contributes significantly to the rate of nucleation [Kondo 1969].

Measurements indicate that the kinetics of hydration are influenced by several factors, including the fineness and composition of the cement, temperature and water-cement ratio of the paste, and admixtures. It is important to note here that if any of these factors increases the specific rate of hydration, then, simultaneously, the same change intensifies the deceleration of the hydration to a greater degree. Thus, the hydration will begin more strongly but will also level off sooner. The kinetics are important not only because they control the quantity of the hydration products at early ages but also because they influence to a certain extent the quality of hydration products as well.

This inverse relationship between early strengths and final strengths is not quite understood yet. The hydration products formed, say, during low-pressure steam curing do not differ greatly from those formed at 20°C. The elevated temperature modifies somewhat the morphology of the calcium silicate hydrates, but this effect does not appear to be large enough to have a major effect on the final strength. Thus, the main reason for the relatively low final strengths of steam-cured and other accelerated concretes seems to be mechanical: The rapid hydration may produce higher final porosity and more microcracks in the gel [Butt 1969, Venuat 1974].

Equation 3.7 with the appropriate a parameters appears suitable for the kinetics, that is, for the calculation of strength and other characteristics of

the hydration process in terms of compound composition fineness and age as was discussed in Chap. 2
and earlier in this chapter.

3.5 STRUCTURE OF THE CEMENT PASTE

3.5.1   Fresh Paste

The initial structure of the fresh cement paste
depends on the volume fraction, particle-size
distribution, and chemical composition of the cement
particles. [Cement and Concr. Ass. 1976] Although the
cement particles are more or less dispersed by the
water, the high concentration of the cement particles
and the resulting van der Waals forces of interparticle attraction produce flocculation and ensure
that the paste as a whole acts as a single flock
[Powers 1945b]. Thus, fresh cement paste can be
considered either as a concentrated suspension of
cement particles dispersed in an aqueous solution, or
as a weak porous permeable solid containing continuous capillaries filled with the aqueous solution
[Powers 1964]. The flocculation in fresh cement
pastes was directly observed by Uchikava by SEM with
sample-freezing, back scattered electron image
method. [Uchikava 1987]
    The interparticle forces can be modified, for
instance, by changing the water content or by certain
surface-active or other agents (admixtures), as
discussed in Chapter 6.

3.5.2 Hardened Paste

When all the cement particles have hydrated
completely, about half of the cement gel occupies the
sites that were originally occupied by cement
particles and the other half occupies space outside
the original boundaries of the particles. Since the
external volume remains relatively constant, the
doubling of volume of the cement grains cannot take
place by a symmetrical growth of cement gel. That is,
the process by which unhydrated cement becomes cement
gel must be such that the cement gel is produced only

where there is sufficient space to accommodate it. Moreover, the cement gel evidently conforms to the shape of whatever space is available to it.

Studies of hardened portland cement paste with the electron microscope [Copeland 1962, 1967, Grudemo 1964, Mills 1968, Richartz 1969, Hadley 1972, Krokosky 1970, Midgley 1971, Hale 1971, Diamond 1972] as well as adsorption studies have revealed that the cement gel consists of exceedingly illformed colloidal products with a layered structure of no definite number and of irregular configuration. Water constitutes a structural part between layers [Feldman 1968, 1971b, 1972]; therefore it does not behave as normal free water. The colloidal material first appears as needlelike or lathlike bodies that bridge the pores between cement grains. In the course of hydration they grow laterally and become irregular, still very small, thin sheets, which frequently agglomerate into conical forms, fibers, and rosettes. In this gel are embedded several more or less well-crystallized hydrates, mainly calcium hydroxide, and unhydrated cement particles [Kondo 1974, Copeland 1974]. Crystalline calcium hydroxide usually constitutes 20-30% of the weight of dry cement gel. This calcium hydroxide is a weak link in the hardened paste because of its chemical vulnerability and low specific surface. It can be improved by binding it chemically into a gellike calcium silicate hydrate. Most commonly, pozzolanic materials are used for this purpose at regular temperatures, and ground quartz in the case of autoclave curing.

3.5.3 Porosity

As has been pointed out, the hardened cement paste always contains a considerable amount of pores of different sizes. There is no general agreement about the shapes of these pores [Brunauer 1970]; more and more evidence has been published recently to indicate that they are essentially interlayer spaces rather than narrow-necked pores [Verbeck 1969, Feldman 1968, 1971a, 1971b], although the layered structure has not yet been demonstrated directly. Porosity of cement paste can be defined as a more or less dispersed

phase the strength of which is negligible as compared
to the strength of the other phases in the composite.
In the case of a hardened cement paste, this is,
somewhat arbitrarily, the fraction of the volume of a
saturated specimen occupied by evaporable water
[Powers 1964]. The system formed by the smallest
pores is called *gel pores* by Powers, and their
average width was estimated originally as 15-30 Å.
This size is too small to allow nucleation of new
hydration products. Powers also estimates that the
minimum possible gel porosity of the completely
hydrated portland cement paste is about 28%, and that
the usual gel porosity is between 40 and 55%.

In addition, there is another void system in the
hardened paste made up of that part of the original
water-filled space that has not become filled with
the porous cement gel. These pores are larger at
least by an order of magnitude than the gel pores and
are called *capillary pores* [Powers 1956]. When the
original water-cement ratio of paste is higher than
about 0.35 by weight, the volume of the cement gel is
not sufficient to fill completely all the original
water space in the hardening paste, even after
complete hydration. Therefore, the volume of capillary pores increases rapidly with an increase in
water-cement ratio [Locher 1976c], also demonstrated
in Fig. 3.8 [Powers 1954b]. It is also affected by
the particle size distribution of the cement since
cement powders with mixed particle size have lower
voids content than one-size cements. Incomplete
hydration would decrease the volume of gel pores and
increase the volume of capillary pores propor-
tionately [Verbeck 1966]. The evaporable water
present in the gel pores may be called *gel water*,
whereas the water in the capillary pores is *capillary
water* (Figs.3.6 and 3.7). The *bound water* is combined
chemically with the gel [Hansen 1963b]. The gel water
is not capable of reacting with unhydrated cement,
but the capillary water is.

A third pore system is formed by *air voids*
resulting either from incomplete compacting, or from
intentionally entrained air, or from both. These
voids are substantially larger than the capillary
pores. The strength of a hardened cement paste as

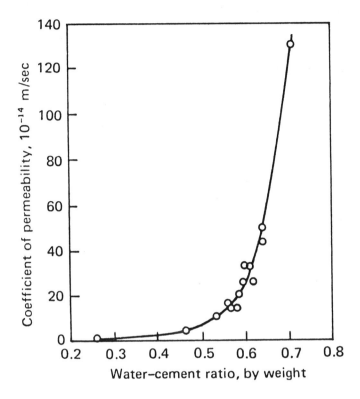

Figure 3.8: Higher water-cement ratios increase the permeability of the hardened cement paste primarily through increased capillary porosity [Powers, Copeland, Hayes, and Mann 1954b].

well as many other technically important properties depend primarily on its capillary porosity and air void content, that is, on its degree of space filling [Locher 1976a].

The amount of capillary pores in a hardened cement paste can be characterized in several ways, such as the ratio of the amount of chemically bound water to the amount of mixing water, or the ratio of the amount of mixing water to the amount of the hydrated cement. One of the often mentioned methods is the *gel-space ratio*, introduced by Powers and Brownyard [Powers 1947-48] and refined several times since [Powers 1949a, 1956]. According to the last defini-tion, the *gel-space ratio* is the ratio of the gel volume to the prevailing volume of the capillary porosity, where the latter term means the volume of water-filled space originally present in the paste, plus the space made vacant by the hydration of cement grains. The *modified gel-space* ratio is defined as the ratio of gel volume to the combined volume of the prevailing capillary pores plus air voids. If the specific gravity of the cement is assumed to be 3.15, and 1 cm^3 (solid volume) of portland cement is to produce a little more than 2 cm^3 of gel, then a simple calculation provides the modified gel-space ratio in terms of the original water and air contents and degree of hydration, as follows:

$$X_F = \frac{0.65}{(w_o + A)/\alpha c + 0.32}$$ (3.3)

where X_F = modified gel-space ratio

α = the fraction of cement that has become hydrated

c = weight of cement in the mixture

w_o = volume of mixing water in the mixture

A = volume of air in the mixture.

Another possibility is to express the porosity as the combined amount of capillary and air voids as the fraction, or percentage, of the total volume of the paste. By using the same assumptions as for Eq. 3.3, the porosity P of a hardening paste in terms of composition and degree of hydration is

$$P = \frac{(w_o + A)/c - 0.34\alpha}{(w_o + A)/c + 0.32} \qquad (3.4)$$

where the symbols are the same as in Eq. 3.3.

The two methods used most frequently for the direct measurements of the distribution of pore sizes in hardened cement pastes are mercury porosimetry and capillary condensation. Unfortunately, there are discrepancies between the measurements of these two values, the mercury porosimetry providing coarser distributions. Evidence available from other techniques (scanning electron microscopy, etc.) is in good agreement with results obtained by mercury porosimetry [Diamond 1971].

Results of recent direct measurements are in substantial agreement with the concept of void systems as discussed above. Kroone and Crook found [Kroone 1961], for instance, that curing at *100%* humidity reduces the total volume of larger pores in the cement paste but increases the amount of small pores.

Verbeck and Helmuth report data [Verbeck 1969] concerning pore-size distributions of 11-year moist-cured cement pastes at three water-cement ratios (Fig.3.9). The data were obtained with a high-pressure mercury porosimeter. The measured pore sizes span more than three orders of magnitude, from about 9×10^4 Å to about 40 Å. The pore-size distribution curves show that there is a broad and continuous spectrum of pore sizes rather than a single size for the gel pores and another for the capillary pores. Nevertheless, the two pore systems can be somewhat arbitrarily separated considering that the curves display two clearly defined maxima corresponding, in a sense, to the gel pore and capillary pore concepts, respectively. The curves also show that the sizes of the gel pores, and particularly the sizes of the capillary pores, increase with an increase in the water-cement ratio. For the paste with 0.35 water-cement ratio, maxima occur at pore diameters of 45

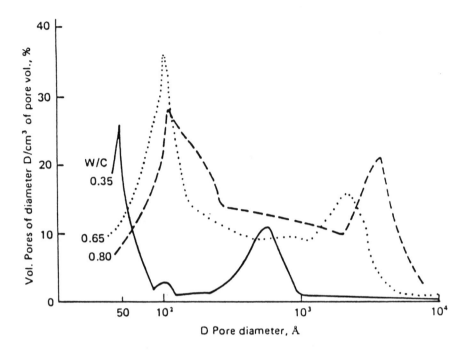

Figure 3.9: Pore-size distribution in cement paste moist-cured for 11 years. After Verbeck and Helmuth [1969].

and 550 Å; for the 0.65 paste, maxima occur at 105
and 2100 Å; for the 0.80 paste, maxima occur at 110
and 3500 Å. Helmuth and Verbeck also report pore-size
distributions calculated from adsorption isotherms
that are in fair agreement with comparable data
obtained with a porosimeter. It should be noted, however, that more recent measurements indicate neither
capillary nor gel spaces but merely spaces between
fine individual particles of cement hydration
products [Winslow 1970]. The reason for this contradiction is unclear.

The permeability of cement paste to water (Fig.
3.8) demonstrates that the pore system has limited
continuity. In fresh paste the water-fllled space
consists of continuous capillaries, but in mature
paste the capillaries may become discontinuous.
Development of cement gel reduces capillary sizes,
and, unless the original water-cement ratio exceeds
about 0.7 by weight (the exact value depending on the
fineness and particle-size distribution of the
cement), cement gel gradually reduces the continuity
of the capillaries [Powers 1959]. Elimination of
capillary continuity, however, does not destroy the
continuity of the pore system as a whole because
water flow can occur through the gel pores.

3.5.4 Specific Surface

A direct consequence of the small sizes of cement gel
particles and the large amount of porosity is that
the specific surface of the CSH gel is very large.
This is perhaps the most significant characteristic
of the gel, because the size of the specific surface
influences many technically important properties of
the cement paste. Thus, the discovery of the magnitude of the specific surface by Powers and
Brownyard [Powers 1947-48] represents a turning point
in the science of cement and concrete. About 75% of
the colloidal fraction of the hardened portland
cement paste (the so-called cement gel) is CSH gel,
and at least 80% of the specific surface of the
hydrated paste is CSH gel surface.

As has been mentioned, physical or chemical factors that speed up the hydration, such as low water-

cement ratio, high temperature, and certain admixtures, first produce cement gel with somewhat
higher specific surface, that is, with finer texture
[Lachaud 1972]. However, the same factors decrease
the specific surface at later ages, as pertinent
experiments by Kantro et al. [Kantro 1961] show in
Fig. 3.10.

It is well known that crystal formation follows
the same tendency as mentioned above. The effect of
water-cement ratio on the specific surface may also
originate in the packing of cement grains in the
fresh paste: The more closely they are packed, that
is, the lower the water-cement ratio, the less chance
there is to develop larger gel particles.

It has also been noticed that changes in the
specific surface are accompanied by changes of
opposite sign in the lime-silica ratio of the CSH
gel. A comparison of Fig. 3.11 and Fig. 3.3
demonstrates this [Kantro 1962]. Note, however, that
although all these variations in the specific surface
seem definite, they are still relatively minor. Thus,
one may say that the specific surface of cement gel
is not much influenced by differences in paste
density, fineness and chemical composition of the
cement, or admixtures. Powers considers this feature
another outstanding characteristic of the cement gel
[Powers 1966]. A notable exception is the case of
high-pressure steam curing, which causes a profound
alteration in the structure. The reactions are
fundamentally different from those occurring during
hydration at temperatures under 100°C; in contrast,
they have much in common with those taking place in
the production of the autoclaved lime-silica products
[Taylor 1964c, Kalousek 1969]. The principal effect
of this curing on the structure is elimination of
most of the colloidal attributes of the hydration
products, and therefore a lowering of the specific
surface.

Although low-angle x-ray scattering can be applied [Winslow 1973, 1974], the most commonly used
methods for measuring the specific surface of a CSH
gel are based on the theory of adsorption. [Powers
1958a, 1958b] The essence of this is that a sample of
dry, hardened cement paste, when exposed to moist

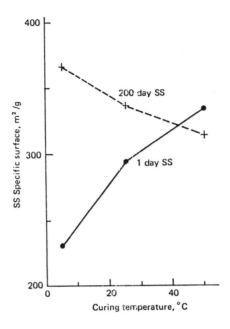

Figure 3.10: The specific surface of CSH gel increases with increasing temperature at early ages but decreases at later ages [Popovics 1972, 1976a]. Data from Kantro [1961].

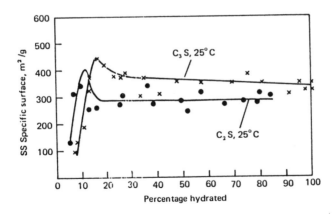

Figure 3.11: Specific surface area of tobermorite gel as a function of percentage hydration. Reproduced from Kantro, Brunsuer, and Weise [1962]. Reprinted with permission from *Journal of Physical Chemistry,* vol. 66, no. 10, October 1962. Copyright by the American Chemical Society.

air, will take up water in definite amounts, which amounts are a function of the relative humidity of the air and the magnitude of the "internal surface" exposed to air. Thus, one can calculate the magnitude of this internal surface, usually expressed as specific surface, from adsorption measurements with suitable vapors or gases. A pertinent method of calculation was developed by Brunauer, Emmett, and Teller [Brunauer 1938], called the *BET method*. Accordingly, gases and vapors are adsorbed not only on a solid surface but also on the layers adsorbed earlier. The energy of adsorption is equal to the heat of evaporation in the second, third, and further adsorbed layers, but it is greater in the first layer. From this an equation can be derived, the so-called BET equation, which provides a factor that is proportional to the specific surface of the porous material tested.

Many gases and vapors are suitable for such adsorption measurements. Theoretically, and also in many practical cases, the specific surface determined from adsorption is independent of the adsorbed material, that is, of the adsorbate applied. Unfortunately, the hardened cement gel is an exception: the specific surface determined by water vapor as the adsorbate is usually over 200 m^2/g, or over 550 m^2/cm^3 of cement gel, which is approximately twice or three times as large as the specific surface determined by nitrogen adsorption. Thus, the question is, which value represents the true specific surface of the hydrated cement gel. Powers and Brunauer, among others, accept the value obtained with water vapor on the basis that the N_2 molecule is too large to penetrate the gel pores of the cement gel, thus it cannot measure all the specific surface, whereas the smaller vapor molecules can. Therefore, they believe that nitrogen adsorption provides a specific surface that is less than the actual value [Brunauer 1970, Powers 1960].

On the other hand, Canadian researchers (Feldman, Sereda, etc.) prefer the specific surface value measured by nitrogen adsorption. Their argument is the following [Feldman 1968, 1969, 1971a]: Water escapes from the spaces between the layers of the CSH gel, that is, from the gel pores, during the drastic drying procedure preceding the adsorption; then, during the water adsorption, not only is the water adsorbed on the gel surface but, in addition, an extra amount reenters the interlayer spaces. Therefore, if one calculates the specific surface from the assumption that the total quantity of water was adsorbed according to the BET theory, the calculated specific surface will be much larger than the true value.

Although recent measurements of helium flow through the hardened paste indicate that the interlayer spaces collapse after the water has been driven off by drastic drying [Feldman 1971b,1972], and thus support the Canadian point of view, there is no definite proof at present concerning which of the adsorption methods provides the most reliable value for the specific surface of the cement gel. The importance of this question is more than academic, since the answer may provide information concerning the internal structure of the cement gel: If the actual specific surface of the gel is greater than 200 m^2/g, the hypothesis may be acceptable that the strength of a hardened cement paste originates from weak physical (or secondary, or van der Waals) bonds acting between the gel particles. If, however, the actual specific surface is much less, the presence of much stronger, chemical (primary), mostly covalent bonds should be assumed as the main source of strength.

Three different models representing the conflicting views concerning the structure of the hardened cement gel are presented.

3.5.5 Models for the Structure of Cement Gel

The *Powers-Brunauer model* of cement gel assumes a large (water vapor) specific surface, and accepts the overwhelming role of van der Waals forces in the strength of the paste. Although these bonds are weak, they can provide high compressive strengths for the paste when they combine with a large specific surface. The fact that a mature paste becomes stronger when it is dried slowly are said to support this proposition. It is also assumed, however, that a small fraction of the boundary of a gel particle is chemically bonded (cross-linked) to neighboring particles. The evidence for this is that penetrating water cannot break the bonds holding hardened cement paste together [Powers 1956, 1958b, Brunauer 1970]. The Powers-Brunauer model, with certain simplifying assumptions, is applicable for the calculation of the volumetric composition of hardened portland cement pastes. [Hansen 1986]

The *Feldman-Sereda model* assumes a layered structure for the cement gel having pores as spaces between these layers, and a smaller (nitrogen) specific surface. [Ramachandran 1981] The model is based again on the proposition that mostly physical, that is, weak bonds between gel particles are responsible for the strength of the paste [Feldman 1968]. They provided the following experimental evidence for the role of the physical bonds [Soroka 1968, 1969]: Cement pastes were prepared in the conventional manner but were compacted into specimens only several hours later. It was found that the delayed compaction did not change the modulus of elasticity-versus-porosity relationship. This was said to be strong evidence for the lack of chemical (primary) bonds, because if such bonds had been present, the delayed compaction certainly would have broken them, thus would have weakened the specimens.

A third model developed recently by Tamas [Tamas 1973, 1975] provides the smallest specific surface for the gel, determined by argon adsorption, and suggests that the primary reason for the strength of cement paste is chemical bonds resulting from a limited polymerization of the calcium silicates, although van der Waals forces and hydrogen bonds also contribute to a certain extent.

The type of the reactions between the calcium silicates and water in the Tamas' model is condensation polymerization, where short-chain silicate anion polymers, so-called oligomers, are formed through binding the SiO_4^{4-} tetrahedrons of the portland cement clinker as monomers together by bridging oxygen ions; the by-product is CaO, or more precisely, its hydrated $[Ca(OH)_2]$, and/or carbonated ($CaCO_3$)

forms. In other words, the hardened CSH gel in this model is a conglomeration of silicate anion oligomers of differing degrees of polymerization which form an irregular, three-dimensional net, and in which there are also cross bonds between oligomers. The exact size distribution of the oligomers is not known yet, but it appears that the average oligomer contains 8-10 Si atoms. This would correspond to a 0.4-0.5 average degree of polymerization. The existence of silicate anion oligomers in the hardened portland cement paste has been proved by a chemical method called silylation [Tamas 1976].

3.6 EFFECT OF CEMENT COMPOSITION ON THE STRENGTH DEVELOPMENT - MATHEMATICAL MODELS

The effects of compound composition on the strength development were mentioned in the previous section but only in qualitative terms. Here and in the following section mathematical *cement models* (formulas) will be discussed that attempt to describe these apparent effects quantitatively.

The term "cement model" means a simplified, hypothetical cement in which many of the factors influencing the hydration and hardening of an actual cement are disregarded; the remaining few variables are combined in a form that can reproduce quantitatively one, specific property, such as strength development. The mathematical form of the strength development of this hypothetical cement is called the *mathematical cement model*.

Modeling is an important tool for the cement manufacturers because reliable relationships between the fineness and composition of a portland cement and its properties help them produce cements of specified properties. For instance, a model can help the cement manufacturer optimize the cement composition, and/or minimize the energy consumption for a cement produced for a specific purpose (steam curing, etc.). It can also be helpful for the concrete engineer to optimize the heat curing process for a cement and/or concrete of a given composition; or, to trace the mechanism of

the effect of an admixture on the strength
development.

As was discussed in the previous sections, the
strength of a portland cement paste originates from
the primary and secondary bonds as well as from the
reduction of porosity in the hardening cement paste.
Therefore, it is safe to say that the fineness and
chemical composition of the cement affects the rate
and magnitude of the developing strength only
indirectly, that is, through their effects on the
quality and quantity of the developing hydration
products. Since, however, the relationships between
fineness and hydration on the one hand and bonds and
porosity on the other had not been established yet,
except in a rudimentary attempt [Jons 1982], direct
relationships between the composition of portland
cement and its properties can be established only at
the phenomenological level, for instance through more
or less empirical models. Nevertheless, such models
do have a rational background since the compound
composition of the cement and/or its fineness determines the rate and the final extent of hydration,
consequently the bonds and the porosity in the hardening cement paste.

The apparent effects of the compound composition
and fineness of cement on many of the technically
important properties of the cement paste, including
the strength development, have been recognized for
more than fifty years. The character of these effects
has also been recognized even earlier in qualitative
terms, for instance, that the early strengths are
influenced primarily by C_3S and C_3A, while the late
strength development by C_2S. The next step was the
attempt to establish numerical relationships between
the compound composition of cement and its strength
development. [Bates 1917, Woods 1932]

3.6.1   The Linear Model

The first attempt to establish a quantitative
relationship between the compound composition of a
cement and the strength it develops at various ages
was the working hypothesis, or hardening models,
stating that each cement compound contributes its

intrinsic strength at a given age in proportion to the percentage of that compound present. The mathematical form of this model for the four main compounds is linear as follows:

$$f = \text{strength} = a(C_3S) + b(C_2S) + c(C_3A) + d(C_4AF) \qquad (3.5)$$

where the symbols in parentheses represent the calculated (Bogue) percentages by weight of the compounds, and a, b, c, and d are empirical coefficients (parameters) representing the contribution of one percent of the corresponding compound to the strength of the hardening mixture at a given age under the given circumstances (temperature, etc.).

A set of the coefficients in Eq. 3.5 for a given set of conditions can be determined by regression analysis from results of such strength test series where the compound composition of the cement is the sole variable; that is, where the fineness and gypsum content of the various tested cements, age, air content and composition of the strength specimens, curing and testing methods, etc. are practically identical. Such a test series were performed by [Gonnerman 1934] where, among others, standard Ottawa-sand mortars were made with a variety of portland cements of differing compound composition and tested for compressive and tensile strengths at ages from 1 day to 2 years. For instance, for the 7-day strength of Ottawa-sand mortars of 1:2.75 mix proportion by weight, he obtained the following coefficients:

$$a = 40.0; \quad b = -5.1; \quad c = 58.4; \quad \text{and } d = -0.2$$

in psi.

Example 3.1

The composition of Cement No. 24 in Gonnerman's investigation was:

$$C_3S = 41\%; \quad C_2S = 37\%; \quad C_3A = 7\%; \quad C_4AF = 12\%.$$

Thus, with the coefficients above, the estimated compressive strength of this cement at the age of 7 days in a 2-in (50-mm) standard mortar cube is:

$$f = 40.0 \times 41 - 5.1 \times 37 + 58.4 \times 7 - 0.2 \times 12 =$$

$$= 1860 \text{ psi } (12.83 \text{ MPa})$$

Note that any change in the conditions concerning the cement (fineness, SO_3 con- tent, etc.) or testing the strength (composition of the mortar, curing, etc.) would result in a different set of parameters.

Equation 3.5, in general, gives a fair agreement between calculated and observed strengths with the appropriate parameters for cements of usual compositions. Nevertheless, this equation is objectionable for both technical and statistical points of view, primarily because it does not reflect the effect of C_3A on the strength development properly [Schramli 1978].

Several attempts have been made to eliminate or reduce these objections above while keeping the linearity of the model but neither the inclusion of many additional variables [Blaine 1968, Von Euw 1970] nor the reduction of the number of independent variables to two [Alexander 1969, 1972] have improved the linear model substantially, the conclusion may be drawn that the linear form is not suitable to express the strength development adequately in terms of the cement composition at various ages because the fundamental assumption is invalid. Therefore, several non-linear models will be discussed below.

3.6.2   The   Exponential   Model   for   Relative Strength

This model depicts the hardening process of portland cement more realistically, although in relative terms (percent), than the linear hardening models. The importance of the relative strength formulas, such as Eq. 3.6, is that they are applicable for the analysis of the hardening process of cement paste. In addition, they can be extended to provide the strength in a stress unit.

The approach of the exponential model for the reconciliation of the greater than linear effect of C_3A on hardening is the working hypothesis that the portland cement consists of two hardening components, essentially the two calcium silicates; the C_3A acts as a catalyst on these components during hydration. [Popovics 1974a, 1976a] A numerical form of this hypothesis is the following [Popovics 1967b, 1968b]:

$$f_{rel} = 100 \frac{f}{f_{28}} = 100 \frac{C_3(1-e^{-a_1 t}) + (100-C_3)(1-e^{-a_2 t})}{C_3(1-e^{-28a_1}) + (100-C_3)(1-e^{-28a_2})} = \tag{3.6}$$

$$= 100 \frac{100 - C_3 e^{-a_1 t} - (100-C_3)e^{-a_2 t}}{100 - C_3 e^{-28a_1} - (100-C_3)e^{-28a_2}} \tag{3.7}$$

where

f_{rel} = relative strength of portland cement paste, mortar, or concrete, percent of the 28-day strength

t = age of the specimen at testing, days

C_3 = computed C_3S content of the portland cement, percent

a_1 and a_2 = rate parameters of the two hardening components which are independent of the strength, age, and C_3S content, but may be a function of the temperature, C_3A content, and any other factor that influences the course of hydration (fineness, gypsum content, admixtures, water-cement ratio, curing and testing method,etc.), 1/day.

Equations 3.6 and 3.7 are the mathematical forms of a cement model that consists of only two hardening components. The first component is the C_3S, the second component is the quasi-homogeneous mixture of the other cement ingredients, mostly C_2S. The hardening of this cement model starts at t = 0 time and proceeds continuously with the same hydration mechanism controlled by the rates of reactions of the two components at all ages. This is the mechanism that acts during the "first stage" of the strength development of an actual portland cement. [Popovics 1968d] The model disregards the effect of diffusion process (Section 3.4) on the strength development.

Numerically, the a parameters can be obtained by applying a suitable statistical method to pertinent experimental data. For instance, by applying Equation 3.7 to the previously mentioned test results by Gonnerman [1934], the following a_1 and a_2 values were obtained by the method of least squares for the compressive strength of standard 1:2.75 Ottawa sand mortars with wet curing [Popovics 1967b]:

$$a_1 = 0.0067 \ C_3A + 0.10 \tag{3.8}$$

and

$$a_2 = 0.0018 \ C_3A + 0.005 \tag{3.9}$$

where C_3A represents the potential tricalcium aluminate in the portland cement computed according the the Bogue method (Eqs. 2.4 through 2.7) expressed in percent.

The application of the presented formulas are illustrated below.

Example 3.2

The compositions of three of Gonnerman's cements, including No. 24, are shown in Figure 3.12. The computed C_3S contents of all three cements are practically the same but the C_3A contents differ. Illustrate the effect of C_3A content on the relative strength development at various ages by using Eq. 3.7 along with Eqs. 3.8 and 3.9. The a parameters of cement No. 24 are calculated by Equation 3.8 and 3.9:

$a_1 = 0.14$ 1/day; and $a_2 = 0.018$ 1/day.

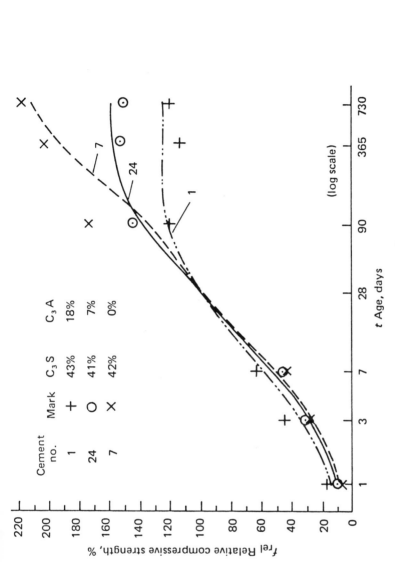

Figure 3.12: Comparison of experimental and computed values to illustrate the effect of C_3A content on the kinetics of the hardening of portland cement in 1:2.75 mortars. Experimental values are represented by points, and values calculated by Eq. (3.7) with $a_1 = 0.0067C_3A + 0.10$ and $a_2 = 0.0018C_3A + 0.005$, by lines [Popovics 1967b, 1968b].

By substituting these values into Eq. 3.7 the
following is obtained:

$$f_{rel,24} = 100 \, \frac{100 - 41e^{-0.14t} - 59e^{-0.018t}}{100 - 41e^{-3.92} - 59e^{-0.50}} =$$

$$= 100(1.58 - 0.65e^{-0.14t} - 0.93e^{-0.018t})$$

Similarly, the equations of the curves for
relative compressive strength versus age for the
cements Nos. 1 and 7, respectively, are

$$f_{rel,1} = 100(1.25 - 0.54e^{-0.22t} - 0.71e^{-0.037t}),$$

and

$$f_{rel,7} = 100 \, (2.12 - 0.89e^{-0.10t} - 1.23e^{-0.005t}).$$

Values calculated by these formulas are presented
in Figure 3.12 by lines along with observed
results of points from Gonnerman's experiments.

Two further illustrations of the use of
exponential model are presented below.

1. The effect of the *strength type* on the strength
development is illustrated in Figure 3.13. Here
experimentally obtained values with standard
Ottawa-sand mortars (points) and calculated
values (lines) of the relative tensile, flexural
and compressive strengths are presented. The
experimental values were published by Klieger
[1957], the calculated values were obtained by Eq.
3.7 with appropriate values of a_1 and a_2
[Popovics 1967b]. Similar phenomenon was
observed on hardening cement pastes. [Hsu 1963]

2. Eq. 3.7 can be used for the estimation of the
ultimate strength, that is, the strength obtained
after a very long wet curing, expressed as

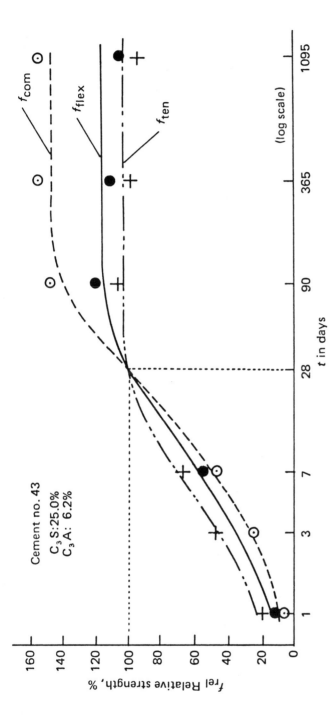

Figure 3.13: Comparison of experimental and computed values to illustrate the effect of test method on the kinetics of the hardening of portland cement in mortars. Experimental values are represented by points, values calculated by Eq. (3.7) by lines [Popovics 1967b, 1968b].

percentage of the 28-day strength. This equation takes on the following form for great t values [Popovics 1972]:

$$f^o_{rel} = 100 \ \frac{100}{100 - C_3 e^{-28a_1} - (100-C_3)e^{-28a_2}} \tag{3.10}$$

$$= 100 \ \frac{100}{98 - (100-C_3)e^{-28a_2}} \tag{3.11}$$

where f^o_{rel} = relative ultimate strength, percent of the 28-day strength. The other symbols are the same as the symbols of Eq. 3.7.

As a numerical illustration for the detrimental effect of the higher C_3A content on the ultimate strength, Figure 3.14 presents the relationship between the compound composition of portland cement and the relative ultimate compressive strength. Points represent experimental values reported by Gonnerman [1934] for cements of approximately 40 and 60 percent C_3S contents, respectively, with constant (1.8%) SO_3 content, and the same fineness (1440 cm^2/g by air permeability). The lines designate values that were calculated from Eq. 3.11 for 40 and 60 percent C_3S contents, respectively, with the a_2 value of Eq. 3.9. Considering the inherent fluctuation of the results in such kind of experiments, the goodness of fit appears reasonably good in Figure 3.14.

In the comparisons above the calculated strength development is controlled solely by the rates of reactions at all ages.

The primary advantages of the exponential model for relative strengths, that is Eq. 3.7, are that (a) it is simple since it requires only two parameters, and (b) it works; that is, it is supported by experimental results within wide limits. On the other hand, Eq. 3.7 provides only the relative strengths. However, this weakness has been eliminated, as shown

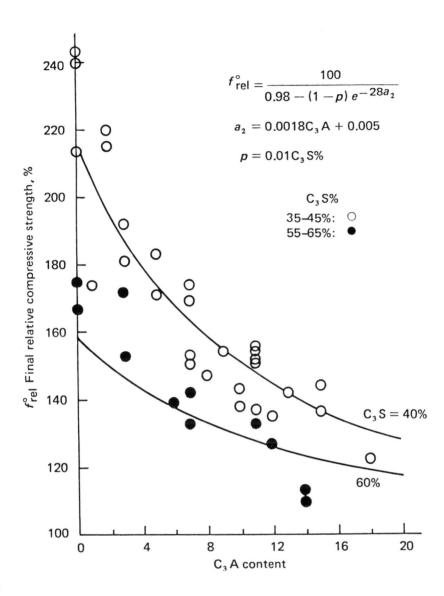

$$f^{\circ}_{rel} = \frac{100}{0.98 - (1-p)\,e^{-28a_2}}$$

$$a_2 = 0.0018C_3A + 0.005$$

$$p = 0.01C_3S\%$$

Figure 3.14: Final relative compressive strength of portland cements in mortars as a function of the compound composition [Popovics 1972, 1976a].

below.

3.6.3   Exponential   Model   for   Strengths   in Stress Units

There are several ways how an exponential model can be used for the estimation of the strength which is expressed in a stress unit instead of percent. One possibility is to combine Eq. 3.7 with another equation that predicts the 28-day strength, such as Eq. 3.5. Three other possibilities are presented below.

A simple approach is the transformation of Eq. 3.7 into the following formula:

$$f_t = f_{90} \frac{100 - C_3 e^{-b_1 t} - (100-C_3)e^{-b_2 t}}{100 - C_3 e^{-90 b_1} - (100-C_3)e^{-90 b_2}} \tag{3.12}$$

where

f_t = estimated mortar or concrete strength at the age of t days, psi or MPa or kg/cm^2

f_{90} = strength of mortar or concrete at the age of 90 days, in the same stress unit as f_t.

The satisfactory goodness of fit of Eq. 3.12 with appropriate *b* parameters [Popovics 1981a] to experimental data is illustrated in Figure 3.15 with five of the LTS cements, one of each standard type. Eq.3.12 shows also good fit to standard mortar tensile strengths obtained with the same five cements.

The exponential model can also be used for the estimation of *concrete* strength. One such an application is its combination with a strength vs. water-cement ratio formula [Popovics 1981b], another is for the prediction of concrete strength from the strength of the same concrete obtained experimentally at different time.

3.6.4   Generalization   of   the   Exponential Model for Curing Temperature (Popovics 1987a)

The exponential model was further generalized

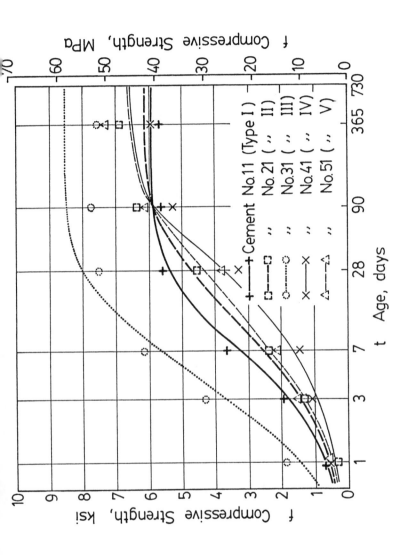

Figure 3.15: Compressive strengths of the five standard portland cement types in 1:2.75 Ottawa-sand mortars in moist-air curing. Lines represent Eq. 3.12, points represent experimental values reported by Klieger [Popovics 1981a].

recently. This new model can produce not only the
concrete strength with various portland cements after
standard curing but also it provides quantitative
information about several important characteristics of
the kinetics of hydration and hardening as a function
of curing temperature. These characteristics are: the
time of the end of setting, that is, the beginning of
the hardening; rates of hardening; the time when the
diffusion control of hydration starts; how much the
strength contribution of this is; and the final
strength potential of various portland cements. Experimental data support this cement model at ages from
1 day to 1 year for concretes cured at different
temperatures, as shown below.

3.6.4.1 The Generalization The generalization
of the exponential model was based on the recognition
that the strength development of a portland cement is
affected by the curing temperature in four different
ways; namely, a lowering (raising) of the curing
temperature [Popovics 1987a]:

1. delays (speeds up) the beginning of the
   hardening;
2. slows down (increases) the rate of chemical
   reactions between cement and water, and even
   more so the deceleration of these reactions;
3. delays (speeds up) the time of transition from the
   first stage of strength development (controlled by
   the rate of chemical reactions) to the second
   stage;
4. increases (decreases) the final strength.

The so-extended cement model provides quantitative
information about these effects on the model cement
separately. The proper combination of these effects
produces the overall strength development of portland
cements at various ages and curing temperatures.
Especially important is the separation of the two stages
in the strength development because the hydration during
the first stage is influenced intensively by the curing
temperature whereas during the second stage it is not.
    The dividing point of time between the two stages in
the model is marked as t_d. (Fig. 3.16) In other words,

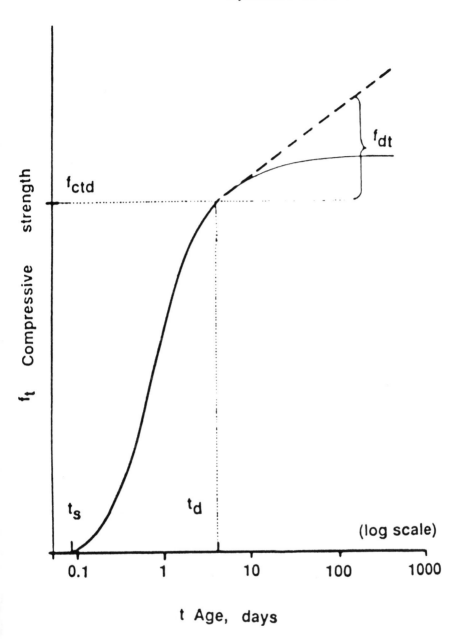

Figure 3.16: Schematic representation of strength development of the model cement as a function of age. Continuous line: strength values representing Eq. (3.13); broken line: strength values representing Eq. (3.14) and (3.17). The thick lines represent the calculated f_t [Popovics 1987a].

the first stage of the strength development, f_{ct}, takes place from the time of beginning of hardening, t_s, up to t_d, whereas the second-stage strength, f_{dt}, develops after the beginning of the diffusion control, t_d. Figure 3.16 shows that the total strength of the concrete, f_t, during the first stage is equal to f_{ct}, whereas the total strength of the concrete in the second stage is equal to the sum of f_{ct} at the age of t_d, which is f_{ctd} and f_{dt}. This figure also shows that f_{dt} is approximated by a logarithmic function of the age t in the model.

In addition to these two stages of strength development, several other factors are included in the model as variables that influence the kinetics of strength development. These are:

compound composition and fineness of the cement; age; and curing temperature.

The mathematical form of this new cement model is as follows [Popovics 1987a]:

1. During the first stage of strength development, (when the hydration is controlled by the rate of chemical reactions), that is when $t \leq t_d$, and therefore, $f_t = f_{ct}$:

$$f_t = f_{ct} = f_o f_{28} \frac{1-C_3 \exp(-a_1(t-t_s)(S/300)) - (1-C_3)\exp(-a_2(t-t_s)(S/300))}{1-C_3 \exp(-a_1(28-t_s)(S/300)) - (1-C_3)\exp(-a_2(28-t_s)(S/300))}$$

(3.13)

where

f_t = strength of portland cement paste, mortar or concrete at the age of t, cured at any constant temperature between 4 and 50°C, in the same unit as f_{28}

f_{28} = strength of the same mixture but with Type I portland cement and cured for 28 days in the standard manner at 22.78°C (73°F), in the same unit as f_{ct}

f_o = relative final strength of the cement paste, mortar or concrete, that is, strength after a very long proper curing, percent/100 of the 28-day strength

t_s = age when the hardening process starts, days

t_d = age when the first stage of strength development ends, and the second stage starts, days

S = Blaine specific surface of the cement, m^2/kg.

The other symbols are identical with the symbols of Eq. 3.7

2. During the second stage of hardening, that is when $t > t_d$:

$$f_t = f_{ctd} + f_{dt} =$$

$$= f_o f_{28} \frac{1 - C_3 \exp(-a_1(t_d - t_s)(S/300)) - (1 - C_3)\exp(-a_2(t_d - t_s)(S/300))}{1 - C_3 \exp(-a_1(28 - t_s)(S/300)) - (1 - C_3)\exp(-a_2(28 - t_s)(S/300))} + f_{dt} \quad (3.14)$$

where

f_{ctd} = strength of portland cement paste, mortar or concrete at the age of t_d, cured at any temperature between 4 (25) and 50°C (120°F), in the same unit as f_{28}

f_{dt} = strength developed after the age of t_d, that is during the second stage, in the same unit as f_{28}. (Fig. 3.16)

The other symbols are identical with the symbols of Eqs. 3.7 and 3.13.

Using appropriate a_1 and a_2 values, strengths can be calculated from these equations for various cement types at various ages for different but constant curing temperatures.

3.6.4.2 Comparison to Experimentally Obtained Strengths The values calculated from Eq. 3.13 or 3.14 were compared to pertinent experimental concrete strengths. This comparison is shown for Type I cement in Table 3.1. The other two cement types showed the

Table 3.1: Calculated and Experimental Strengths Obtained at Various Curing Temperatures [Popovics 1987a].

Cement: Type I

Curing temp. °C	Source of strength	a₁ 1/d	a₂ 1/d	t_d day	f_ctd MPa	f_o %/100	Compressive Strength, MPa day					
							1	3	7	28	90	365
4.44	Calc.	0.08	0.014	98.6	55.03	1.49	0.20	5.61	14.36	36.73	53.88	58.85
	Exp.						0.56	5.11	14.98	36.26	49.84	54.11
12.78	Calc.	0.20	0.060	68.1	48.00	1.14	4.12	13.71	25.77	43.83	48.81	52.90
	Exp.						4.06	13.51	26.25	43.75	52.85	55.93
22.78	Calc.	0.48	0.11	43.8	42.56	1.00	9.42	21.52	31.59	41.85	44.67	48.75
	Exp.						9.87	21.84	31.43	42.35	44.52	50.33
22.78	Calc.	0.43	0.12	43.8	38.32	1.00	7.67	18.43	28.25	37.68	40.40	44.51
	Exp.						8.05	18.20	28.42	38.08	42.00	45.15
32.22	Calc.	0.70	0.12	28.8	37.43	0.99	10.65	20.96	28.67	37.37	40.75	44.84
	Exp.						10.43	21.63	29.12	36.68	39.90	43.68
40.56	Calc.	0.95	0.13	19.9	34.72	0.95	12.27	21.38	27.98	35.71	39.12	43.20
	Exp.						12.95	22.40	26.95	33.04	36.75	41.02
48.86	Calc.	3.0	0.20	13.8	27.31	0.74	14.53	19.50	24.44	29.38	32.79	36.87
	Exp.						14.28	19.88	24.08	28.77	33.11	37.24

1 MPa = 145 psi

same goodness of fit, as illustrated in Figures 3.17 and 3.18. The experimental data were taken from a paper by Klieger [1958] which also contains all the pertinent experimental details.

It can be seen that the goodness of fit is quite good between the calculated and experimental strengths.

3.6.4.3   Characteristics   of   the   Kinetics   of Hydration From fitting Eq. 3.14 to experimental strengths (Table 3.1, Figs. 3.17 and 3.18) specific formulas can be obtained for several characteristics of the kinetics of strength development of the model cement as a function of curing temperature and cement type. These are, for the three cement types used by Klieger, presented in Table 3.2. Table 3.1 also contains illustrative values for other characteristics of the hydration kinetics of Type I cement for several temperatures.

3.6.4.4 Interpretation of the Formulas
The above-presented equations for the kinetics of three types of model cements (Table 3.2) indicate the following:

1. As expected, the rate of hardening of the C_3S, that is, the a_1 parameter is the largest regularly   for the Type   III   cement   and   the smallest for Type II. The a_2 values are the largest for Types I and III intermittently and the smallest for Type II. Note also that these rates increase rapidly with increasing curing temperature. This increase is more than an order of magnitude for each cement type as the curing temperature increases from 4 (25) to 48°C (120°F).

2. The starting time of hardening in the cement model, t_s, is naturally related to, although regularly less than the standard initial setting time of an actual portland cement. (Fig. 3.16) At higher temperatures it can be even negative. t_s is a fictive time for the model, a relative indicator of the beginning of hardening at a given temperature as compared to that at

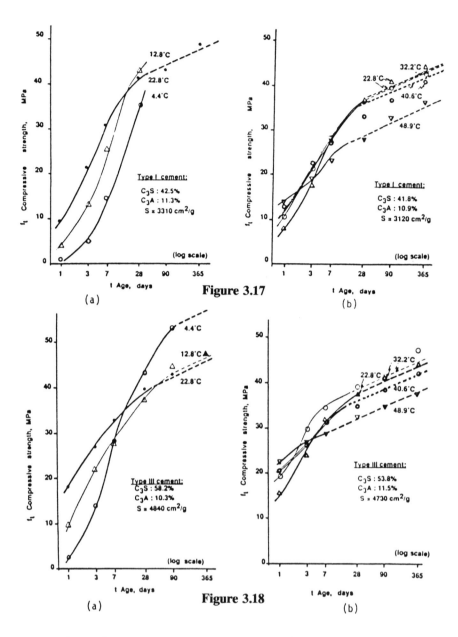

Comparison of concrete strengths calculated from the cement model (lines) to experimentally obtained values (points) for Type I cements in Figure 3.17 and Type III cements in Figure 3.18. Continuous lines represent strength development controlled by the rate of reactions; broken lines represent diffusion control [Popovics 1987a].

Table 3.2: Predictive Equations for Characteristics of the Kinetics of Strength Development of Portland Cements [Popovics 1987a]

Hydration characteristics		Equations					
Name	Symbol	For Type I cement Equation	Eq. No.	For Type II cement Equation	Eq. No.	For Type III cement Equation	Eq. No.
Beginning of hardening	t_s day	$\dfrac{215}{(t^o+10)^2} - 0.10$	3.15	$\dfrac{190}{(t^o+10)^2} - 0.12$	3.19	$\dfrac{165}{(t^o+10)^2} - 0.10$	3.23
Beginning of the diffusion control	t_d day	$120 \times 10^{-0.0192 t^o}$	3.16	$115 \times 10^{-0.0156 t^o}$	3.20	$122 \times 10^{-0.0235 t^o}$	3.24
Compressive strength developed during the control of rates of chemical reactions	f_{ct} MPa	See. Eq. 3.13	3.13	See Eq. 3.13	3.13	See Eq.3.13	3.13
Compressive strength developed during the diffusion control	f_{dt} MPa	$6.62 \log (t/t_d)$	3.17	$7.10 \log (t/t_d)$	3.21	$5.93 \log (t/t_d)$	3.25
Rel. final compressive strength	f_o %/100	$1-33 \times 10^{-6}(t^o-29)^3$	3.18	$1.1-30 \times 10^{-6}(t^o-29)^3$	3.22	$1-30.2 \times 10^{-6}(t^o-29)^3$	3.26

t^o = constant curing temperature, C^o. The log is of 10 base. The other symbols are identical with the symbols of Eq. 3.13 and 3.14. 1 MPa = 145 psi.

standard curing temperature. Note that when the curing temperature is -10°C (12°F) or less, the model cement does not start hardening. The value of t_s primarily affects the very early strengths.

3. The time of transition in the cement model, t_d, from the first stage of hardening to the second stage (Fig. 3.16) shortens exponentially with increasing curing temperatures, as shown by Eqs. 3.16, 3.20 and 3.24 in Table 3.2. This reduction is the largest in the case of Type III cement and the smallest for Type II. Consequently, the hardening is rate-controlled in Type II cement longer than in Type I, which, however, is still longer than in Type III for every temperature. This trend has been recognized earlier but only in qualitative terms. t_d is also controlled, negatively, with the speed of hardening, that is, C_3S and C_3A contents, fineness, etc. of the cement. Similar tendency appears with the strength development of the various cements during the second stage; that is, the higher the rate of hardening, the smaller the strength increase. Note that the strength development in the second stage, f_{dt}, is less sensitive to the temperature of curing than the chemical reaction-controlled strength development. The t_d and f_{dt} values affect the strength at later ages.

4. Table 3.1 demonstrates that the concrete strength at the age of t_d, that is f_{ctd}, decreases with increasing curing temperature, although to a much lesser extent than t_d. Apart from the two extreme temperatures, the value of f_{ctd} is between 35 (5,000) and 45 MPa (6,500 psi) for each temperature regardless of the cement type for the tested concretes.

5. The final strength f_o decreases with increasing curing temperatures approximately with identical rates for each cement type, as the pertinent equations in Table 3.2 as well as the numerical values in Table 3.1 demonstrate. Numerically, f_o is, again, a function of the speed of hardening.

The faster the hardening, the lower the final
strength. For instance, f_0 is the highest for Type
II cement and lowest for Type III cement indicating the inverse effect of C_3S and C_3A contents
on the final strength which has also been demonstrated experimentally. (Fig. 3.14) The f_0 equations in Table 3.2 also show that the final
strengths change little as long as the curing
temperature is not too far from 29°C (85°F).
However, the changes increase rapidly (with the
third power) with further increases or decreases
in the curing temperature.
The equations show that a low value of f_0 re-
produces numerically the strength-reducing
effect of the higher (early) curing temperature
on the f_{tc} portion of the concrete strength at
later ages.
The value of f_0 primarily affects the strength
at later ages.

Eqs. 3.15 through 3.26 are approximations with
limits of validity restricted to the domain of the
analyzed experimental data. Nevertheless, all
these formulas seem to be useful tools between 0 and
50°C (32 and 122°F) for the estimation of several
characteristics of the kinetics of hardening
numerically. [Popovics 1987a]

4

Hydraulic Cements Other Than Standard Portland

SUMMARY

Although the five standard types of portland cement
can be used for a wide variety of purposes, there are
technical and/or economical reasons to use other cementing materials for special purposes. Some of these
cements are portland cements modified for fulfilling
the requirements for a special purpose (white portland cement, oil-well cement, regulated-set cement,
hydrophobic cement, barium and strontium cements,
etc.). Other such cements have compositions such that
they cannot be considered as portland cement (high-
alumina cement, expansive cement, hydraulic lime,
natural cement, masonry cement, supersulfated cement,
slag cement, etc.).

High-alumina cements consist primarily of
calcium aluminates. They have several outstanding
properties, such as high early strength, excellent
refractoriness, good resistance against certain
chemical attacks, and so on. Nevertheless, the
structural applicability of these cements is
restricted because even slightly elevated
temperatures can produce permanent strength
reductions in high-alumina cement concrete when
moisture is present. This is attributed to an
increase in the internal porosity of the hardened
paste of high-alumina cement and reduction of bonds
which are the consequence of the spontaneous
conversion of the hexagonal crystals of the hydration

products (mainly calcium aluminate hydrates) into the more compact cubic form.

Expansive cements undergo increases in volume during hydration but do not become unsound, and develop satisfactory strength. Small volume increases may be provided to compensate for shrinkage. Larger increases may be provided in concrete for the development of self-stressing. There are three types of expansive cement: types M (Soviet), K (Klein), and S (Portland Cement Association). The volume increase in all three types is based on the formation of sulfoaluminate hydrates which compound produces expansions also in other cementitious mixtures. The strength is contributed by a high-alumina cement, or a portland cement component. Self-stressing and especially shrinkage-compensating concretes have been used successfully in numerous structures and products.

The names of the various *special portland cements* usually provide information about the major feature or application of the cement. Of the nonportland hydraulic cementing materials, *the supersulfated cement* has the best structural properties. *Hydraulic lime, natural cement,* and *slag cement* have low strengths, thus limited applicability. The *latent hydraulic materials* (granulated blast-furnace slag, natural pozzolans, fly ash, etc.) develop cementitious properties only in the presence of lime. They are used in rather large quantities because (1) they may provide a reduction in the cost of concrete by saving in the amount of cement used; (2) they reduce the heat of hydration in the concrete, increase its resistivity against sulfate and certain other che- mical attacks, and reduce the aggregate-alkali reaction; and (3) there are environmental benefits from the consumption of slag and, especially, fly ash.

4.1   INTRODUCTION

The cementitious materials discussed in this chapter are either modifications of standard portland cements to obtain special properties, or hydraulic cements the compositions and properties of which differ from those of the standard portland cements discussed earlier. There is a market for such cements because standard portland cements do not satisfy   all the needs of the concrete industry.

Additional information about these special cements can be found in the literature. [Ramachandran 1981]

4.2 HIGH-ALUMINA CEMENT

4.2.1   History, Manufacture, Hydration

High-alumina cement (HAC) is used in small quantities as compared to portland cement [Robson 1962]. Its technical importance originates from the special properties, such as high early strength, good refractory, and so forth. The term high-alumina cement refers to the class of hydraulic cements the essential constituents of which are calcium aluminates. The limits of the usual compositions are presented in Fig. 2.4. High-alumina cement is more expensive than portland cement.

Although the hydraulic properties of calcium aluminates have been recognized since about 1850, the first suitable method for the commercial manufacture of high-alumina cement was developed by Jules Bied and patented in 1908 in France. The commercial manufacture of such cements did not begin in the United States until 1924. Presently high-alumina cement is produced in many countries. It is called *ciment alumineux* or *ciment fondu* in France, and *Tonerdezement* or *Tonerdeschmelzzement* in Germany.

The raw materials customarily used for the manufacture of ordinary high-alumina cements are *limestone* and *bauxite*. (Bauxite is a residual deposit formed by the weathering of rocks of high aluminum content.) The bauxite is ground with limestone, then the dry mixture is fed into a kiln where it is heated until it melts (fusion), which process takes place about 2900°F (1600°C). Then the molten clinker is cooled down, for instance by spreading water on it, and the granulated material is ground to cement fineness with or without the addition of a small amount of gypsum [Eitel 1966].

The clinker of a high-alumina cement consists of a series of calcium aluminates, iron-containing compounds, beta-dicalcium silicate, and other, minor constituents [Lea 1956, Bogue 1955b, Robson 1969, Lhopitallier 1960]. Considerably less is known about the compound composition of high-alumina cements than that of portland cements. It is clear, however, that the monocalcium aluminate (CaO.Al$_2$O$_3$) is by far the most important compound, since it is primarily responsible for the particular cementitious behavior found in this kind of cement. This compound is slow-setting but subsequently hardens with great rapidity. High-alumina cements that provide the highest early strength usually contain less than 5% silica.

The hydration of high-alumina cement produces again small crystals and gel. These consist mainly of a series of calciumaluminate hydrates, the most important of which is CaO.Al$_2$O$_3$.10H$_2$O [Lhopitallier 1960, Turriziani 1964b, Chatterji 1966a,b, Majumdar 1966, Kuhl 1961b, Double 1977]. The hydrates first form hexagonal crystals which, however, later tend to transform into the more stable 3CaO.Al$_2$O$_3$.6H$_2$O of cubic form, particularly at temperatures higher than about 85°F (30°C), when water is present [Talaber 1974]. It is also important that no free calcium hydroxide develops during hydration which is probably the reason that the pH of high-alumina cement paste is lower (around 10) than that of portland cement paste. The water of hydration of high-alumina cements, that is, the amount of water that combines chemically with the anhydrous cement during hydration, is about 50% of the weight of the cement [Graf 1960]. This is about twice as much as the water required for the hydration of portland cement. Otherwise the mechanism of hydration, origin of strength, role of pores, and so on, appear to be similar to those in portland cement pastes.

4.2.2   Properties and Applications

Standard specifications for high-alumina cements

have been established in several countries, for instance in Great Britain (B.S. 915), which require a minimum alumina content of 32% and an alumina-lime ratio between 0.85 and 1.3. There is no standard yet in the United States that would cover high-alumina cement for structural concrete.

Despite the chemical differences, many of the technically important properties of high-alumina cements (color, specific gravity, fineness, time of initial setting, total amount of heat of hydration, storability, etc.) are similar, although not always identical, to those of portland cements. The workability, frost resistance, impermeability, creep as well as shrinkage properties, and thermal expansion of high-alumina cement concretes are also satisfactory. There are also minor differences; for instance: high-alumina cements require less water for normal consistency [Robson 1962]; the usual interval between initial and final settings is short, about 30-45 min. in high-alumina cements; their times of final setting are less sensitive to the water-cement ratio [Graf 1960]; the initial rate of heat development is high, around 9 cal/g.hr; and unsoundness does not occur in high-alumina cements. More important is that high-alumina cements are greatly superior to portland cements in sulfate and weak acid resistance, a property that is attributed to the absence of free calcium hydroxide in the hydrated high-alumina cement. These cements, especially the white alumina cements and the high-alumina cements with considerable calcium dialuminate contents, are also superior in refractory bonding, that is, in suitability for prolonged service at high temperatures [Robson 1964, Hansen W.C. 1955], for instance in furnaces or kilns, up to 2900°F (1600°C). Concretes with high-alumina cement lose strength during exposure to high temperature, just as portland cement concretes do, but from about 1300°F (700°C) upward they begin to undergo solid reactions that create a ceramic bond to the aggregate.

An additional difference is in the early strength development. Although there are considerable variations in the initial rate of hardening shown by

high-alumina cements of different origins, all such
cements are rapid-hardening. The low-silica cements
can give with a water-cement ratio of 0.5 by weight,
concrete strength of around 4000 psi (27 mPa) at 6
hrs. after placing. Another example is shown in Fig.
4.1, where the early strength development of a
high-alumina cement (low-silica type) is compared to
portland cements. The strength of the same portland
cement concrete is about half of the high-alumina
cement concrete at the age of 90 days. The supe-
riority of the strength developing ability of
high-alumina cements is even more conspicuous when
the applied curing temperature is close to the
freezing point.

Despite these important advantages, the
structural applicability of high-alumina cements is
restricted because even slightly elevated
temperatures can cause a permanent strength reduction
in high-alumina cement concrete when water is
present. A typical example is given in Fig. 4.2,
which shows clearly the detrimental effect of the
curing temperature of 100°F (38°C) under various
conditions. The weaker the concrete, the greater the
relative loss in strength. The strength reduction is
attributed to an increase in the internal porosity of
the hardened paste of high-alumina cement; that is,
the consequence of the spontaneous conversion of the
hexagonal crystals of the calcium aluminate hydrates
into the cubic form since the latter has a smaller
volume [Talaber 1974, Revay 1974, Neville 1975]. This
can occur at any time in the life of the concrete
when the conditions are sufficiently hot and wet, as
was recognized by Bolomey as early as 1927.
Completely dry concrete is not subject to conversion,
but if the concrete still contains some mixing water
in a free state, the conversion may easily take place
[Talaber 1962, Neville 1980, 1981, Mehta 1964]. The
conversion takes place at a rate that depends on a
number of factors, including the temperature,
water-cement ratio, and others. The strength reduction of the concrete seems to increase with an
increase in this rate and with an increase in the
water-cement ratio [Midgley 1975, Building Res. E.
1978, Quon 1982]. The temperature conditions,

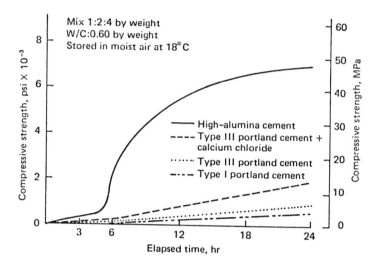

Figure 4.1: Early strength development of concretes made with a high-alumina cement and portland cements. From Robson [1962].

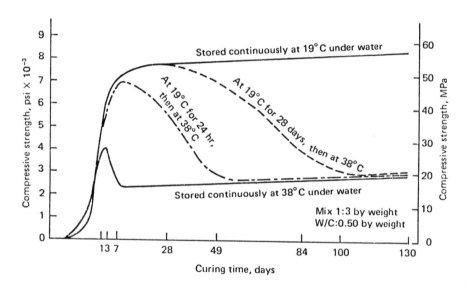

Figure 4.2: Reduction in compressive strength of a mortar made with high-alumina cement when cured in water of 38°C (100°F). From Robson [1962].

especially during the first 24 hr, are also extremely important for strength at later ages [Teychenne 1975, 1978].

The extent of the conversion can be determined experimentally, for instance by DTA, but these results do not show good correlation with strength reduction in the concrete. Results of rebound hammer tests are not reliable either, since badly converted high-alumina cement concrete can, and usually does, have a sound outer layer [Mayfield 1974]. Ultrasonic pulse testing is more suitable, although it provides information about the average condition of the concrete, not about the weakest spots. Thus, only cube or cylinder test results provide strength with acceptable reliability [Bretz 1969].

A recommended procedure for the application of high-alumina cement is to make concrete of low water-cement ratio, using ice water, keeping the concrete temperature below 74°F (23°C), and placing and compacting it, in thin layers, when the ambient temperature is below 85°F (30°C). Otherwise the proportioning procedure is based on the same principles as that of portland cement concrete [Shacklock 1974]. Intensive curing must be employed during the first 24 hr after placement. The storage of high-alumina cements at temperatures higher than 85°F (30°C) is also undesirable [Anon. 1973].

Heat curing of high-alumina cement concretes is usually forbidden. It should be mentioned, however, that such concretes with low water-cement ratios produced compressive strengths up to 6000 psi (41 MPa) in 15 to 20 min. at 212°F (100°C) curing temperature [French 1971].

Properties of high-alumina cement concretes may or may not be modified by admixtures, such as by water reducing, or air entraining, or some other admixtures. [ACI 1985b] For instance, superplasticizers have been shown less effective with these cements than with portland cements.[Quon 1981] Mixtures of high-alumina and portland cements usually have very short setting times and low strengths; therefore the use of such mixtures in normal constructions is undesirable [Lea 1956, Orchard 1973b].

4.3 EXPANSIVE CEMENT

4.3.1   Composition

The term *expansive cement* or *expanding cement* means a hydraulic cement containing a constituent (or constituents) that, in normal processes of hydration, setting, and hardening, undergo increases in volume, but do not become unsound, and develop a satisfactory strength. Small volume increases may be provided to compensate for decreases resulting from shrinkage; larger potential increases may be provided in concrete placed under restraint to develop a prestressed condition, called *self-stressing* or *chemical prestressing* [Mather 1965, Halstead 1964].

The earliest mention of deliberate attempts to produce cements that would cause concrete to expand after setting occurred in 1936 in a paper by Lossier in France. Cements of similar type have since been investigated in the USSR by Budnikov and Kosyreva, and Mikhailov in the 1950s, and in the United States by Klein and co-workers beginning in 1958 [North 1965a]. Further historical background and other details can be gained from a bibliography on expansive cements prepared by the Portland Cement Association [Lyon 1965].

Expansive cements can, in principle, be made either by manufacturing a special clinker containing an expansive constituent, or by mixing an additive containing such a constituent with a regular cement. The latter method seems to be simpler.

Several different possibilities have been recommended for the production of expansive cement, the use of MgO for instance, but so far only the cements based on the formation of sulfoaluminate hydrates have proved to be satisfactory. Incidentally, this compound produces expansions also in other cementitious mixtures [Popovics 1990a, 1990d].

The *Lossier expansive cement* consisted essentially of three components, (1) portland cement, which imparts its particular properties to the final product; (2) sulfoaluminate clinker, which is the expanding agent, and is produced by burning together gypsum, bauxite, and chalk; and (3) blast-furnace slag, the stabilizing agent, the function of which is slowly to take up the excess calcium sulfate and bring the expansion to an end.

One of the *Mikhailov cements* is of low expansion for shrinkage compensation and is based on high-alumina cement with added gypsum and calcium aluminate hydrate. The other is a strongly expansive cement for self-stressing and is based on a mixture of 60-70% regular portland cement with 30-40% of an expansive component consisting of a mixture of high-alumina cement and gypsum [Mikhailov 1960]. The recommended name for this latter kind of cement is *expansive cement, type M*. The properties of these and other Russian-made expansive cements are discussed by Budnikov and Kravchenko [Budnikov 1969].

Klein's cement appears to have been based on the application of a stable expanding compound obtained by heat treating a mixture of bauxite, chalk, and gypsum at about 2400°F (1330°C). Although the ingradients are quite similar to those used in the Lossier cement, the material selection and clinkering conditions probably contribute to the formation of a different material, an anhydrous calcium sulfoaluminate compound identified by X-ray diffraction [Klein 1958, 1961, ACI 1970, Mehta 1966]. The recommended name for this kind of cement is *expansive cement, type K*. This cement type proportioned for shrinkage compensating concretes contains approximately 10-15% expansive component.

Another kind of cement is the *expansive cement, type S*, developed by the Portland Cement Association. This is a portland cement having a large C_3A content and modified by an excess of calcium sulfate above the usual amount found in regular portland cement.

Type K cement has been available commercially since 1963. The vast majority of structures utilizing expansive cement concrete have incorporated this type. Type S cement was first made available commercially in 1968. Application of these cements has been largely restricted to production of shrinkage-compensating concrete. Type M cement is not presently available commercially in the United States.

4.3.2   Hydration and Properties

The basic reaction of hydration in the expansive cements discussed above is the formation of calcium sulfoaluminate hydrates [Mehta 1966, 1967]. The tri-sulfate hydrate has the composition $C_3A \cdot 3CaSO_4 \cdot 31H_2O$, and is called *cement bacillus* in the European literature, after Michaelis, or *ettringite*, which is the name of a naturally occurring mineral of the same composition and structure. The trisulfate can be the result of the reaction between C_3A and gypsum or some other sulfate in the presence of water. Note that the volume of the hydration product is about eight times the volume of the C_3A used in its formation because of its high bond-water content. The trisulfate is thus a material capable of acting as a cement and also of causing expansion in concrete [Halstead 1964, Bentur 1974, Ish-Shalom 1975]. Another possible expansion mechanism is that the ettringite swells in a colloidal form, since the combination of the high specific surface and the characteristic microcrystal structure of this collide, capable of attracting water, enables the imbibition of large quantities of water [Mehta 1974]. In any case, the expansion continues only as long as the concrete is kept moist.

Note that a major part of ettringite must form after attainment of a certain degree of strength [Bentur 1975], otherwise the expansive force will dissipate in deformation of a still plastic or semi-plastic concrete placing no stress on the restraint provided. Also, if the sulfoaluminate reaction continues for too long a period, it results in unsoundness, which can destroy concrete, as has been known for a long time. For instance, the formation of cement bacillus has been recognized as the cause of excessive expansion, cracking, and even complete disintegration of concrete subjected to intensive sulfate attack.

ASTM C 845-87 is the pertinent American specification for expansive cements. Several countries have also established specifications, for instance the USSR.

The only marked difference in physical properties of mortar or concrete made with an expansive cement and those made with regular cement appears to be the expansion. One should realize, however, that the paste of an expansive cement is subject to drying shrinkage in the same way as paste of a regular cement. The use of expansive cement produces a volume increase exceeding the amount of the anticipated shrinkage, after which the subsequent contraction on drying reduces or offsets the original swelling.

Cements having greatly different expanding abilities can easily be manufactured. The upper limit of linear expansion appears to be 2% without the concrete becoming unsound, that is, disrupted. About 60-80% of the total expansion occurs in the first 24-36 hr.

The expansion is influenced by numerous factors. Some of these are controlled by the producer of the cement, others by the producer of the expansive cement concrete. The expansion is increased by: (1) increasing the amount of the hydratable aluminates and the amount of calcium sulfate present in the cement; (2) increasing the amount of the expansive component; (3) decreasing the fineness of the cement; (4) increasing the water-cement ratio at constant cement content; (5) increasing the ambient humidity and/or temperature; (6) decreasing the specimen size; (7) decreasing the degree of restraint of specimen; (8) decreasing the mixing time; and (9) using structural lightweight aggregate [Budnikov 1969, ACI 1970, 1983, Monfore 1964, Klein 1963, Klieger 1971, Polivka 1976, Epps 1970].

By using blends of highly expansive cement and normal portland cement, mortars and concretes can also be made that have a wide range of expansive potentials. With proper selection of the ingredients and proportioning, the expansive potential of a mixture can be tailored to satisfy most job-expansion requirements. For instance, high levels of expansion with low cement contents are a possibility. Pumpable mixtures can also be produced [Hoff 1974].

Air-entraining, water-reducing, and water-reducing-retarding admixtures can be used with

expansive cements, but this may result in reduced
expansions as well as higher slump losses [Call
1979]. The best way to determine the effects of
admixtures is by trial mixes.

Shrinkage-compensating concretes develop compressive, tensile, and flexural strengths equivalent
in rate and magnitude to Type I or II portland cement
concretes. The strength falls off, however, as the
degree of unrestrained expansion increases [Klein
1968]. If the concrete is restrained during the
expansion and hardening, it has a much higher
strength than it would have had if the restraint had
been removed before the completion of the expansion.

Type K cement concretes require about 2-in
(50-mm) more slump than regular concretes to provide
an equivalent slump when a 30-min haul is necessary.
The minimum cement content generally recommended for
shrinkage-compensating concretes is about 500 lb/yd^3
(about 300 kg/m^3), and the minimum reinforcement for
restraint is approximately 0.15% steel. Such a
concrete when moist-cured will provide 0.03-0.1%
uniaxial restraint expansion which, in turn, will
induce about 25-100 psi (170-680 kPa) compressive
strengths in the concrete. If, however, the concrete
is appropriately reinforced, a large enough expansion
can induce tensile stresses in the steel greater than
120,000 psi (820 MPa), and compressive stresses in
the concrete up to 800 psi (5.5 MPa). This implies
that the concrete must gain strength at rapid rate to
a value sufficient to establish adequate bond to the
reinforcement before the expansion becomes of too
high a magnitude.

Self-stressing concrete applications include
precast concrete pipes, precast architectural panels,
highway pavements, sidewalks, and tunnel linings.
Applications of shrinkage-compensating concretes
include industrial floor slabs, highway and airport
pavements, parking decks, many types of water-holding
structures, food-processing and pharmaceutical buildings, and other storage areas where crack-free
surfaces are needed for functional or hygienic
reasons. [Mehta 1980b]

Short-term experiments on the durability and
permeability of expansive concretes have provided

favorable results [Gutman 1967, Li 1965, Palotas 1967]. Long-term experiences are not available yet. Other aspects of the behavior of shrinkage-compensating concretes in structures and pavements are discussed in an ACI report [ACI 1970]. In the same report, results of numerous laboratory tests are summarized with structural elements of self-stressing concrete.

The composition and manufacture of expansive cements, the fabrication of expansive concrete members, and chemical prestressing are still in the research-and-development stage. It seems, however, that the advantages of shrinkage compensating cements can be utilized in the field with traditional concreting techniques and a few additional measures. For instance, expansive cement should be protected from moisture during storage even more than a regular portland cement, and the storage period should be short. The more intensive slump loss of a fresh expansive concrete can be overcome either by special mixing procedures or by the use of icy water for mixing. Prolonged mixing and retempering reduce the potential expansion [Gaynor 1973]. In general, quality control of high order is essential in obtaining the desired expansion.

The increasing number of recent studies concerning expansive cements indicates the significance of this material [ACI 1970, 1973, Fifth Intern. Symp. 1969].

4.4 SPECIAL PORTLAND CEMENTS

The first rule in using any special cement is to follow the manufacturer's instructions. Improper or careless application can be costly.

4.4.1 White Portland Cement

The composition of a *white portland cement* is characterized by low iron and manganese contents, which are the compounds primarily responsible for the dark color of regular, gray portland cement. Usually the Fe_2O_3 content of white portland cements is limited to

0.5% maximum, whereas it is not surprising to find the Fe_2O_3 content of Type II cements as high as 3.5%. White portland cement is ground more finely and has lower specific gravity than gray cement, but in its other chemical properties as well as in physical properties it is essentially the same as regular portland cement. Because of the considerably higher burning temperature and the less efficient iron-free grinding, white portland cement costs nearly twice as much as standard cements [Blanks 1955]. Pertinent details can be found in the literature [North 1965b].

White portland cement is the ideal base when color pigments and/or colored aggregates are to be used for decorative purposes in cast stone, terazzo, and so on. To obtain good color, white concrete of rich mix proportion is generally used with low water-cement ratio. It may be applied in making portland cement based paints for greatest pigmentation effects. It may also be used unpigmented for light reflectance and high visibility in traffic marking, lines and curvings for highways and airport runways.

It may be mentioned here that high-alumina cements can also be produced in white. These cements are used mostly for refractory concretes such as in furnace and kiln constructions.

The use of white cements requires careful attention to cleanliness on the job site. Form oils and curing media should be of a type that do not cause discoloration. To produce uniformity over the entire surface, mix proportions, mixing time, as well as placing and finishing procedures for the batches have to be identical [French 1971].

4.4.2 Colored Cements

The base material for *colored cements* is portland cement with which 3-10% of chemically inert pigment, usually some metallic oxide, has been interground [Lea 1956]. Strong pigments can be added to regular portland cement, but for the light colors white portland cement has to be used as basis. Alternatively, the pigment can be added when mixing the concrete, but this is not a recommended practice because it is almost impossible to maintain uniformity of the color of such concrete. Satisfactory black concretes are very difficult to attain.

Colored cements are used in concretes for decorative purposes. Such concretes frequently fade even when they are made with pigments of approved types. This fading is due to the formation of a white calcium carbonate film over the surface resulting from the calcium hydroxide liberated during the hydration of the cement.

4.4.3 Oil-Well Cements

Although slightly modified (retarded) Types I, II, and III portland cements can be used for cementing around steel castings of gas and oil wells having depths not exceeding 6000 ft (1800 m), deeper wells require special slow-setting cements, so called oil-well cements.

The cement slurry has to be pumped into position before it sets under conditions of higher temperature and pressure prevailing in such depths. Also an oil-well cement must harden quickly after setting. The usual mix proportion of the slurry is about 16 lb of cement per gallon of water (2 kg/liter).

Oil-well cements may differ in composition from regular portland cement in that they have low C_3A contents, are coarsely ground, may contain friction-reducing additives and special retarders (starch, sugar, etc.) in addition to or in place of gypsum, and have been interground or blended with them during manufacture. API Standard 10A of the American Petroleum Institute specification for oil-well cements sets requirements for six classes. One of these, the class G cement, has no special admixtures, thus the addition for retarders, viscosity-control agents, and other chemicals is the responsibility of the user [Harder 1967].

4.4.4 Rapid-Setting Portland Cements

Rapid-setting portland cements have times of initial setting much shorter than the standard period of 45 min. The times of final setting are also short.

Rapid-setting cements can be manufactured by reducing the gypsum content, increasing the C_3A content, and increasing the fineness of grinding. Rapid setting can also be produced (1) by adding an accelerating admixture in a sufficiently large quantity when mixing the concrete, especially when Type III cement is used; (2) by adding portland cement, up to 15%, to high-alumina cement; and (3) by adding high-alumina cement, up to 40%, to portland cement. Mixtures of latter type are preferable, partly because of an easier control and partly because of a better strength development of such mixtures [Orchard 1973b, Robson 1952]. As a rule , however, the strengths of rapid-setting cements are considerably lower than those attainable with cements of normal setting times.

Rapid-setting cements are employed in emergency cases, such as sealing of water leaks. Rapid- and very rapid-setting mixtures have been used successfully by the shotcrete method of conveying.

4.4.5 Regulated-Set Cement

Regulated-set cement is a special rapid-setting portland cement containing an additional rapid-setting and rapid-hardening ingredient. The portland cement has a low C_2S content and no C_3A. Under patents issued to the Portland Cement Association, this latter has been replaced in the quantity of 1-30% by calcium haloaluminate having the formula $11CaO \cdot 7Al_2O_3 \cdot CaX_2$ in which X is a halogen. The halogen can be fluorine, chlorine, iodine, or bromine, but fluorine is preferred. Ordinarily the sulfate content of such cements is 1-12% SO_3 as calcium sulfate [Hoff 1975a]. Various formulations provide controllable setting times that range from a couple of minutes to 0.75 hr. Citric acid has been found to be an effective retarder in the quantity of 0.1%. The correspondingly rapid strength-development levels are also adjustable, being more or less proportional to the amount of fluoroaluminate contained in the cement. Compressive strengths of mortars at 1-hr age can exceed 1000 psi (6.9 MPa) when the calcium fluoroaluminate content is 12-13%.

The hydration products observed in the hardened paste of regulated-set cements are mainly ettringite and monosulfate hydrate at the early ages [Uchikava 1973].

Associated with the development of high early strength is the liberation of large quantities of heat. Apart from these characteristics at very early age, the long-term strength and other physical properties of regulated-set cement concretes are comparable to those of concretes made with Type I or Type III portland cements.

Curing procedures for this cement are the same as those for standard portland cement concretes. Handling, placing, consolidating, and finishing of fresh mortars and concretes must be completed within the handling time available. Revibration or rework after the initial hardening is not feasible.

Suggested applications include the manufacture of products such as block, pipe, and prestressed, precast, or extruded elements; paving for airports; patching and resurfacing of highways and bridge decks; vertical slip forming; fireproofing columns and beams; shotcrete; winter concreting; lightweight insulating concrete; and so on [French 1971, Hoff 1975b].

4.4.6 Waterproofed cement

Waterproofed cement is a portland cement interground with a small portion of a water-repellent material, such as calcium stearate, with the intention of reducing the water permeability of the mortar or concrete made with such a cement. There has been controversy as to whether or not these cements actually fulfill this aim. Alternatively, it is possible to add a permeability-reducing admixture to the concrete during mixing. In either case, the use of a chemical does not replace the normal rules for producing concrete with negligible permeability.

4.4.7 Hydrophobic Cement

A *hydrophobic cement* is a portland cement interground with a small amount of hydrophobic (water-repellent) material, such as oleic acid, with the intention of reducing the harmful effects (lumping, strength loss, etc.) of prolonged storage on the cement. The hydrophobic material forms a water-repellent film around each cement grain, which inhibits hydration by humidity until this protection is broken down when the concrete is mixed. As little as 0.1% oleic acid cuts significantly the strength losses that a cement may suffer from atmospheric moisture during long storage [Stoll 1958]. The odor of such cements is distinctly musty. The water-repellent agents usually act also as air-entraining agents. Concretes made with hydrophobic cement may show smaller strengths, especially at early ages, as well as smaller water absorptions, but this latter effect reduces with time. Hydrophobic cements are particularly suitable for transport in bulk [Kosina 1959]. They are used in soil stabilization and in concrete mixtures to be pumped.

4.4.8 Antibacterial Cement

An *antibacterial cement* is a portland cement inter-ground with an antibacterial agent with the intention of sublimating microbiological metabolism in the hardened concrete. This minimizes deterioration caused by fermentation, always present in the concrete floors of food-processing plants, dairies, breweries, and sugar plants. Such cements are also recommended for swimming pools, bath houses, and so on.

4.4.9 Barium Cements and Strontium Cements

Barium cements and strontium cements are ordinary portland or high-alumina cements in which calcium oxide is replaced completely or partially by barium oxide or strontium oxide, respectively. *Barium portland cements* can produce good strength combined with excellent sea-water resistivity. They have also good absorption capability of x-ray and gamma-ray, thus they are suitable for shielding concretes. Barium portland cements can produce excellent white cement, too [Eitel 1966]. *Strontium portland cements* develop strength through the hydration of tri-strontium silicate contaminated by small amounts of minor constituents. This silicate is similar to belite.

Considerable improvement of refractory properties is achieved with *barium aluminous* and *strontium aluminous cements.* Although some of these cements are not hydraulic but only air-hardening, they are refractory above 2850°F (1580°C) if they are low in iron oxide. Some of them can withstand temperatures up to 3600°F (2000°C) when combined with a suitable aggregate [Braniski 1962a 1962b]. Barium aluminous cements are also recommended for shielding concretes.

There is no ASTM specification for these special portland and other cements.

4.5 OTHER HYDRAULIC CEMENTING MATERIALS

4.5.1 Hydraulic Lime

Hydraulic lime is the predecessor and primitive version of portland cement. It is produced from the burning of impure limestone containing clayey matter. Unlike common lime, hydraulic lime will harden under water by taking up water and forming hydrated compounds. It is burned at relatively low temperatures, about 2200°F (1200°C), characteristically has a high free-lime content, low strength, and varying quality. The free lime is usually hydrated during the manufacturing process, but the calcium silicates are not. It was first produced and used in 1756 in England by John Smeaton [Lea 1956]. The significance of hydraulic limes nowadays is negligible.

4.5.2 Natural Cement

The term *natural cement* is defined by ASTM C 219-84 as hydraulic cement produced by calcining a naturally occurring argillaceous limestone at a temperature below the sintering point and then grinding to a fine powder. The same specification covers two types:

Type N: Natural cement for use with portland
cement in general concrete construction

Type NA: Air-entraining natural cement for the
same uses as type N.

The free-lime contents of natural cements are
less than those of hydraulic limes. Requirements
concerning loss of ignition and insoluble residue as
well as fineness, soundness, time of setting, air
entrainment, and compressive strength are also
included in the ASTM standard.

When the lime content is sufficiently high,
dicalcium silicate is the principal hydraulic
compound in natural cements, but no tricalcium
silicate is present. Otherwise these cements vary in
composition, a result largely of differences in
magnesia content. The quality of a typical natural
cement is intermediate between hydraulic lime and
portland cement. Although natural cements were used
in the United States as early as the construction of
the Brooklyn Bridge (around 1875), they are still
being used to a limited extend, mainly for economic
reasons, for masonry, autoclaved foamed concrete,
blended with portland cement for hydraulic structures, and so on.

4.5.3 Masonry Cement

The term *masonry cement* is applied either to
interground mixtures of portland cement clinker,
limestone, gypsum, and air-entraining agent; or
mixtures of portland cement, air-entraining agent,
and lime or ground slag or diatomaceous earth. The
latter group develops lower strengths. These cements
may also contain portland-pozzolan cement, natural
cement, slag cement, hydraulic lime, hydrated lime,
chalk, talc, pozzolan, or clay. Masonry cements are
used to make mortars for cementing together masonry
units, such as clay brick, concrete bricks, concrete
blocks, glass blocks, cut stones, and so on. The
purpose of these cements is to provide more plastic
mortar than portland cement. They also exhibit better

water retention. These goals are fulfilled primarily by the large amount of entrained air. They are often produced with high fineness. The strength of masonry cements is lower than that of portland cements, but this is usually irrelevant or even advantageous. Requirements concerning fineness, soundness, time of setting, air entrainment, and compressive strength of masonry cements are given in ASTM C 91-87a.

The low-strength masonry cement mortar usually consists of 1 volume of portland cement or 1 or 2 volumes of hydrated lime or other blends with which sand is mixed up to 5 volumes. A better mortar is obtained by mixing 1 volume of high-strength masonry cement to 2 or 3 volumes of sand. Sands having a fineness modulus of about 2.3 are usually used for this purpose.

4.5.4 Supersulfated Cement

Supersulfated cement is made by intimately inter-grinding a mixture of granulated blast-furnace slag (75% or more), calcium sulfate (in the form of anhydrite), and a small amount of lime, portland cement, or portland cement clinker (not more than 5%) as a source of calcium hydroxide. It is so named because the equivalent content of sulfate exceeds that for portland blast-furnace slag cement [ACI 1985d]. The properties of blast-furnace slag will be discussed later under "Latent Hydraulic Materials."

The activation of blast-furnace cement for hydration by calcium sulfate was discovered by Kuhl more than 60 years ago. The mechanism of this hydration differs from the hydrations of other cements. Here the early strengths appear to be due primarily to formation of ettringite and the later strengths to formation of the tobermorite phase [Lea 1956, Kuhl 1961b, Czernin 1962, Nurse 1964]. Such cements are used mainly in Europe, where suitable slags with high alumina and lime contents are available. They are called *Sulfathuttenzement* in Germany, and *ciment metallurgique sursulfate* or *ciment sursulfate* in Belgium and France.

Supersulfated cements, ground very finely, show as good strength development at normal temperatures as portland cements [Orchard 1973b]. In spite of this, their heat of hydration is low: about 40-45 cal/g at 7 days and 45-50 cal/g at 28 days. Such cements are highly resistant to attack by sodium and calcium sulfates, but they do not equal the resistance of sulfate-resistant portland cement and other special cements to attack by magnesium sulfate and some acid solutions [Thomas 1969, Evans 1960]. The hardening of supersulfated cements is quite sensitive to cold weather [Hummel 1955]. At higher curing temperatures, above about 100°F (38°C), the strength development may again drop seriously. Additional details can be found in the literature [Schroder 1969].

Supersulfated cement can be used in many cases in concrete and reinforced concrete where portland cement is applicable [Graf 1960]. It is particularly suitable for harbor structures, foundations, mass concrete, floors and containers resisting attack by diluted acids, and sewers [Blondiau 1959,1960]. It is unsuitable for steam curing or for carbonation treatment. Also, it should not be mixed with other cements. Mixtures leaner than 1:6 by weight and water-cement ratios less than 0.5 by weight are not recommended. Concretes made with supersulfated cement require careful and extended wet curing.

4.5.5 Slag Cement

Slag cement is a finely divided material consisting essentially of an intimate and uniform blend of granulated blast-furnace slag and hydrated lime in which the slag constituent makes up more than 60%. ASTM C 595-86 covers two types as follows:

> Type S: Slag cement for use in combination with portland cement in making concrete and in combination with hydrated lime in making masonry mortar

> Type SA: Air-entraining slag cement for the same use as type S.

The standard requires higher specific surface

and permits higher autoclave expansion as well as lower strengths for slag cements than for blended portland cements. The hydration products in pastes of slag cement activated with lime are, among others, tobermoritelike phases, ettringite, and hexagonal calcium aluminate hydrate [Nurse 1964].

Slag cements are hardly used for any other purpose than mentioned in the standard because of their low strength [Lea 1956]. Efforts have been made to improve the strength development by adding accelerating agents, such as fly ash with an alkali salt. [Nagai 1960]

4.5.6   MgO-Based Cements

Magnesium oxychloride cement consists of MgO powder and $MgCl_2.6H_2O$ solution to which aggregate(s) and pigment are added. The usual aggregates are: sawdust, wool, leather, cork, sand, gravel and crushed stone. It was discovered by Sorel, therefore it is also called *Sorel cement*. It has many good properties, as a flooring material, such as: high compressive and flexural strengths; good resistance to abrasion; good fire resistance; it is light therefore has low thermal conductivity; it has excellent bonding capability to many organic and inorganic solids; and it is resilient. It is also fairly resistant to attacks by grease, oils and paints, and to deterioration by alkalies, sulfates and organic solvents.

On the other hand, it is unstable in water, and loses strength on prolonged exposure to it. It can also be damaged by acids and some salts, and is also corrosive to aluminum and steel. [Ramachandran 1981]

Magnesium oxysulphate cement is also a good binding material. It consists of MgO powder and aqueous solution of $MgSO_4.7H_2O$. Although it is usually weaker than the magnesium oxychloride cement, the abrasion resistance is also lower, many properties of the magnesium oxysulphate cement are similar to those of magnesium oxychloride, including instability in water. However, the oxysulfate cement is less sensitive to higher temperatures which makes it useful for certain applications.

4.5.7 Phosphate Cements

These cements can be formulated to develop strength very rapidly along with short setting times. Two types are discussed below.

An *aluminum-phosphate* cement has been produced on an experimental basis. It consists of a granular solid cementing component, which is a mixture of MgO and flyash, and a liquid component which is a 50 percent by weight water solution of $Al(H_2PO_4)$. The magnesium oxide reacts chemically with the liquid component producing rapid strength and heat development.

Aluminum-phosphate cement mortars of stiff-plastic consistency showed [Popovics 1987e]

initial setting time	15 to 20 min.
final setting time	20 to 25 min.
compressive strength	250 to 550 psi
at the age of 1 hour	(1.75 to 3.85 MPa)
compressive strength	4000 to 7000 psi
at the age of 24 hours	(28 to 49 MPa).

The strength development of *magnesium-phosphate* based cements can be even faster. Several such cements are available commercially. A typical composition is a granular blend of MgO and $NH_4H_2PO_4$ with a small amount of fly ash. These react with water rapidly producing $NH_4MgPO_4.H_2O$ and $NH_4MgPO_4.6H_2O$ [Popovics 1987f]. The following properties were measured on mortars made with a commercially available magnesium-phosphate cement [Popovics 1987e]:

initial setting time	6 to 40 min.
final setting time	9 to 47 min.
compressive strength	1000 to 10,000 psi
at the age of 1 hour	(7 to 70 MPa)
compressive strength	7500 to 12,000 psi
at the age of 24 hours	(52.5 to 84 MPa).

Due to the special properties, these very rapidly hardening cements require modifications of most of the test method standardized for portland cements. [Popovics 1988b]

4.6   LATENT HYDRAULIC MATERIALS

4.6.1   General

Latent hydraulic materials are materials that in themselves possess little or no cementitious value but will, in finely divided form and in presence of water, react chemically with calcium hydroxide at ordinary temperatures to form compounds possessing cementitious properties. In other words, the hardening energy is dormant and becomes active only under the influence of an activator, such as calcium hydroxide or some other strong alkaline compound. When a latent hydraulic material is blended with portland cement and water, it becomes activated by the calcium hydroxide developed during the hydration of the cement.

Latent hydraulic materials can be used to make cementitious materials in the form of blended cements in different ways, as has been discussed. They can also be added to the concrete mixer together with portland cement and aggregate as finely divided mineral admixtures [Mather 1958]. The *economic* justification for using latent hydraulic materials can be economical and/or technical. It may provide the possibility of reducing cost by saving in the amount of cement used. A concrete containing such a material, if properly and economically proportioned, will usually include a smaller amount of portland cement than would otherwise be required. Therefore these materials may be used as "replacements" or "substitutes" (Ersatz) for part of the portland cement. Since the specific gravity of most of our latent hydraulic materials is about 2.5 and that of portland cement is about 3.15, the absolute volume of a replacement by weight will be 20-25% greater than that of the cement which is replaced.

The *technical* advantages of using latent hydraulic materials are usually more important than the economic reasons. They may serve in the fresh concrete as correctives for mixtures deficient in fine materials. If such a deficiency exists, which is typical for lean concretes, then the proper use of a latent hydraulic material improves the workability, and reduces the tendency for segregation and bleeding. If, however, the concrete does contain an adequate amount of fines, which is typical for rich concretes, then the addition of a such a material, as a rule, increases the water requirement, or impairs the workability [Powers 1932]. Also, the advent or air entrainment, with its marked effects on workability, bleeding, and segregation, has in great measure displaced latent hydraulic materials as a means for improving the physical texture of the fresh concrete. The use of air entrainment together with the such materials is possible. It should be kept in mind, however, that the amount of entrained air is usually reduced by the latent hydraulic material, particularly if the latter is fly ash [Larson 1964a].

The ecological benefit is equally important since the consumption of such materials as slag and, especially, fly ash reduces the air and water pollution which result from the storage of these waste materials in large quantities.

Note that the same latent hydraulic material may behave in concrete differently with different portland cements even when the cement type remains the same. Generally, improved strengths are obtained with cements that have higher contents of alkali and C_3A, and higher fineness.

The two most important groups of latent hydraulic materials are the *granulated blast-furnace slags*, and the *pozzolans*. The usual composition limits of these materials are shown in Fig. 2.4.

Other details concerning the effects of latent hydraulic materials on various properties of concrete can be found in the literature [ASTM 1950, Malhotra 1983a,b, 1987, 1989a,b].

4.6.2 Granulated Blast-Furnace Slags

ASTM C 219-84 defines the term *ground granulated blast-furnace (GGBF) slag* as a glassy nonmetallic product consisting essentially of silicates and aluminosilicates of calcium that is developed simultaneously with iron in a blast furnace and is granulated by quenching the molten material in water or steam and air. The composition of blast-furnace slag is determined by that of the ores, fluxing stone, and impurities in the coke charged into the blast-furnace. The major oxides - silica, alumina, lime, and magnesia - constitute 95 percent or more of the total oxides.

ASTM C 989-89 provides for three grades of ground granulated blast-furnace slags, depending on their respective mortar strengths when blended with an equal mass of portland cement. The classifications are Grades 120, 100, and 80, based on the *slag-activity index* (SAI) expressed as:

$$SAI = 100SP/P \qquad\qquad (4.1)$$

where
$\qquad$ SP = average compressive strength of slag-
$\qquad\qquad$ reference cement mortar cubes
$\qquad$ P = average compressive strength of reference
$\qquad\qquad$ cement mortar cubes.

The greater the grade, the higher the required slag-activity index of the GGBF slag. For instance, any slag sample of Grade 120 should have a SAI at least 110 at the age of 28 days.

In addition to strength requirements, the C 989-89 limits the residue on a 45-µm (No. 325) sieve to 20 percent and the air content of a mortar containing only GGFB slag to a maximum of 12 percent. The Canadian standard for granulated blast-furnace slag as a latent hydraulic material is CSA A 363.

Blended cements, in which GGBF slags are combined with portland cement, are covered by ASTM C 595. (Table 2.7)

There are only empirical formulas for the composition of slag with good hardening properties. Some of these are included in specifications, for instance in the German and Soviet standards. High lime and alumina contents are usually advantageous [Lea 1956, Kramer 1960, Kholin 1960]. Note, however, that the chemical composition of the slag is just one of the factors determining the cementitious properties of a slag. The others are: (a) alkali concentration of the reacting system; and (b) glass content and fineness of the the slag.[ACI 1987b]

The process of alkali activation is complex. When GGBF slag is mixed with water, the initial hydration is much slower than that of a portland cement paste. Therefore, portland cement or alkali salts are used to increase the rate of hydration and strength development. Hydration of GGBF slag in the presence of portland cement depends largely on breakdown and dissolution of the glassy slag structure by hydroxyl ions released during the hydration of the portland cement. On the whole, the compounds of the slag appear to form the usual products of portland cement hydration except that the slag hydrates without the liberation of calcium hydroxide [Kuhl 1961b, Fifth Intern. Conf. 1969, Czernin 1962, Nurse 1964].

Effects of chemical admixtures on the properties of concrete containing GGBF slag are similar to, although not necessarily identical with those for concretes made with portland cement as the only cement. For instance, the amount of high-range water-reducing admixtures required to produce flowing concrete is usually 25 percent less than that used in concretes not containing GGBF slag. In any case, the effects of chemical admixtures on concretes containing a slag should be checked by trial mixes. [ACI 1987b]

In most cases, GGBF slags have been used in proportions of 25 to 70 percent by mass of the total cementitious material. As the percentage of GGBF slag increases, a slower rate of strength gain should be expected, particularly at early ages. The proportioning techniques for concretes incorporating slags are similar to those used in proportioning concretes made with portland cement or blended cements.

The workability and placability of *fresh concretes* are usually improved by the presence of GGBF slag. The time of initial setting is typically extended by the slag one-half to one hour at temperatures around 73°F (23°C). The effect on bleeding depends on its fineness.

When compared to portland cement concrete, compressive and flexural strengths of *hardened concretes* containing GGBF slag of Grade 120 are usually lower at early ages and higher at 7 days and beyond. Other grades tend to impart reduced strengths at all ages. Concretes containing GGBF slag have been found to respond very well under elevated temperature curing. Conversely, larger relative strength reduction at early ages is expected with slag concretes when cured at normal or low temperatures. Other effects of incorporation of GGBF slag in concrete are, as follows:

> reduction in the temperature rise in mass concrete;
> reduction in the permeability of mature concretes;
> improvement in the sulfate resistance;
> reduction of expansion due to alkali-silica reaction.

Other properties of hardened concrete containing GGBF slag, such as, resistance to freezing and thawing or deicing chemicals, protection of reinforcement against corrosion, modulus of elasticity, creep, remain unaffected by the presence of slag.[Mielenz 1983, Nilsson 1988] Storage and handling of, as well as batching with GGBF slag are essentially the same as those for portland cement. [ACI 1987b]

4.6.3 Pozzolans

A combination of pozzolan and lime was historically the first hydraulic binder. This was developed primarily by the Romans more than 2000 years ago for various concrete structures. Parts of such structures still stand today. One of their quarries was near the town Pozzuoli, which provided the name pozzolan for this group of latent hydraulic materials [Lea 1956, Haegerman 1964].

Pozzolans may be natural in origin or artificial. Naturally occurring pozzolans include volcanic tuffs and pumices, trasses, diatomaceous earths, opaline cherts, and some shales.[Mielenz 1983, Price 1975] These form pozzolan Class N according to ASTM C 618-89. The most important man-made pozzolan is *fly ash* (pozzolan class F), which is the finely divided residue resulting from the combustion of ground or powdered coal. This pozzolan is discussed in the next chapter.

The composition of a pozzolan or pozzolanic material is siliceous and aluminous, and varies widely. ASTM (C 618-89), the Bureau of Reclamation, and the Corps of Engineers, as well as many foreign countries, have all established specifications for chemical and physical properties of pozzolans. The chemical requirements usually call for a minimum of 70-75% of the pozzolan to be composed of SiO_2 + Al_2O_3 + Fe_2O_3. Unfortunately, the latent hydraulic ability of a pozzolan depends primarily on its reactive, or soluble, silica, alumina, and iron oxide contents; these are not identical — not even proportional — to the total silica, alumina, and iron oxide contents. The siliceous ingredient is in an amorphous state in a good pozzolan. Crystalline siliceous materials, such as quartz, combine with lime very slowly, except under curing at high temperatures [Higginson 1966]. Also, calcination, that is, heating the pozzolan at temperatures around 1000°C (1800°F) for a short time, or chemical treatments may increase the reactivity of many latent hydraulic materials but not all of them. [Mielenz 1950, Alexander 1955a, 1955b, Turriziani 1964a]

There have been several chemical and physical methods recommended for the determination of the amounts of reactive ingredients in pozzolans, that is, for the prediction of how a pozzolanic material will behave in a mortar or concrete, but none of them is completely satisfactory yet [Alexander 1955c, Massazza 1974].

Two methods are specified by ASTM C 618-89 and C 311-89 for the estimation of pozzolanic activity index:

1.With portland cement by making 2-in. (50-mm) Ottawa-sand mortar cubes where 35%, by absolute volume, of the cement is replaced by pozzolan. The obtained compressive strengths are compared to the compressive strengths of comparable control cubes made with 100% portland cement. At the age of 28 days the index, that is, the strength of the fly ash mortar, should be min. 75% of the control mortar.

2.With lime by making 2-in. (50-mm) cubes that contain 1 part hydrated lime, 9 parts of standard Ottawa sand, by weight, and pozzolan equal to twice the weight of the lime multiplied by a factor obtained by dividing the specific gravity of the pozzolan by the specific gravity of the lime. After a special curing of 7 days the compressive strength of these cylinders should be at least 800 psi (5.5 MPA).

Strengths of a concrete with pozzolanic materials are typically lower at early ages, and higher at later ages, than are obtained with portland cement alone [McIntosh 1960, Kokobu 1969, 1974]. Simultaneously, the temperature rise is decreased, resulting from heat of hydration of the cement. The setting times are also increased by the addition of pozzolanic materials, although the values are usually still within the normal specification limits. Use of pozzolanic material with other than sulfate-resisting portland cements generally increases resistance of the concrete to aggressive attack of sea water, sulfate solutions, and natural acid waters. It has also been indicated by laboratory tests that certain pozzolans are capable of reducing the expansion caused by alkali-aggregate reaction (see Chap. 9). Improved impermeability of the concrete frequently accompanies the use of pozzolan, especially in lean mixtures. In general, the creep of concrete as well as the drying shrinkage are greater when pozzolans are used than when they are not. Freezing and thawing resistance of non-air-entrained concretes has also been observed to be impaired by the use of pozzolans [Davis 1950, Abdun-Nur 1961]; this resistance of air-entrained concretes seems to be influenced mainly through the effect of the pozzolan on the amount of entrained air.

Considering all these, the chief use of pozzolans is in mass concrete, harbor works, sewers, and when chemical reasons justify it.

The mechanism of the improvements by pozzolans in harsh *fresh concretes* consists of an increase of the paste content, an improvement of the lubricating ability of paste in the concrete, and an increase in the cohesiveness of the mixture. In these respects the fineness and the particle shape of the mineral admixture are significant; the chemical characteristics are only of secondary importance. However, the carbon content of a fly ash has a definite role in the reduction of the amount of entrained air in the fresh mixture, because it has a capacity for the adsorption of air-entraining agents.

The main source of the effects of pozzolans on the *hardened concrete* is the chemical reactions between the reactive siliceous and aluminous compounds of the pozzolan and the calcium hydroxide liberated from the portland cement during hydration. This process forms mechanically and chemically stable cementing substances of similar types as those derived from portland cement clinker hydration. [Lea 1956, Malquori 1960, Ludwig 1960, Jambor 1963, Turizziani 1964a, Terrier 1966, Fifth Interntl. Symp. 1969]. The process of binding the hydrated lime is intensified by the presence of alkalis. Nevertheless, it is fairly slow but continues for a long period if the water needed for the reaction is available. This explains not only the low strengths at early ages and the high strengths at later ages, but also the improved chemical resistance of pozzolan concretes, since the free hydrated lime represents a weak spot in the hardened cement paste.

Pozzolans and other latent hydraulic materials should be handled, conveyed, and stored in the same manner as cement.

4.7   FLY   ASH   AND   SILICA   FUME

4.7.1 General

According to ASTM C 618-89, *fly ash*, or *pulverized-fuel ash (PFA)* in the U. K., is a "finely divided residue that results from the combustion of ground or powdered coal." It is primarily the inorganic portion of the source coal in a particulate form.

Two kinds of fly ash are produced from the combustion of coal:

Class C - High, more than 10%, calcium content produced from sub-bituminous coal
Class F - Low, less than 10%, calcium content produced from bituminous coal.

Fly ash is available in most areas of the United States. Approximately 50 million tons of fly ash are produced annually in America of which 10% is used in concrete. Unused fly ash in large quantities represent environmental hazard and its storage can be expensive.

The amount of literature concerning fly ash is considerable, including an ASTM standard (C 311-89) for sampling and testing fly ash for use as an admixture in portland cement concrete.[Abdun-Nur 1961, Faber 1967, 1974, Proc. Fifth Interntl. Symp., Part IV 1969, Kokobu 1969, 1974, Helmuth 1987, Bakker 1982, Berry 1986, Halstead 1986, Dunstan 1983, American Coal 1987, Malhotra 1983a,b, 1987, 1989a,b].

4.7.2 Fly Ash Properties

A number of standards exist which specify the desired properties of the fly ash. In the United States, ASTM C-618 is the standard. There, certain requirements concerning both the physical and chemical properties as well as performance requirements of the two classes of ash are specified. It is usually agreed upon that the following properties should be considered when assessing the suitability for fly ash use in concrete: pozzolanic activity, particle size or surface area (sieving is preferable to surface area measurements for this), carbon content or lose on ignition or LOI (to minimize air entrainment variability), moisture content (protect purchaser from buying wet ash), and various chemical parameters (sulfates, magnesium, and alkali contents.) Important performance tests include pozzolanic activity, alkali reactivity, soundness and shrinkage, etc.

Also, the hydraulic behavior of a fly ash is influenced by (1) its carbon content, which should be as low as possible; (2) its silica content, which should be finely divided and as high as possible; and (3) its fineness, which should be as high as possible [Orchard 1973b]. In addition, both the strength contribution and the water reducing capacity can be variable with different portland cements even if a constant ash is used.

Although the constituents of fly ash are typically not present as oxides, it is customary to represent them in this manner. The major oxide components are usually SiO_2, Al_2O_3, Fe_2O_3, CaO, MgO, and SO_3. If the sum of the first three ingredients is 70% or greater, the fly ash is technically considered as Class F. For a Class C fly ash, the sum of the first three components is greater than only 50%. The principal active ingredient in Class F ash is siliceous or aluminosilicate glass. For Class C, calcium alumino-silicate glass is the active ingredient. The glassy material in Class C ash is often more reactive than the glass in Class F ash. In addition, Class C ash may contain additional reactive components such as free lime, anhydrate, C_3A, C_3S, etc. [ACI 1987c] It has been noted that the reactivity of a glass is influenced by the characteristics of the particular cement it is combined with.

The physical properties of a fly ash also contribute to improvement of concrete quality. The majority of fly ash particles are spherical in shape although this may depend on the type of ash, burning conditions, collection conditions, etc. Modern ashes are typically finer than ashes collected in the past. Today, most particles pass the No. 325 sieve. The particle size distribution of the ash usually stays

constant with fuel type and processes. Tests have
shown that higher ash fineness improves the
properties of the concrete. Specifically, it was
noted that a large percentage of particles finer than
10μm had a positive influence on the strength. The
specific gravity of Class C ashes (2.4-2.8) tend to
be higher than Class F ash.

The chemical specifications of fly ash require
that the minimum for the sum of the oxides (reactive
glasses) is 70% for Class F and 50% for Class C ash.
However, this definition has been criticized.
Alternatively, Class C ash can be defined as that ash
having a minimum calcium oxide content of 10% and
self cementing capabilities. For both Classes of ash,
however, an upper limit of SO_3 content is usually set
at 5%. This is because the compound has an effect on
setting time and other properties as well as making
the concrete more susceptible to sulfate attack. A
limit of moisture content (3% max.) is necessary to
insure handling characteristics as well as to prevent
premature hydration of Class C ashes. For both types
of ash, the maximum allowable LOI is 6% although
there is speculation that this value should be lower,
particularly for air entrained concretes. Similarly,
a maximum content for alkalis is set at 1.5% for both
classes of ash.

The primary physical requirement of both classes
of fly ash is that not more than 34% should be
retained on the No. 325 (45 μm) sieve by wet sieving.
This is an important specification because coarser
fly ash fails to react pozzolanically. Other physical
requirements include water requirement, soundness,
variability limits in terms of specific gravity,
fineness, and effect on AEA, increase in drying
shrinkage (uncommon), and reduction of alkali-
aggregate reaction.

4.7.3   Test Methods

Test and sampling methods for fly ash are outlined in
ASTM and elsewhere. In a RILEM report the methodology
for various fly ash tests are outlined. Aside from
general comments and sample preparation, the
following fly ash tests are discussed in detail:

moisture content, loss on ignition, silicon oxide content, soluble silicon oxide content, ferric oxide content, aluminum oxide content, calcium oxide content, magnesium oxide content, sulfuric anhydride content, chloride content, free calcium oxide content, total alkali oxides content, ammonium content, glass phase content (by X-ray diffraction spectroscopy and by determination of insolvable residue in HCl methods), particle density determination (by pycnometer bottle and Le Chatelier flask methods), and fineness determination (by dry sieving, wet sieving, Blaine air permeability and laser methods). The following tests for fly ash-cement pastes, mortars, or concretes are outlined: soundness (expansion test), water requirement (expressed as water content of test specimen divided by water content of control specimen to achieve equal specified consistencies), preparation and curing of specimens, determination of compressive strength (28 days), and the pozzolanic activity index at a quoted age and type of curing (activity is defined as ratio of test strength to control mix strength).[RILEM 1989] In general, it is agreed that these performance tests are an inadequate means to predict fly ash performance in concrete. However, they may be used as a means to detect gross variations in fly ash reactivity.

Tests for ash should provide the following needs: ensuring that the ash will not be harmful to the desired properties of the concrete; indicating the potential performance of the ash in concrete; and monitoring key properties to ensure uniformity.. For reasonably consistent Class F ash, the most important ash properties are fineness, LOI, and autoclave expansion. Significant properties of Class C fly ash that affect performance in concrete include fineness, LOI, autoclave expansion, SO_3, CaO, and alkali content. Variability in ash color should also be monitored.

It is important to check for contaminants, in the form of start-up oil or stack additives, in the fly ash. These additives among others may have an adverse effect on the suitability of using the ash in concrete.

The traditional chemical method for determining fly ash reactivity is based on measuring the amount of lime which combines with the ash when mixed with hydrated lime or portland cement. Variations of this test method are still frequently used although some investigators find little correlation between lime absorption and strength development. Another notable chemical method is the measurement of acid soluble oxides in an ash. However, it is generally accepted that chemical methods are unlikely to be of value in predicting long term strengths. The reason for this can be attributed to the fact that most fly ashes have different responses to curing conditions.

Several physical methods have been used to assess the reactivity of fly ashes including measuring the heat of solution in acid, differential thermal analysis, petrographic analysis, and various specific surface measurements. At present however, these methods do not appear able to reliably predict the reactivity of fly ashes.

The reactivity of a fly ash can be determined by a method developed by Alexander that measured compressive strength of fly ash-lime mortars under standard curing conditions. Because of the sensitivity of the pozzolanic reaction to curing temperature, the tested specimens can be cured at different temperatures and the results averaged out and the apparent reactivity can be estimated. However, many investigators concluded that such test methods do not reliably predict the reactivity of fly ashes, particularly of blended cements.

The most direct and reliable method for determining the reactivity of a fly ash in a portland cement concrete mixture is to make trial mixtures for a specific ash and cement. However, the correlation between long term strength and the strength of accelerated test specimens is often poor because of a specific fly ash's sensitivity to curing temperature. [Clifton 1977]

The tests for determining an index of pozzolanic activity can be divided into three types: chemical, physical, and mechanical. The primary chemical test is the Fratini method where the reactivity of an ash is assessed by determining the quantity of lime which combines with the pozzolana after curing a specified time and temperature in lime solution. The Flourentine method assesses the reactivity by measuring the amount of SiO_2 rendered soluble in cold HCl by the reaction of pozzolana and lime solution. Jambor proposed a physical test for fly ash reactivity which measures the heat evolved during dissolution of the pozzolana in a mixture of nitric and hydoflouric acid together with a determination of the insoluble residue. (A measurement of the dissolved silica.) ASTM defines pozzolanic reactivity in terms of two mechanical tests. The pozzolanic activity index with lime is represented by the strength of a specified lime-pozzolan mortar cured for 1 day at 23°C and 6 days at 55°C under sealed conditions. The index with portland cement is the ratio of the strength of a specified cement-pozzolan mortar to that of a control mortar. The sealed specimens are cured for 1 day at 23°C and 27 days at 38°C. The cement-pozzolan mortar is proportioned by 35% cement replacement by volume and the water content is adjusted to give a constant flow value. However, neither of these mechanical tests is considered to give a value of the fly ash concrete. Reactivity tests for blended cements are similar to the above mentioned tests for fly ash alone. In general, tests on pozzolanic material do not show good relationships with mature cement-pozzolan systems; the use of these tests is limited to rejecting unsuitable materials. In addition, there is at present no generally accepted method for assessing the effect a given fly ash will have upon the strength of a mature fly ash concrete.

Lea developed a method for evaluating pozzolanic cements that took into account the temperature response of the strength of pozzolanic concrete with respect to standard concrete. The values obtained for this difference were shown to correlate with 180 and 365 day strengths of normally cured concretes. These differences were claimed to be a characteristic of the pozzolana used. The British Standard test (BS 3892) for fly ash pozzolanicity is based on this technique.

In order to test the reliability of using the

"Lea ratio" as a British Standard, Lea prepared some test specimens with various kinds of fly ashes and portland cements. All specimens had the same mix proportions (30% cement replacement, etc.) and several types of curing were employed. It was found that the strength ratio, the ratio of the strength of the blended cement mortar to that of the portland cement mortar, seemed to depend upon the type of portland cement that was used. It was found that these differences were inversely related to the strength of the ordinary portland cement mortar cubes. It was also found that the strength ratios were poorly correlated to the "Lea difference" because each portland cement behaved differently. (The Lea difference is the difference in strength of blended cement mortar after Lea cycle curing and after 7 day 20°C water curing.) However, it was found that the "Lea ratio" correlated very well for all cement and fly ash types with the strength ratios because this ratio is less dependent upon cement type. (Lea ratio is the ratio of the strength of blended cement mortar to that of the corresponding ordinary portland cement mortar cured the same way.) Hence, this method is accepted as a British Standard to estimate pozzolanic reactivity.

When comparing the Lea ratio with 90 and 180 day strength ratios, it was noted that the relationship was dependent on the type of portland cement used. The linear regressions for the individual cements have a similar slope although they are displaced from the common regressions by an amount depending upon the type of portland cement. This means that using the Lea ratio to estimate long term strength on known ash but unknown portland cement is unreliable. In other words, tests must be undertaken on a specific cement, but not necessarily on a specific fly ash with this method.

4.7.4   Strength Development and Hydration

The contribution to the strength of the concrete, by the flyashes, can be determined by assuming the ash to be an inert filler and then calculating the expected strength loss due to reduced cement content. The comparison of the expected strength to the actual strength gives an indication of the contribution of the fly ash to the strength [Popovics 1991].It was found again that the strength contribution varied with the type of fly ash as well as the type of cement. In particular, it was found that the strength contribution of the ashes was inversely proportional to the 45μm sieve residue while little correlation was noticed between strength contribution and specific surface or chemical analysis. ASTM found that concrete strength was influenced by the type of cement that was used rather than the type of ash in most cases. The strength contributions of the ash also seem to be dependent upon the curing conditions. At early ages, it was noted that the presence of fly ash in concrete may actually take away from the strength; again the actual strength was compared to calculated strength assuming the ash to be inert filler. Some researcher demonstrated that the presence of ash retards C_3A and C_4AF hydration to a greater extent than gypsum during early hydration. Conversely, the presence of a "fine powder" increases the effectiveness of the calcium silicate's hydration. This would suggest that the effectiveness of the ash at early ages is improved for high calcium silicate cements and lowered for high aluminate cement. Test results support this. It was also noted that the metal alkali content of the cement was directly proportional to the strength contribution of concrete cured under a Lea cycle. The differences between the ash types for a given cement are attributed to their fineness, again the "fine powder effect" that increases the effectiveness as reaction products precipitation sites. However, it is interesting to note that the 180 day strengths of the blended cement mortars are largely independent of the cement used, so that, provided a suitable ash is selected, the choice of portland cement type is unimportant.

The greater the reactivity of an ash, the greater it is accelerated by increasing curing temperature; hence, simple maturity factors as used in portland cement systems cannot be used.[Dalziel 1983]

The principal product of the reaction of fly ash and lime is CSH, although there are some differences with respect to portland cement CSH. The reaction of fly ash depends largely upon the breakdown of the glassy structures by hydroxide ions and heat evolved from cement hydration. It appears that at early ages the performance of ash in concrete is more dependent on its physical and morphological properties (fineness, etc.) than on chemical composition, but the ultimate reactivity depends on the total silica and alumina available. Some suggest that the reaction of fly ash in concrete is a two step process. During early curing, the primary reaction is with alkali hydroxides. Subsequently, the main reaction is with calcium hydroxide. When these reactions commence has been disputed however, with test results ranging from immediate to two week delayed reaction. Therefore, there are no general quantitative relationships between the progress of the fly ash reaction and development of strength. Some researchers theorize that the pozzolanic reactions may be initiated by the absorption of calcium hydroxide on the surface of the fly ash particles. The reaction products are then precipitated in a ring around the reacting ash particle.

The effects of intrinsic factors on the reactivity of a fly ash are difficult to quantify because of the wide range of chemical and mineralogical composition. In addition, most studies reveal that no well defined relations between the inorganic constituents and the development of strength have been found. The glassy phase (silica, alumina, and ferrous oxides) of the ash is recognized to be the most reactive phase in fly ash. However, it has been found that the physical condition of the glassy particles also influence their reactivity. Non-porous glassy particles have been found to be more reactive than sponge-like particles. According to Rosauer, the factor which has the greatest influence on fly ash reactivity is the extent of strain in the glass particles, the more highly strained particles being more reactive. Strain can be developed by rapid cooling of the ash. Other intrinsic properties that may have an influence on ash reactivity are particle density, size distribution, fineness, and specific surface although some investigators found little correlation between these properties and ash reactivity.

Several investigators have reported that the strength development of fly ash-cement mixtures depends on physiochemical properties of the cement as well as the ash. Most fly ashes react more rapidly with cements having higher C_3S and C_2S contents. The contribution of fly ash to the strength development of concrete differs greatly depending on the curing temperature. During the first several months, the strength development generally increases with higher temperatures. At later ages, the role of curing temperature decreases. Similar to standard concrete, moisture content has an effect on the strength development of fly ash concrete. The reactivity of an ash may be improved by a number of methods including quick quenching, grinding, and "dressing" unacceptable ashes with metallurgic ashes.[Clifton 1977]

4.7.5   Effects of Fly Ash on Fresh Concrete

The addition of fly ash to concrete has a considerable effect on the properties of fresh concrete. There is agreement that low calcium ashes show some retarding influence on the mix. This may be due to the fact that the cement is becoming more "diluted."

The effects of fly ash on fresh concrete are well known. Workability and pumpability of concrete is improved with the addition of ash because of the increase in paste content, increase in the amount of fines, and the spherical shape of the fly ash particles. Note that this improvement in workability may not be true for coarse, high carbon fly ashes. The use of fly ash may retard the time of setting of concrete. This is especially true of Class F ashes. Class C ash may or may not extend setting time and there are results that show reduction of setting time.

Fly ash, in contrast to other pozzolans, reduces the water requirement of a concrete mix. It has been

suggested that the major influencing factor in the
plasticizing effect of fly ash is the addition of
very fine, spherical particles. In fact, it has been
shown that as the particle size increases, the
plasticizing effect decreases. This indicates that
some fly ashes do not improve workability. The
rheology of fly ash cement pastes has been shown to
behave as a Bingham model. Finally, the inclusion of
some fly ashes in a mix reduce bleeding and
segregation while improving finishibility. This again
can be attributed to the increased amount of fines in
the mix and lower water requirement.

It is reported that the use of some fly ashes
causes an increase in the amount of air entraining
admixture required in concrete. It is proposed that
carbon in the fly ash absorbs the AEA therefore
requiring more to be used as an active role in the
mix. In general class C fly ashes require less AEA
than class F ashes.
Also, there may be an increased rate of air content
loss with manipulation if this ash is used. [ACI
1987c]

4.7.6   Effects of Fly Ash on Hardened
Concrete

4.7.6.1 Strength The largest effect that fly ash
addition may have on hardened concrete is its effect
on *strength development*. The most important variables
that effect the strength development in fly ash concrete are the properties of the fly ash such as chemical composition, particle size, and reactivity. The
curing conditions also have an effect. In general,
the rate of strength development in concrete seem
only to be marginally affected by high calcium fly
ashes whereas it is drastically affected by class F.
The particle size of the fly ash may or may not have
an effect on the strength development, although research indicates that the finer the ash, the more
intensive the pozzolanic activity. Also, some benefits of intergrinding fly ash and cement have been
noted. Finally, it appears that fly ash concretes are
more sensitive to curing temperature than standard
portland cement concretes. This means that curing at

elevated temperatures   increases (in excess of   30°C)
the ultimate strength of   fly ash concretes and that
large quantities   of fly ash may be used successfully
in high temperature applications. The rate of reaction of fly ash-cement systems is clearly increased by
temperature, as is the case   for portland cement concrete. Yet, the products of   hydration do not exhibit
the poor mechanical properties associated with curing
portland cement at elevated temperatures.   This would
suggest   that the   products of   fly ash-cement hydration, their   relative proportions or   morphology, are
significantly   different than   from those formed from
thermally   accelerated hydration   of portland   cement
alone.

    Of the properties of hardened concrete that are
affected, the one that is usually   given most   attention   is   the compressive   strength   and the   rate of
strength gain. Compared with standard concrete, a mix
containing   typical   Class F   ash may   develop   lower
strengths   at   early   ages(up to 7 days).   However,
equivalent   strengths   at early   ages   with   fly ash
mixes can be achieved   with attention   to mix   design
and   use of admixtures such as accelerators or superplasticizers. However, fly   ash   concrete with   equivalent   or lower   strength at   early   ages   may   have
equivalent or higher strengths at later ages   because
of   continued   pozzolanic   activity. Class   C   ashes
often   exhibit   a higher rate   of reaction   at early
ages than Class F. However, both   ash types   seem   to
contribute   to   the   late strength,   particularly   if
the ash is fine. Some tests   show that certain   Class
C   ashes   may   not show   the   later strength typical
of   Class F fly ash. In addition,   changes in   cement
source may effect concrete strengths   as much as 20%.

4.7.6.2   Other Properties   Since   fly ash has   an
effect on the compressive strength of concrete,   it
naturally has an effect on associated properties such
as modulus of elasticity,   creep   strain rate,   and
impact resistance.   In general, the addition of fly
ash to concrete has   no effect on bond to steel, bond
to   old   concrete,   drying   shrinkage,   or   abrasion
resistance.

    The replacement   of cement   with low calcium fly

ash results in a significant reduction in the *heat of hydration*. Of course, temperature rise depends on other factors such as thermal properties of the concrete, rate of heat loss, dimension of the element, etc. However, high calcium fly ashes may or may not reduce the heat of hydration.

The heat of hydration of a mass concrete structure can be reduced by fly ash replacement. Generally, as the amount of cement that is replaced is increased, the heat of hydration of the concrete is reduced. However, some Class C ashes contribute to the early temperature rise.

Fly ash content may increase the required AEA dosage to achieve a given *air content* and may also increase the rate of entrainment loss of prolonged mixing of fresh concrete. When hardened concretes of similar entrained air content and strength are compared however, there are no apparent differences in freeze-thaw durability between plain and fly ash concretes. Of course, it should be kept in mind that fly ash concretes develop strength slower than plain concrete, and this may have an effect on the durability of young concrete.

Published data indicate that fly ash has little influence on the *elastic properties* of concrete. In general the modulus of elasticity parallels the strength development in concrete; that is, at early ages there is a reduction and at late ages an increase in E. However, the elastic properties of portland cement concrete and cement-fly ash concrete are similar.

Similarly, the incorporation of fly ash appears not to influence the *abrasion* or *erosion resistance* of a concrete although sufficient information in this area is lacking.

Data on the effect of fly ash content on *creep* is limited. However, preliminary data shows that concretes with large ash contents creep more than low ash or no ash concretes.

Some studies have shown that use of fly ash in concrete may reduce *drying shrinkage* to some degree. However, these improvements are usually not significant.

A number of investigations have been made

concerning the effect of fly ash on concrete *permeability*. From the results, it is clear that the permeability, at a given porosity, is directly related to the quantity of hydrated cementlike material. Thus at early ages, fly ash concrete is more permeable than standard concrete because there is less hydrated material. At later ages, as the pozzolanic reactions occur to produce more hydration products, the permeability is significantly reduced. That is, there is a transformation from relatively large pores to small pores as a result of the pozzolanic reaction. Similarly, it has been found that slag and fly ash cements are more effective in limiting chloride ion diffusion in pastes.

Because of the pozzolanic reactions, fly ash chemically combines with calcium, potassium, and sodium hydroxide to produce CSH thus reducing the risk of leaching calcium hydroxide. As a result, the permeability is reduced.

Interestingly, the addition of fly ash to concrete had varying effects on the *sulfate resistance*; some ashes had a significant effect in improving resistance while others had no effect or adverse effects. Studies have been contradictory; thus, no definitive light has been shed on the controlling mechanisms of sulfate resistance of pozzolanic concrete. Some researchers believe that the CaO and Fe_2O_3 content in the ash are the main contributors to the resistance or susceptibility of fly ash concretes to sulfates. Specifically, the sulfate resistance was found to increase as the CaO content decreased and the Fe_2O_3 content increased. This seems to suggest that, in general, Class F ashes have better resistance to sulfates than Class C ashes assuming equal maturity. However, other studies contradict these findings by proposing that pozzolans of high fineness, high silica content, and highly amorphous silica is the most effective pozzolan for reducing sulfate expansion in concrete. The resistance of fly ash concretes to *marine environments* is an unresearched area although it can be speculated that the addition of fly ash may improve the concretes' behavior in such environments.

Some tests have been performed which indicate

that the presence of fly ash in a concrete may decrease depth of *carbonation*, especially at later ages. In general, it can be stated that good quality fly ash concrete is comparable to plain concrete in its resistance to carbonation.

The incorporation of fly ash appears not to influence the behavior of concrete at *elevated temperatures*. Loss of strength and changes in other structural properties occur at about the same temperatures for both concretes.

A variety of natural and artificial pozzolans have been found to be effective in reducing the damage caused by *alkali-aggregate reactions* (AAR). There are substantial published data to indicate that Class F ashes are effective in reducing expansion when used at replacement levels of 25-30%. However, this is limited to the alkali-silicate reactions. It is also widely reported that larger volumes of fly ash substitution can reduce AAR expansion even more significantly. It is questionable, however, whether early strength losses caused by 40-50% replacement with Class F ash would be tolerable. Similar levels of replacement with Class C ash may be acceptable.

The source of the soluble alkali is unimportant; soluble alkalis from fly ash (some Class C ashes with high alkali sulfate contents) are just as detrimental as those from cements, and some Class C ashes are known to be ineffective or detrimental in relation to alkali reactivity. However, other researchers concluded that only small differences were found between the effectiveness of different ashes. These differences could best be correlated to the differing pozzolanicity of the ashes rather then the alkali contents of the ashes although there was some correlation. In addition, the available alkali content of the ashes gave no better correlation with the observed expansion than did the total alkali content. However, in all cases the addition of ash reduced expansion more than would be expected for a simple dilution of the alkali content. Interestingly, some tests show that small additions of fly ash to mortars containing reactive aggregate may increase expansion until a critical amount of ash was used where the expansion would decrease. Researchers have concluded that this pessimum point, that is, the amount of ash that results in maximum expansion, may be related to the CaO content of the ash. As the CaO content increases, so does the the fly ash replacement value that results in maximum expansion. This would seem to indicate that Class C ashes would have to be used in greater quantities than Class F to achieve the same amount of AAR protection. [Carrasquillo 1987]

It is generally accepted that fly ash concrete does not decrease the *corrosion protection* of steel members in concrete when compared to normal concrete. However, if the late age permeability of a fly ash concrete is lower than a standard concrete, this may affect the corrosion rate of the steel. There is no evidence that the inclusion of fly ash reduces the concrete "protective" alkalinity as this alkalinity is controlled by the dissolved alkali ions rather than quantity of $Ca(OH)_2$ present. [Malhotra 1987]

4.7.7   Proportioning

Several methods of fly ash proportioning in concrete have been established. These include simple replacement of portland cement by fly ash, simple addition, modified replacement, and variations of these. A reasonable concept is to preestimate the relative amount of fly ash to be used in the mixture, and then follow a proportioning technique similar to that used in proportioning concretes made with portland cement or blended cement. Methods for the preestimation of the optimum amount of fly ash are available. [Popovics 1982b] It is possible to design a fly ash concrete of comparable strength to standard concrete at any age. However, factors such as curing condition sensitivity, sensitivity to particular cement, etc., must be considered.

The interaction between fly ash and other admixtures and by-products in concrete must be studied. In general, there are four classes, other than air entraining admixtures, of concrete admixtures that are used along with fly ash: accelerators, retarders, water reducers, and superplasticizers. Results show that all types of accelerators are effective in the presence of fly ash. However, they do not completely compensate for the slow rate of strength gain that results with ash substitution. Similarly, fly ash does not seem to consistently affect the performance of water reducing admixtures or superplasticizers. In the case of superplastcizers however, some test results show that their effectiveness may be reduced by the presence of fly ash. The simultaneous use of fly ash and other mineral admixtures was investigated. It was found that fly ash and condensed silica fume or fly ash and ground slag can be used together in concrete. When used with a modest amount of silica fume, the early strength of the fly ash concrete improved considerably. In addition, the later strength also improved dramatically. However, the use of silica fume increases the water requirement of a mix, and this has to be accounted for.

4.7.8   Applications

Fly ash is typically introduced into concrete mixes to reduce cement costs, reduce heat of hydration, improve workability, reduce the effects of aggregate reactivity, or attain required long term strength levels, and for ecological reasons.

Fly-ash modified concrete can have some specific applications in construction. Ash is often used in the production of high strength concrete because the addition of ash may reduce the water requirement without a loss in workability and because fly ash produces increased strength at late ages that cannot be achieved with additional portland cement.[Popovics 1990b] Fly ash is also often used in the production of roller compacted concrete (RCC). Here, fly ash may be used in two different applications: lean concretes with lower ash contents or rich concretes with high ash content. [ACI 1987c] In RCC, the paste content must be kept high enought to assure compaction; fly ash does this.

Because the performance of a given fly ash vary inherently and with different cements, trial mixes are recommended for the prediction of the performance of fly ash concrete mixtures.

4.7.9   Silica Fume

Silica fume is a by-product from the operation of electric arc furnaces used to reduce high-purity quartz with coal in the production of elemental silicon and ferrosilicon alloys. It consists of extremely small spherical particles of amorphous silicon dioxide that are collected from the gases escaping from the furnaces. Its specific surface is around $20,000$ kg/m^2, as determined by nitrogen adsorption method which is about 50 times that of typical cements. The specific gravity of a typical silica fume is about 2.2, its bulk loose density is of the order of 250 to 300 kg/m^3 (15.6 to 18.7 lb/ft^3) as compared with about 1200 kg/m^3 (75 lb/ft^3) for normal portland cement. Its color varies from light to dark gray depending mainly on the carbon content.[ACI 1987a] ASTM C 618-88 does not cover silica fume.

The use of silica fume in concrete as a pozzolanic material, that is, as a partial replacement of portland cement in concrete, is a relatively new development.[ACI 1985b]

Silica fume is very reactive with portland cement because of its extreme fineness and high silica content. The incorporation of silica fume in portland cement pastes contributes to the hydration reactions possibly by providing nucleation sites for $Ca(OH)_2$, and also by reacting with alkali and Ca^{++} ions.[Detwiler 1989] The calcium silicate hydrate that is formed by the reaction of silica fume may be different from that formed by hydration of the calcium silicate phases in portland cement but the difference is small.

In general applications, part of the cement may be replaced by a smaller quantity of silica fume without loss of strength. For instance, one pound of silica fume can replace 3 to 4 pounds of cement under the right circumstances. The amount of silica fume used is generally less than about 10 percent by mass of cement. Because of its high specific surface, it increases the water demand of the concrete, therefore frequently high-range water reducers are used with silica fume to maintain mixing water requirements at acceptably low levels and to control drying shrinkage.

The effects of silica fume on properties of the *fresh concrete* include improvement of the cohesiveness and reduction of bleeding. The dosage of air-entraining admixture to produce a required volume of air in concrete increases markedly with increasing amounts of silica fume. This is due again to the very high specific surface of silica fume and to the effect of carbon when the latter is present.

The main contribution of the silica fume to the strength development in *hardened concrete* at normal curing temperatures takes place from about 3 days on. At 28 days the strengths of silica-fume concrete (compressive, flexural, splitting) are always higher than the strength of the comparable portland cement concrete. Silica fume has been successfully used to produce: concretes of 70 MPa (10,000 psi) or greater compressive strength; low water and chloride permeability [Berke 1988, Berntsson 1990]; and chemically resistant concrete. Such concretes contain up to 25 percent silica fume by mass of cement. The loss in slump that would be caused by such use is compensated for by the use of a suitable high-range water reducing admixture. Like other pozzolans, silica fume can be used to prevent deleterious expansion due to alkali-silica reaction in concrete. The added advantage is that only small quantities may be needed as compared to other pozzolans. [ACI 1987a]

Silica fume's extreme fineness and loose density creates *handling problems*. This can be overcome with properly designed loading, transport, storage and batching systems. Such systems include pelletized, compacted or densified forms as well as water slurry. All the product forms have positive and negative aspects as presented in Table 4.1.

Health hazards from exposure to silica fume are very small. Nevertheless, it is recommended that workers handling silica fume use protective gear and systems which minimize the generation of dust.

Table 4.1: Performance Characteristics of Different Silica Fume Forms (from ACI Comm. 226, 1987a)

Silica fume form	Positive characteristics	Negative characteristics
Uncompacted	* Highest pozzolanic activity, technical performance, and efficiency * Temperature protection not required for transport, storage, or batching	* Dusty and moisture sensitive * High transportation cost * Limited pneumatic transport length
Pelletized/ densified	* Lowest transportation cost * Lowest storage and transport volume requirements * Decreased dust	* Lowest pozzolanic reactivity performance and efficiency * Moisture sensitive
Water slurry	* No dust * Simplest transport and batching systems	* Inaccuracy in silica dosage due to sedimentation * Difficult to control concrete water content * Storage and dispensing tanks must be constantly agitated to keep silica fume in suspension * Must control pH to prevent premature gel formation * Temperature control of transport, storage, and batching to prevent freezing

4.8   SELECTION OF CEMENTS

Since cement is the most active component of concrete and usually has the greatest unit cost, its selection and proper use is important in obtaining the balance of properties and cost desired for a particular concrete mixture. The selection should take into account the properties of the available cements and the performance required of the concrete. A general guideline for the application of various portland cement is presented in Table 2.3.

Usually a variety of cements is available to the user. Most cements will provide adequate levels of strength and durability for general use. Some provide higher levels of certain properties than are needed in specific applications. For some applications, such as those requiring increased resistance of sulfate attack, reduced heat evolution, or use with aggregates susceptible to alkali-aggregate reaction, special requirements should be imposed in the purchase specifications. While failure to impose these requirements may have serious consequences, imposing these requirements unnecessarily is not only uneconomical but it may degrade other more important performance characteristics. [ACI 1985b]

The goal of the selector or specifier is to recommend the type, and perhaps quantity, of cement that meets the structural and durability requirements of the structure - no more, no less. Nevertheless, for a long time there have been virtually no economic penalties to discourage users and others from overspecifying cement characteristics. For example, eventhough a fully satisfactory Type I cement has been available, users have often chosen to specify a Type II cement or a low-alkali cement on the basis that it could do no harm, and its special characteristics might be beneficial. This, however, is not always true. As an example, it is not a reasonable idea to specify Type II cement, if increased sulfate resistance is not needed, for certain plant-manufactured structural elements which require rapid strength gain in the production process; the cement composition that imparts sulfate resistance tend to reduce the rate of strength development which hurts the production. Also, the effects of increased attention to pollution abatement and energy conservation are changing the availability and comparative cost of all types of cement. This brings about a need for greater understanding of factors affecting cement performance than was previously necessary.

It is usually satisfactory, and advisable, to use a general purpose cement that is readily obtained locally. Because such a cement is manufactured in large quantity, it is likely to be uniform and its performance under local conditions will be known. A decision to obtain a special type of cement may result in the improvement of one aspect of performance at the expense of others. For this reason, a strong justification is usually needed to seek a cement other than a commonly available Type I or Type II portland cement, or corresponding blended cement.

In recent years, admixtures have become an essential part of concrete specification and production. The effects of admixtures on the performance of concrete are usually intricately linked to the particular cement-admixture combinations used. Since these effects may not be fully understood, the prediction of the behavior of an admixture with various cements, especially with cements other than portland cements, is uncertain. There are general guidelines, such as that a greater quantity of air-entraining admixture is required to produce a given air content when a blended cement, or fly ash is used, than when portland cements. Nevertheless, the use of an admixture makes it almost mandatory to base the cement selection on previous field experience with the specific combination, or on results of trial mixtures simulating job materials, conditions and procedures.

4.9 FUTURE OF CEMENTS

A better understanding of the future of cements, coupled with improvements in cement manufacturing techniques, will make it possible to produce cements

designed for particular purposes. Some of the desirable possibilities are: cements with reduced slump
loss, nonbleeding cements, cements for the elimination of "cold joints" in concrete construction,
cements of reduced thermal coefficient, increased
tensile strength, modified elastic properties,
reduced permeability and shrinkage, and increased
chemical resistivity [ACI 1971b].

5

Water

SUMMARY

Water is used in concrete making (1) as mixing water; (2) for curing of concrete; and (3) for washing. The amount of impurities in the water is restricted for each case.

Impurities may be either dissolved in the water or present in the form of suspensions. Some of these impurities such as sugar, tannic acid, vegetable matter, oil, and sulfates, may interfere with the hydration of the cement, thus delaying setting and reducing the strength of the concrete. These effects vary markedly with the brand and type of cement used as well as with the richness of the mixture. Specifications for tolerable maximum amounts for impurities in mixing water are available.

The rule of thumb is that if a water is potable, it is suitable as mixing water for concrete. However, many waters unsuitable for drinking are still satisfactory for mixing. Also, the appearance of a suspicious water is not a reliable basis on which to judge its fitness for mixing water. Therefore, if there is any question about the concrete-making quality of a water, a sample should be tested, for instance with trial mixes if times is available, or be submitted to a laboratory for testing and recommendations.

The requirements for curing water are less stringent than those for mixing water, but the

214

permissible maximum amounts of impurities are still restricted. Water for washing aggregate or concrete mixers should not contain materials in quantities large enough to produce harmful films or coatings on the surface of aggregate particles or concrete mixers.

5.1 INTRODUCTION

Water is used in concrete making for three different purposes: (1) as mixing water; (2) for curing of concrete; and (3) for washing. The quality requirements for the water depend on the type of the use. Summaries of the physical and physicochemical properties of water are available in the literature [Duriez 1962, Forslind 1954, Portland Cem. A. 1968].

5.2 MIXING WATER

The mixing water, that is, the free water encountered in freshly mixed concrete, has three main functions: (1) it reacts with the cement powder, thus producing hydration; (2) it acts as a lubricant, contributing to the workability of the fresh mixture; and (3) it secures the necessary space in the paste for the development of hydration products. The amount of water needed for adequate workability is practically always greater than that needed for complete hydration of the cement. The very important quantitative aspect of the mixing water, that is, the problem of how much water should be added to a batch, will not be discussed here. Only the qualitative aspects will be analyzed, namely, the questions of what kinds and what quantities of impurities may make a water less suitable or unsuitable for concrete making.

Impurities can be either dissolved in the water or present in the form of suspensions. Some of these impurities, above certain levels, may interfere with the hydration of the cement, thus changing setting times, and reducing the strength of the concrete. Impurities may also cause staining on the concrete surface or increase the risk of corrosion of the reinforcement. It is important to note that these effects of impure water on concrete vary markedly with the type of cement used and the richness of the mixture [Steinour 1960a, Graf 1960]. Nevertheless, there is a simple rule concerning the acceptability of a mixing water: If a water is potable, that is, fit for human consumption with the exception of certain mineral waters, and waters containing sugar, then it is also suitable for concrete making. In other words, if a water does not have any particular taste, odor, or color, and does not fizz or foam when shaken, then there is no reason to assume that such a water will hurt the concrete when used properly as a mixing water. On the other hand, many waters unsuitable for drinking are still satisfactory for concrete making, as Abrams's study reporting comparative compressive strengths of concretes at ages up to 2 1/3 years [Abrams 1924], and other subsequent publications, indicate [Steinour 1960a, Anon., 1963]. This is so because waters containing no more impurities than 2000 ppm of common ions (that is, 0.2% of the weight of water) are generally acceptable as mixing waters, whereas for drinking waters 1000 ppm is the permissible upper limit in some specifications. Nevertheless, use of untreated domestic waste water is not advocated as mixing water. It is important for the field worker to know that the appearance of a suspicious water is not a reliable basis on which to judge its fitness for mixing water. Therefore, if there is any question about the concrete-making quality of the water, a sample should be tested or submitted to a laboratory for testing and recommendation. Sometimes it may be sufficient to test the setting time of cement paste made with the doubtful water as well as that with a water proved satisfactory [Lea 1956]. It is better, when time is available, to test specimens or concrete or mortar containing the water for comparative strength and durability with respect to control specimens made with water known to be satisfactory. In such cases, a tolerance between results of about 10-15% is usually permitted. ASTM C 94-89b specifies similar

values.
The waters that are considered to be harmful are
mostly those that contain excessive amounts of
sugar, tannic acid, vegetable matter, oil, humic
acid, sulfates, free carbonic acid, alkali salts,
and effluents from sewage, gas, paint, and ferti-
lizer works.[Portland C. A. 1975] Some of these are
natural impurities, others occur in sewage or waters
carrying in- dustrial wastes. Presently available
information is inadequate to set up rigid
specifications for tolerable maximum amounts for
these impurities in mixing water. A typical
specification is shown below only as an informative
example as quoted by Waddell [Waddell 1962]. This
requires that the water should be clean, clear, free
from sugar, and should not contain acid, alkali,
salts, or organic matter in excess of the following
amounts when tested in accordance with AASHTO
Designation: T 26-79:

1. Acidity and alkalinity:
    (a) Acidity - 0.1 N NaOH, 2ml max. to neutralize
        200-ml sample
    (b) Alkalinity - 0.1 N HCl, 10ml max. to neutra-
        lize 200-ml sample.

    A similar pertinent requirement may be that the
    pH value of the mixing water should not be less
    than 6 or greater than 9.

2. Total solids:
    (a) Organic   0.02% max.
    (b) Inorganic   0.30% max.
    (c) Sulfuric   anhydride (SO_3)   0.04% max.
    (d) Alkali chloride as sodium
        chloride (NaCl)   0.10% max.

    As a guide, a turbidity maximum limit of 2000
    ppm can be used.

When standard 1:3 mortar briquettes made with
cement, sand, and water from the sample are compared
with briquettes made with the same cement and sand
and distilled water, there should be no indication

of unsoundness, marked change in time of set, or variation of more than 10% in strength.

Experience has shown that waters with impurity contents higher than those above may be used (for nonprestressed concrete), but use may produce a reduction in concrete strength or other harmful effects [US. Bureau of Recl. 1966]. For instance, sea-water with an average salt content of 3.5%, mainly sodium chloride and magnesium sulfate, has been used to manufacture concrete, particularly plain concrete, in many areas, and the service record has been good, although strength reduction up to 15% may have occurred [Abrams 1924, Dewar 1963, McCoy 1966, Griffin 1963]; or the alkali content of water may contribute to alkali-aggregate reaction. Efflorescence may also occur [Griffin 1961], and corrosion of the reinforcement can become a problem [Shalon 1959], especially with chlorides present. Seawater and other waters with high salt contents must never be mixed with high-alumina cements, as salt has a very adverse effect on strength.

It is recommended that any deleterious materials in the mineral aggregate to be used be taken into account when the permissible amounts of impurities in a mixing water are determined [Fulton 1964]. The amounts of impurities in a water from a given source are apt to vary with the time, for instance between wet and dry seasons, hence it may be necessary to obtain more than one sample during the progress of a job.

In certain cases, the amounts of impurities can be reduced in the water. For instance, a water carrying excessive silt can be stored in settling basins; or a water charged with gases can be forcibly circulated or aerated until adjustment to the acceptable level is attained. It has also been shown that biologically treated average domestic sewage is indistinguishable from distilled water when used as mixing water. [Cebeci 1989] The health hazard of the use of the latter is minimal because the relatively low pathogenic bacterial activity is substantially reduced immediately after its pH exceeds 12.

It can be seen that some impure waters that

would be destructive if in contact with the hardened concrete over the period of its service life can be used without harmful effects for mixing purposes. The reason for this paradox is that even a highly impure mixing water can carry only a relatively small total amount of deleterious materials into a concrete mixture. For instance, 1000 ppm impurity (that is, 0.1%) in the water corresponds only to 0.06% of cement weight for the water-cement ratio of 0.6 by weight. On the other hand, where the hardened concrete is continuously exposed to an aggressive water, the deleterious material can build up concentrations in the concrete pores much higher than that of the original aggressive water. An exception is again the case of high-alumina cement concrete, which has a high resistance against contacting seawater.

5.3 WATER FOR CURING AND WASHING

The requirements for *curing water* are less stringent than those discussed above, mainly because curing water is in contact with the concrete for only a relatively short time. Such water may contain more inorganic and organic materials, sulfuric anhydride, acids, chlorides, and so on, than an acceptable mixing water, especially when slight discoloration of the concrete surface is not objectionable. Nevertheless, the permissible amounts of the impurities are still restricted. In cases of any doubt, water samples should be sent to a laboratory for testing and recommendations [McCoy 1966].

Water for *washing aggregate* should not contain materials in quantities large enough to produce harmful films or coatings on the surface of aggregate particles [Ujhelyi 1973]. Essentially the same requirement holds when the water is used for *cleaning* concrete mixers and other concreting equipment. Chemical limitations for the impurities in wash water are specified in ASTM C 94-89b.

The disposal of washing water, as for many other types of waste water, may be a serious problem because pollution-control regulations usually prohibit the direct discharge of such impure waters into existing sewers, lakes, rivers, or even open drains. The solution to this problem may be (1) reduction of the amounts of solids and other impurities in the waste water to an acceptable level by sedimentation, chemical methods, or other means before disposal [Ernst 1972, Krenkel 1972, Wakley 1972, Lauwereins 1971]; or (2) recycling the waste water, that is, reducing the amounts of solids and other impurities to such levels that the water becomes acceptable for reuse for mixing, curing, or washing [Ullman 1972, Meininger 1972]. It should be noted, however, that admixture requirements are likely to change if they are batched in recycled waste water to be used as mixing water.

The recycling of the cement slurry of the unused concrete has also been tried for reuse in concrete, but this slurry waste showed little cementitious value [Pistilli 1975].

6

Admixtures

SUMMARY

Concrete admixtures are special chemicals added to the batch before or during mixing. The quantities used are, with few exceptions, very small, nevertheless they can impart certain desirable properties to the concrete that cannot be secured by other methods, or not as economically.

The most frequently used admixtures are: (1) air-entraining admixtures to introduce a system of small air bubbles into the fresh concrete during mixing, usually anionic surface-active agents; (2) accelerators to increase the rate of setting or the rate of hardening at early ages or both, including some of the soluble chlorides (primarily calcium chloride), carbonates, and silicates; (3) water-reducing and set-retarding admixtures to reduce the water requirement of concrete, or to retard the set, or both, including lignosulfonic acids, hydroxylated carboxylic acids, carbohydrates, polyols, and the salts and modifications of these (primarily calcium lignosulfonate); and (4) finely divided mineral admixtures, including natural and artificial pozzolans (primarily fly ash), hydraulic lime, blast-furnace slag, and ground quartz, to improve the properties of fresh concrete, increase the chemical resistance of concrete, reduce the heat of hydration, reduce expansion produced by alkali-aggregate reaction, and so on.

It is important to recognize that admixtures are no substitute for sound concrete-making practices. As a matter of fact, the proper utilization of admixtures requires increased care, for instance, in batching and mixing. The other aspects of the concrete-making procedure should also be kept as constant as feasible. Since many admixtures affect more than one property of concrete, sometimes affecting desirable properties adversely, and since these effects may be dependent on several factors (brand and type of cement, etc.), and since the mechanism of action of most admixtures is not quite understood, in using any admixture careful attention should be given to the instructions provided by the manufacturer of the product. Also, an admixture should be employed only after appropriate evaluation of its effects, if necessary by use of trial mixes with the particular concrete.

6.1   INTRODUCTION

An admixture is defined in the "Standard Definitions of Terms Relating to Concrete and Concrete Aggregates" (ASTM C 125-68) as: "A material other than water, aggregates, and portland cement (including air-entraining portland cement and portland blast-furnace slag cement) that is used as an ingredient of concrete and is added to the batch immediately before or during its mixing." Under the proper circumstances, admixtures present in comparatively small amounts can impart relatively large effects on the properties of fresh and/or hardened concrete that cannot be secured by other methods, or not as economically. In other instances, the desired objectives may be achieved most economically by changes in composition of the concrete mixture rather than by the use of an admixture. [Popovics 1968e, 1969b, 1971a] Admixtures are no substitute for good concreting practices. Admixtures should be required to conform with applicable ASTM or other relevant specifications. In using any admixture, careful attention should be given to the instructions provided by the manufacturer of the product.

Many admixtures affect more than one property of concrete, sometimes affecting desirable properties adversely. The optimum use of certain admixtures may require reproportioning the concrete. The specific effects of an admixture may be dependent on a number of variables, such as the type and amount of admixture, the compound composition, gypsum content, kind, brand, and amount of cement, factors influencing the kinetics of hydration, the time of addition of the admixture to the fresh mixture, and so on. Therefore, an admixture should be employed only after appropriate evaluation of its effects, if necessary by use with the particular concrete and under the conditions of use intended.

Admixtures that may have immediate reaction with the cement should be solved or dispersed in the mixing water, or in a portion of it, and added to the concrete in this diluted form.

6.2 CLASSIFICATION

A pertinent ACI report [ACI 1989] classifies the *chemical admixtures* for concrete into the following groups:

    air-entraining admixtures
    accelerating admixtures
    water-reducing and set-controlling admixtures
    admixtures for flowing concrete
    miscellaneous admixtures.

In addition to these, *polymers* as admixtures and *finely divided mineral admixtures* are discussed below. Properties of most of the chemical admixtures are specified in ASTM C 494-86.

Admixtures may contain materials that individually would belong in two or more of these groups. For example, a water-reducing admixture may be combined with an air-entraining admixture or with a set-controlling admixture. It is also possible to add two or more separate admixtures to the concrete [Lovewell 1971, Torrans 1968], but this requires even greater caution. For instance, the practice of mixing two admixtures of differing types together prior to addition to the concrete is not recommended, especially without try out.

The successful use of admixtures depends on the use of an appropriate method of preparation and batching. [ACI 1981b] The batching system requires accuracy, reproducibility, and speed. Such a system for use in manual or semiautomatic plants should be capable of volumetric batching within an accuracy of ±3% of the required volume, unless otherwise stipulated [ACI 1989]. The importance of admixtures is indicated by the high number of related publications [Vivian 1960, Kuhl 1961b, Eitel 1966, Fifth Intern. Symp. 1969, RILEM-ABEM 1968, RILEM 1975, Highway Res. B. 1965, 1971, Commissie 1968, Concrete Soc. 1968, Venuat 1967b, Bonzel 1976a, Price 1976, Vavrin 1974, Rixom 1977, 1978, Ramachandran 1981, 1984, Gibson 1987]. A general description of several groups of admixtures is presented below.

6.3   AIR-ENTRAINING ADMIXTURES

6.3.1   Air Entrainment

Air entrainment is the purposeful introduction of a relatively stable system of small air bubbles into the fresh concrete during the mixing operation by means of surface-active chemicals, usually anionic ones, called *air-entraining admixtures* or *agents* (AEAs). [Mielenz 1965, 1969]. Its development goes back to about 50 years [Klieger 1980] and made the production of concretes with improved frost resistance possible.

The entrained air should be distinguished from the entrapped air because only the entrained air improves the frost resistance and workability of the concrete. Entrapped air is the result of incomplete compaction of the concrete. Although it is usually present in concrete in bubbles larger and having more irregular shapes than the entrained air, the distinction between the two types of air is not sharp because (1) all air bubbles carry films produced by AEAs; (2) some of the entrained air is held in the concrete by entrapment; (3) there is no method available for reliably determining the entrained and entrapped air contents separately, partly because (4) there may be an overlapping in the sizes and shapes of the bubbles of the two types of air.

A properly air-entrained concrete is severalfold more resistant to frost action than a non-air-entrained concrete made of the same materials. Air-entrained concrete is always required under conditions of severe natural weathering and where sodium chloride or calcium chloride is used for ice removal on pavements.

6.3.2   The Admixtures

ACI 116R defines an air-entraining admixture or agent (AEA), as "an addition for hydraulic cement or an admixture for concrete or mortar which causes entrained air to be incorporated in the concrete or mortar during mixing, usually to increase its workability and frost resistance." Many materials may be used in preparing air-entraining admixtures. Some materials, such as hydrogen peroxide and powdered aluminum metal, do develop pores, nevertheless they are not considered to be acceptable air-entraining admixtures since they do not produce satisfactory air-void systems for the improvement of frost resistance. Those AEAs that have been traditionally produced in the United States can be classified into the following groups [Jackson 1954, Klieger 1978]:

Group A: Salts of wood resins
Group B: Synthetic detergents
Group C: Salts of sulfonated lignin
Group D: Salts of petroleum acids
Group E: Salts of proteinaceous materials
Group F: Fatty and resinous acids and their salts
Group G: Organic salts of sulfonated hydrocarbons.

Some of these compounds are called anion active, or *anionic*, because the hydrocarbon structure contains negatively charged hydrophilic groups, such as COO^-, SO_3^-, and OSO_3^-, so that large anions are released in water. Conversely, if the hydrocarbon ion is positively charged, the compound is cation active, or *cationic*. Compounds composed of positively and negatively charged organic ions are said to be *amphoteric*. *Nonionic* compounds release no ions into aqueous solutions. In other words, anionic surface-active agents produce bubbles that are negatively charged; cationic agents cause bubbles to be positively charged; and nonionic agents do not induce an appreciable charge on the bubbles. Surface-active agents of all these classes can cause air entrainment in concrete, but their efficiency and the characteristics of the air void system vary widely. These materials are usually insoluble in water, but they can be dissolved through the use of alkalis and other materials.

Anionic surface-active agents are the most widely used as air-entraining admixtures in concrete mixtures for general construction [Halstead 1954]. Salts of wood resins, synthetic detergents, salts of fatty and resinous acids, and organic salts of sulfonated hydrocarbons, which are most widely used as air-entraining agents, typically are employed at rates of 0.002-0.06% by weight of the cement content, based on the active constituents of the admixture. In tests of diverse anionic detergents, including both sodium and organic salts, in concrete with water-cement ratio about 0.45 by weight and slump of 3-4 in (75-100 mm), air content of 5-6%, or a cement paste-air void ratio of about 3.5-6 by volume, was achieved by use of the admixture at rates of 0.002-0.006% by weight. Mielenz also presents data about cationic and nonionic air-entraining admixtures, although these are not available commercially [Mielenz 1969].

Solid particles having a large internal porosity and suitable pore size seem to act in a manner similar to that of air voids. [Litvan 1985] These materials, however, currently are not being used extensively.

6.3.3   Applications

The air-entraining admixture is added to the concrete in or along with the mixing water. Air-entrained concrete can also be made by using *air-entraining portland cement*. (Section 2.5) Of these two possibilities it is usually more desirable to add the air-entraining admixture at the concrete mixer; changing the admixture quantity makes it readily possible to change the air content that may be required to produce the optimum. The use of air-entraining cement, although simple, does not provide a means of adjusting the air content to the job conditions.

Regardless of the method of air entrainment employed in the preparation of air-entrained concrete, the properties of the concrete-making materials, the proportioning of the concrete mixture, and all aspects of the mixing, handling, and placing procedures should be maintained as constant as feasible so that the air content of the concrete will be uniform and within the range specified for the work. [ACI 1989] The air content of the concrete should be checked and controlled during the course of the work. Particular attention should be given to the unusually high amount of air-entraining admixture often required in concrete containing Type III portland cement, portland pozzolan cement, fly ash, or other finely divided powders. If the air content of a concrete is much higher than was intended, and this is discovered before the concrete is placed, the air content can be reduced by the addition of an *air-detraining* admixture. The most widely used such admixture is tributyl phosphate. After placement or compaction, the air content may be reduced by intensive vibration or revibration.

Air entrainment materially alters the properties of both fresh and hardened concrete. Air-entrained concrete is considerably more plastic and workable than non-air-entrained concrete, because the bubbles increase the spacing of the solids in the fresh mixture, thus decrease dilatancy. The better co-hesiveness is the result of increased viscosity and yield stress of the paste (Fig. 6.1). The paste can then be handled and placed with less segregation and there is less tendency for bleeding. The concrete strength may be reduced somewhat by air entrainment, but the durability of the hardened concrete is improved considerably [Klieger 1980] not only directly but also indirectly by increased uniformity, decreased absorption and permeability, and by the elimination of planes of weaknesses at the tops of lifts. These effects are due to the presence of a large number of minute bubbles in the cement paste and mortar.

Because of the unfavorable shape and surface texture of the fine fraction of most lightweight aggregates used for structural concrete, it is usually desirable to use air-entraining admixtures with them. Without entrained air such concrete is generally harsh, has a high bleeding rate, and high water requirement. The greatest improvement is obtained in harsh, lean mixtures. Lightweight concretes containing more than 600 lb/yd^3 (360 kg/m^3) of cement do not always require air entrainment.

6.3.4   The Entrained Air

The air content, specific surface, and size distribution of air voids in air-entrained concrete are influenced by many factors, among the more important of which are (1) the nature and concentration of the air-entraining admixture (Table 6.1 and Fig. 6.2); (2) the nature and proportions of the ingredients of the concrete [Bruere 1961] (Fig. 6.3), including other admixtures present [Bruere 1974]; (3) the type and duration of mixing employed (Fig 6.4); (4) the consistency; (5) temperature and other factors influencing the setting time [Mielenz 1958]; and (6) kind and degree of compaction applied in consolidating the concrete. For instance, a given percentage of admixture entrains progressively more air as (1) the cement content of the concrete decreases; (2) the sand content increases; (3) the proportion of sand passing the No.30 (600-μm) and retained on the No.50 (300-μm) or No. 100 (150-μm) sieves increases; (4) the particle shape of the fine

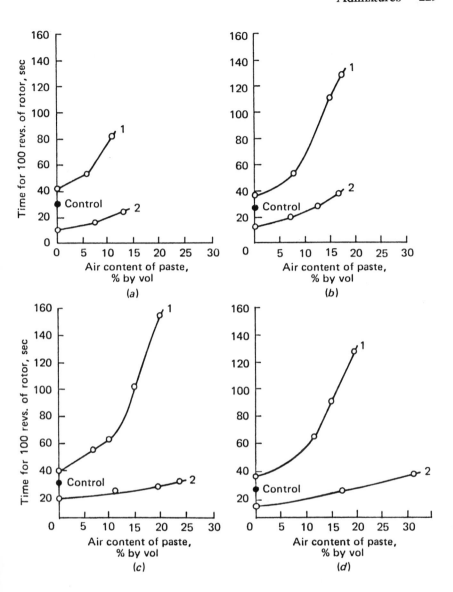

Figure 6.1: Viscosities of portland cement pastes containing fixed concentrations of two surface-active agents in the presence of varying amounts of entrained air [Bruere 1958]. Pastes were mixed by the standard techniques, temperature 68° ± 1.8°F (20° ± 1°C). Surface active agents: 1, sodium abietate, 0.5% by weight of cement; 2, saponin, 0.10% by weight of cement. (a) W/C = 0.36 by weight; weight driving rotor = 550 g. (b) W/C = 0.40 by weight; weight driving rotor = 300 g. (c) W/C = 0.44 by weight; weight driving rotor = 190 g. (d) W/C = 0.48 by weight; weight driving rotor = 135 g.

Table 6.1: Effect of Chemical Type and Concentration of Surface-Active Agent on Bubble Characteristics in Cement Pastes[a]

Surface-active agent	Concentration of Agent (% by wt. of cement)	No. of bubbles intersected per inch of traverse (n)	Air content (fraction of paste volume) (A)	Surface area of bubbles (in^2/in^3) (α)	Computed spacing factor $(\bar{L})$
Sodium sec-octyl sulfate	0	0.6	0.003	800	0.032
	0.100	18.1	0.089	810	0.0082
Sodium dodecyl sulfate	0.0025	28.1	0.065	1730	0.0044
	0.005	64.6	0.140	1850	0.0029
	0.010	84.1	0.175	1920	0.0026
	0.025	105.2	0.213	1970	0.0023
	0.050	131.9	0.272	1940	0.0019
Sodium tetradecyl sulfate	0.010	12.0	0.028	1710	0.0064
	0.020	15.3	0.036	1700	0.0058
Sodium dodecyl benzene sulfonate	0.005	14.2	0.040	1420	0.0066
	0.010	21.9	0.056	1560	0.0053
	0.025	47.3	0.112	1630	0.0037
	0.050	50.7	0.131	1550	0.0036
Neutralized Vinsol resin	0.005	15.3	0.046	1330	0.0067
	0.010	36.2	0.101	1420	0.0044
	0.025	58.7	0.174	1350	0.0037
	0.050	75.6	0.226	1360	0.0033
Darex AEA	0.025	9.7	0.027	1440	0.0077
	0.050	13.6	0.036	1510	0.0065
	0.100	21.2	0.059	1440	0.0055
	0.150	26.8	0.071	1510	0.0049
Igepon T	0.025	19.6	0.036	2180	0.0045
	0.050	28.8	0.055	2090	0.0039
	0.100	40.7	0.076	2170	0.0033
Decyl trimethyl ammonium bromide	0.025	7.8	0.047	660	0.0134
	0.050	13.0	0.062	840	0.0093
	0.100	30.0	0.136	880	0.0063
Dodecyl trimethyl ammonium bromide	0.010	14.4	0.053	1090	0.0077
	0.025	38.5	0.153	1010	0.0052
	0.050	61.7	0.265	930	0.0041
	0.075	72.2	0.310	930	0.0035
Tetradecyl trimethyl ammonium bromide	0.010	9.4	0.038	990	0.0098
	0.018	19.5	0.067	1160	0.0065
	0.025	30.4	0.105	1160	0.0054
	0.050	48.6	0.171	1140	0.0044
	0.075	68.8	0.230	1200	0.0036

(continued)

Table 6.1: (continued)

Surface-active agent	Concentration of Agent (% by wt. of cement)	No. of bubbles intersected per inch of traverse (n)	Air content (fraction of paste volume (A)	Surface area of bubbles (in²/in³) (α)	Computed spacing factor ($\bar{L}$)
Hexadecyl trimethyl ammonium bromide	0.010	3.3	0.014	980	0.0150
	0.025	7.4	0.021	1400	0.0088
	0.050	24.8	0.073	1350	0.0054
	0.100	30.9	0.080	1550	0.0044
Lissapol N300	0.050	2.6	0.015	700	0.020
	0.100	4.9	0.021	930	0.0132
Saponin	0.025	20.1	0.065	1240	0.0062
	0.050	38.8	0.145	1070	0.0050
	0.100	64.2	0.216	1200	0.0037

[a]Mixing temperature 68° to 72°F (20° to 22°C); stirring time 4 minutes; speed of stirrer 1,000 rpm; W/C = 0.45 by weight. From Bruere [1960].

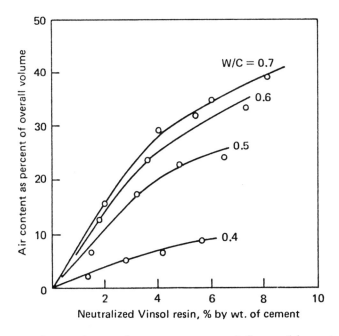

Figure 6.2: Air entrainment in cement paste as influenced by water-cement ratio and dosage of air-entraining agent. From Powers [1954a, 1968].

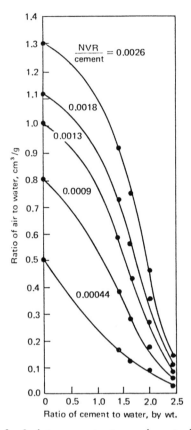

Figure 6.3: Effect of admixture content on air entrainment in mixtures of cement and water for which a high-speed, milkshake type of stirrer was used. NVR stands for neutralized Vinsol resin [Powers 1954a, 1968].

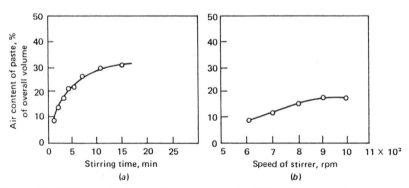

Figure 6.4: Effects of stirring time and speed of the stirrer on the air content of neat paste. Sodium dodecyl sulfate was used as the air-entraining agent, at 0.0125% by weight of the cement. (a) The speed of the stirrer was 1,000 rpm. (b) The stirring time was 3 minutes [Bruere 1955, Powers 1968].

aggregate changes from round to angular; and (5) the concrete temperature decreases, other conditions being equal [Mielenz 1958, Walker 1955].

Because of the large number of these influencing factors, it is highly recommended that trial mixes be used to check the compatibility of the air-entraining admixture with the other concrete components to be used, and to determine the needed amount of AEA under the given circumstances. This is particularly true when air entrainment is used simultaneously with other admixture(s), such as a superplasticizer, since certain air-entraining admixtures have been reported to be more effective with certain superplasticizers than others.

Wherever exposure to severe weather is a consideration, the proportion of air-entraining admixture should be fixed at the minimum required for frost resistance. In other cases the amount should be adjusted as required for workability without excessive reduction in strength. Note that for any given set of materials and conditions, the amount of entrained air is roughly proportional to the quantity of the air-entraining admixture used.

The entrained air is produced by stirring or kneading the fresh mixture. Stirring introduces air as it draws material at the surface of a batch to the interior by a vortex action. The formation, dispersion, and stabilization of discrete air bubbles is fundamentally a process that forms an emulsion distinct from a foam. (An example for the latter is the so-called cellular concrete, in which very high amounts of air or other gas, 60% or even more by volume of concrete, are produced to substitute for all or part of the aggregate.) Materials too stiff or too dilatant to be stirred should be kneaded for air entrainment. On the other hand, such violent agitations, particularly turbulence in the flow pattern, are likely to cause collision between air bubbles, giving a colliding pair of bubbles opportunity to merge into one. Such coalescence is a natural tendency because it results in a reduction of interfacial area and hence decreases the free energy of the system. The principal function of the air-entraining admixture is to prevent such coalescence during mixing, placing, and consolidation. It does so by forming a tough film through adsorption at air-water interfaces, and it may also give an electrostatic charge to the bubbles that gives them some degree of mutual repellency. The mechanism of this is as follows:

Because of the presence of hydrophobic and hydrophilic portions in the molecule of the surface-active compound, an air-entraining admixture is concentrated at air-water interfaces, with the hydrophobic portion preferentially oriented toward the air. Another role that can be played by the surfactant is to reduce the surface tension of water, thus allowing a given shear stress to divide smaller bubbles than could otherwise be divided [Powers 1968].

Properly air-entrained concrete includes air in amounts equivalent to about one-sixth to one-fourth of the volume of cement paste (3-10% by volume of the concrete) [Klieger 1978], primarily in nearly spherical voids ranging in size from about 10 to 1000 µm [Mielenz 1965]. Because of their stability, these air bubbles act in the fresh mixture like water drops or sand particles, therefore air entrainment increases the plasticity if the composition of the mixture remains otherwise unchanged. At constant cement content and constant slump, the entrained air displaces fine aggregate and water; more specifically, the leaner the mixture, the more water and less aggregate is displaced. The resulting reduction in water-cement ratio can offset, partially or entirely, the reduction in concrete strength caused by entrained air. [ACI 1989]

A more complete discussion of air-entraining agents, the mechanics of air entrainment, and air entrainment in water, aggregate, cement paste as well as in concrete has been presented by Powers, and by Mielenz [Mielenz 1969].

6.3.5   Improvement of Frost Resistance

A mechanism of how air voids may prevent the damage caused by freezing in hardened cement paste or concrete was offered by Powers based on *hydraulic pressure* [Powers 1945a, 1949b, 1953a]. The essence of this is that the destruction of concrete by freezing is caused mainly by hydraulic pressure generated by the freezing of water rather than by direct crystal pressure produced through growth of bodies of ice crystals. Development of this hydraulic pressure is a consequence of the expansion of about 9% accompanying the conversion of water to ice because this expansion tends to press the unfrozen water from the freezing zone through capillary pores. As a result, tensile stresses rise in the hardened paste, which causes failure of the cement paste if the stresses exceed the tensile strength. The magnitudes of the hydraulic pressure and the tensile stresses are dependent on several factors. The most important is perhaps the length of the capillary pores, because the pressure is proportional to this. But the length of capillary voids can be reduced by well-distributed air voids through air entrainment, resulting in a relief of hydraulic pressure during progressive freezing. This means that the air voids are more effective the closer they are together. In other words, at a given air content, the protection afforded by the voids against damage by freezing and thawing usually is greater the smaller are the voids, that is, the larger is the number of voids per unit volume of paste. For this reason the entrapped air is of little or no help against frost action because of their larger size. The cement paste in concrete is normally protected against the effects of freezing and thawing if the *spacing factor* of the air void system is 0.008 in (0.02 mm) or less as determined in accordance with ASTM C 457-82a. The same standard defines the term spacing factor as the maximum distance of any point in the cement paste from the periphery of an air void, in inches. The freezing resistance should not be affected adversely by loss of entrained air as a result of vibration, provided that the concrete originally contained an adequate void system, because mainly the larger, less effective bubbles are expelled by the compaction.

6.3.6   Test Methods

ASTM C 260-86 and AASHTO M 154 specify the requirements for air-entraining admixtures whereas the pertinmment test methods are specified in ASTM C 233-89. There are also specified test methods for the measurement of the total air content in fresh concrete (ASTM C 138-81, 231-89a, and 173-78) as well as in hardened concrete along with the determination of the important parameters of the air-void system, such as spacing factor, specific surface, and air-space ratio (ASTM C 457-82a).

6.4 ACCELERATING ADMIXTURES

6.4.1   Acceleration

The purpose of using accelerating admixtures, or accelerators (ASTM C 494-86, Type C and Type E), in concrete is to increase the rate of setting, or the rate of early hardening, or both. Note that reduction of the *time(s) of setting* does not mean necessarily the intensification of the *strength* development of the concrete; but even when it does, higher early strengths usually are not followed by increased concrete strengths at later ages. The probable mechanism of this constantly experienced phenomenon is that most of the chemicals that reduce the setting time do this through the acceleration of the hydration of C_3A and the formation of ettringite; at the same time they tend to retard the hydration of the calcium silicates in the cement. Therefore, with a few exceptions, later strengths are either not influenced considerably by the accelerator, or they are reduced. This means that the *deceleration* of the hardening process is intensified more by the accelerating admixture than the rate increase. Therefore the name "accelerator" is not completely correct.

The accelerating action is caused primarily by increasing the dissolution of certain cement compounds by the presence of accelerator in the cement - water system, although speeding up the rate of crystallization of the hydration products can also play a role. Since there are several different anions in the system, acceleration takes place when the dissolution of the compound(s) is promoted that have low dissolving rate(s) during the early hydration period, such as the silicate ions. [Joisel 1967] Incidentally, it follows from this that if a chemical admixture reduces the dissolution of the cement compounds, more specifically the dissolution of the anion(s) that have high rate of dissolution rate(s) during the early hydration period (e. g., aluminate ions), retardation will occur.

Note that the same chemical admixture can produce retardation in one concentration and acceleration in another one. For instance, the presence of a strong anion in the cement - water system (i. e., Cl$^-$, NO$_3^-$, or SO$_4^-$) reduces the solubility of less strong anions (i. e., silicate and aluminate anions) but tends to accelerate the solubility of calcium ions. In small concentrations (around 0.1 percent by weight of cement) the former effect is dominant (retardation), in large concentrations (1 percent or more) the latter effect becomes dominant (acceleration). On the other hand, the presence of a strong cation in the cement - water system (i. e., K$^+$ or Na$^+$) reduces the solubility of a less strong cation (Ca^{2+} ion) but tends to accelerate the solubility of silicate and aluminate ions. In small concentrations the former effect is dominant (retardation), in large concentrations the latter effect becomes dominant (acceleration). [Mehta 1986]

The benefits of an increase in the *early strengths* may include: (1) earlier removal of forms; (2) reduction of the required period of curing and protection; (3) earlier placement in service of a structure or a repair; and (4) partial or complete compensation for the effects of low temperatures on strength development. The benefits of a reduced *setting time* may include: (1) early finishing of surfaces; (2) reduction of pressure on forms; and (3) more effective plugging of leaks against hydraulic pressure.

6.4.2   The Admixtures

An *accelerating admixture*, or *accelerator*, is a material added to mortar or concrete for the purpose of reducing the time of setting and/or accelerating early strength development. The admixture can be either in powdered or in liquid form. Accelerators used in powdered form primarily for rapid setting include:

> sodium carbonate; sodium aluminate; sodium hydroxide; trivalent iron salts; salts of alpha-hydroxy acids; and aluminum hydroxide.

Accelerators used in liquid form primarily for rapid setting include:

> sodium silicate; sodium aluminate; sodium hydroxide; aluminum hydroxide; and salts of alpha-hydroxy acids.

Admixtures used primarily for the acceleration of strength development include:

> soluble inorganic salts (chlorides, bromides, carbonates, silicates, fluorides, fluoro-silicates hydroxides, etc.); soluble organic compounds (triethanolamine, formates, etc.); carboxylic acids; and miscellaneous solid admixtures (calcium aluminate cements, and addition of finely ground hydrated portland cement, called "seeding" [Duriez 1961].

Water-reducing admixtures, without any direct accelerating or retarding effect, may also produce increase in the early concrete strength by a reduction in the water-cement ratio.

The most common accelerator for concrete is *calcium chloride*, $CaCl_2$, either by itself or as the principal ingredient of most accelerators made under various brand names [Shideler 1952, Ramachandran 1976]. Calcium chloride is available in two forms: (1) solid form ASTM D 98-87 Type S which is flake, pellet, or granular calcium chloride, and comes in

three grades according to the concentration: Grade
1 is 77 % minimum calcium chloride concentration,
Grade 2 is 90 % minimum concentration, and Grade 3
is 94 % minimum concentration; and (2) liquid form
Type L which is water solution in varying
concentrations.

Because of their different physical
characteristics and different chemical purities,
these two types may have different effects on
concrete properties [McCall 1953]. Calcium chloride
can generally be used safely in amounts up to 2% by
weight of the cement in concrete without steel
reinforcement. Larger amounts may be detrimental
and, except in rare instances, provide little
additional advantage.

6.4.3   Effects of Accelerators

The increase in *early strengths* and the reduction of
the *time of setting* are relatively more significant
when the $CaCl_2$ is employed in concrete with a curing
temperature below 70°F (21.1°C). Also, $CaCl_2$ is more
effective when used in rich concretes [Popovics
1968f] (Fig. 6.5). Other effects resulting from the
use of $CaCl_2$ include an increase in *air content* of
the fresh concrete when used with an air-entraining
agent, and, perhaps not independently, a small
increase in workability. An early commencement of
the *stiffening* with some cements has been also
noticed and, accordingly, a reduction in *bleeding*.
Drying *shrinkage* generally, but not always, has been
found to be increased when $CaCl_2$ is used, which may
have a detrimental effect on the flexural and
tensile strengths of the accelerated concrete. *Creep*
is also increased [Hope 1971], which may be due to
morphological changes in the CSH gel, such as
increased specific surface, higher gel porosity, and
lower capillary porosity.

The rate of *heat evolution* is increased
materially, but the total heat liberated is not
changed appreciably by $CaCl_2$. The use of $CaCl_2$ in
warm fresh concrete may result in such rapid
stiffening as to impede placing or finishing.
Calcium chloride generally increases expansion

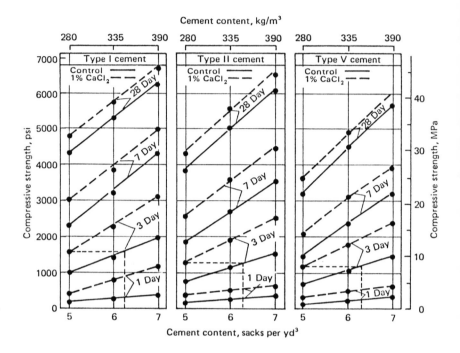

Figure 6.5: Calcium chloride produces a greater strength gain than does an additional sack of cement at early ages and is more effective in increasing the early strength of the richer concretes [Hickey 1953]. (1 American sack weighs 94 lb = 42.6 kg.)

caused by the *alkali-aggregate* reaction, but its effect appears to be unimportant when expansion is controlled by the use of low-alkali cement or pozzolan.

Calcium chloride lowers the resistance of concrete to *sulfate attack* and to *freezing and thawing* at later ages. This may be attributed to the fact that this admixture appears to increase the average size of air voids in concrete when used with air-entraining agents.

The major disadvantage of calcium chloride when used in effective quantities in concrete is its tendency to promote *corrosion* of metals in contact with the concrete. This is due to the presence of chloride ions, moisture and oxygen. In accordance with ACI Committee 222, the maximum acid-soluble chloride contents of 0.08 percent for prestressed concrete, and 0.20 percent for reinforced concrete are suggested to reduce the risk of chloride-induced corrosion. The chloride content should be measured by ASTM C 114 and expressed by weight of the cement. Values for water-soluble chloride ion are given as permissible maxima in ACI 318-83, as follows: 0.06 percent for prestressed concrete; 0.15 percent for reinforced concrete exposed to chloride in service; 1.0 percent for reinforced concrete that will be dry or protected from moisture in service; and 0.30 percent for other reinforced concrete.

The effectiveness of accelerators in producing high-early-strength concrete containing blended cements are similar to those for portland cements, the effects being greater for cements using ground granulated blast-furnace slag than for those using pozzolanic additions. The acceleration seems to be proportional to the amount of portland cement in the mixture. Various effects may be produced when an accelerator is used with other special cements. Thus, the concrete proposed for use should be evaluated with the accelerating admixture by trial mixes.

Calcium chloride does not act as an accelerator with high-alumina cements.

6.4.4   Mechanism of Acceleration by $CaCl_2$

The effects of $CaCl_2$ on the hydration of portland cement have been recognized since 1885 [Highway Res. B. 1965, 1952], but the true mechanism of this "acceleration" is still not clear. [Ramachandran 1980] Most of the previous explanations were based on the chemical reaction between C_3A and $CaCl_2$. According to recent investigations, however, the essence of the hardening intensification is that the $CaCl_2$ as a catalyst acts on the hydration of calcium silicates. For instance, Rosenberg found evidence [Rosenberg 1964] that: (1) although $CaCl_2$ reacts with C_3A, the reaction rate, particularly in the presence of gypsum, is too slow to account for the increase in the rate of setting and hardening; (2) $CaCl_2$ definitely increases the rate of hydration in C_3S. However, it does not react chemically with the C_3S since the CSH gel, although altered in character, does not contain chemically bound chloride; (3) the high ionic character of the $CaCl_2$ solution at the time when the silicates are hydrating might cause the silicate to crystallize out more rapidly in small fibrous crystals. Another line of evidence was obtained by an x-ray method showing that at a time when 68% of C_3S in a portland cement paste had hydrated, 96% had hydrated in the paste also containing 2% $CaCl_2$. Also, 48% hydration of β-C_2S was found as compared with 82% hydration when 2% $CaCl_2$ was present [Copeland 1968].

This mechanism is supported by further investigations, not only for conventional curing conditions [Copeland 1968, Tamas 1966, Tenoutasse 1969, Skalny 1967, Ludwig 1974, Ramachandran 1971] but also when the hydration takes place under low-pressure or high-pressure steam curing [Balazs 1967].

6.4.5   Chloride-Free Accelerators

Because of the corrosion potential of the calcium chloride, non-corrosive accelerating admixtures have been sought. Proprietary nonchloride accelerating admixtures, certain nitrates, formates, nitrites, etc., perhaps blended with other materials, offer users alternatives, although they may be less effective and are more expensive than calcium chloride.[McCurrich 1979, ACI 1989]

One of the most effective chloride-free accelerators is a water-soluble organic material belonging to the *carboxylic acid* group.[Popovics 1983] This acid has a short hydrocarbon chain which is probably the source of its accelerating ability by promoting the solution of calcium ions from the cement compounds. Organic acids with long hydrocarbon chains cannot do that; besides, the long chains may hinder the bond formation among the hydration products, therefore such acids generally acts as retarders. [Mehta 1986]

This carboxylic acid is applicable in a wide variety of cementitious compositions including: portland cement [Popovics 1987c]; non-hydraulic cements, such as epoxies [Popovics 1987b]; finely divided mineral admixtures, such as fly ash [Popovics 1985b]; water-reducing admixtures; or any suitable combinations of these [Popovics 1988a]. For instance, a 2 percent addition of this accelerator to portland cement paste reduced the setting times at room temperature to about half. [Popovics 1990c] The effects of this accelerator on the strength development are illustrated in several figures. It can be seen from these that this accelerator has the tendency

1. to produce greater strength increases with Type III cement than with Type I, and especially with Type V cements (Fig. 6.6);
2. to yield not only higher early but also higher late strengths up to 180 days, than the control concrete (Fig. 6.6);
3. to produce greater strength increases at elevated curing temperatures than at normal or low temperatures. Thus, it is an excellent accelerator to make steam-cured concrete of high late strength (Fig. 6.7);
4. to be compatible with superplasticizer (Fig.

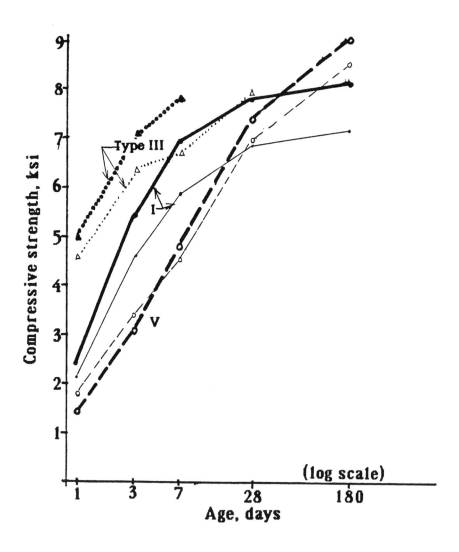

Figure 6.6: Effect of cement type on the changes of mortar strengths produced by 3% accelerator at various ages. Thin line: without accelerator; thick line: with accelerator; 1 ksi = 6.90 MPa [Popovics 1987c].

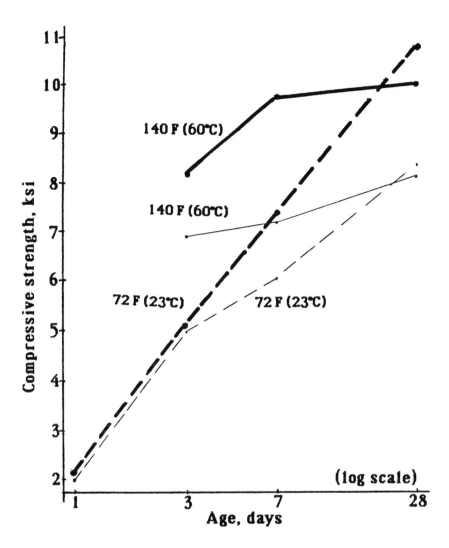

Figure 6.7: Effect of 2% accelerator on the compressive strength of concrete at normal and elevated temperatures. Cement: Type II; thin line: without accelerator; thick line: with accelerator (2%); 1 ksi = 6.90 MPa [Popovics 1987c].

6.8);
5. to produce greater strength increases with
   increasing control strength (Fig. 6.9).

6.5 WATER-REDUCING ADMIXTURES AND SET-CONTROLLING ADMIXTURES

Various results can be expected with a given
water-reducing or set-controlling admixture due to
differences in dosage, cements, aggregates, other
admixtures, and weather conditions. Therefore, if
previous pertinent experience is not available for
the given set of materials, laboratory trial mixes
should be run for the evaluation of the admixture
prior to job use. ASTM C 494-86 is a useful guide
for such tests. Methods for testing these admixtures
are also described in this standard.

6.5.1   Classification

Certain organic compounds or mixtures of organic and
inorganic compounds can be used as admixtures for
several purposes. These are:

1. to reduce the water requirement of a concrete
   for a given slump (ASTM C 494-86, Type A)
2. to retard the set (Type B)
3. both to reduce and retard (Type D)
4. to reduce the water requirement and
   accelerate the set (Type E).

These types are called (conventional)
water-reducing admixtures, or *water reducers*, or
plasticizers, and *set-retarding admixtures* or
retarders, respectively. The same standard also
acknowledges *high range water-reducing admixtures*
(*HRWRA*) as Type F (also Type 1 according to ASTM C
1017-85), and *water-reducing high range, and
retarding admixtures* as Type G (Type 2 according to
ASTM C 1017-85). The high range water-reducing
admixtures are also know as *superplasticizers*.

The materials that are generally available for
such use fall into eight categories [Prior 1960, ACI

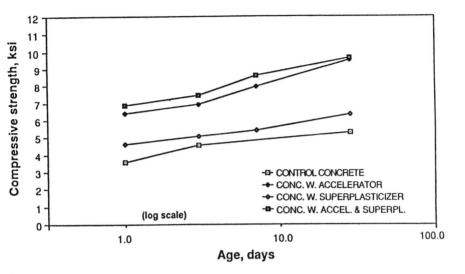

Figure 6.8: Comparison of the strengths of steam-cured concretes with and without the accelerator at various ages. Cement: Type III. 1 ksi = 6.90 MPa [Popovics 1988a, 1990c].

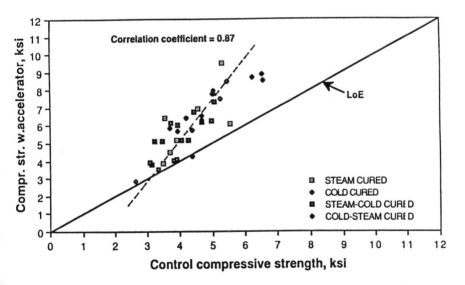

Figure 6.9: Demonstration of the strength-increasing effects of the accelerator on concretes subjected to different curing methods. Cement: Type III. LoE: line of equality. 1 ksi = 6.90 MPa [Popovics 1988a, 1990c].

1989]:

1. lignosulfonic acids, and their salts
2. modifications and derivatives of lignosulfonic acids and their salts
3. hydroxylated carboxylic acids and their salts
4. modifications and derivatives of hydroxylated carboxylic acids and their salts
5. salts of the sulfonated melamine poly-condensation products
6. salts of the high molecular weight condensation product of naphtalene sulfonic acid
7. blends of naphtalene or melamine condensate with other water-reducing or set-controlling materials
8. other materials, which include: (a) inorganic materials, such as zinc salts, borates, phosphates and chlorides; (b) amines and their derivatives; (c) carbohydrates, polysacharides, and sugar acids; (d) certain polymeric compounds, such as cellulose-ethers, melamine derivatives, naphtalene derivatives, silicones, and sulfonated hydrocarbons.

Admixtures of categories 1 and 3 are (conventional) water-reducing, set-retarding admixtures. Admixtures of categories 2 and 4 are (conventional) water-reducing admixtures designed either to have no substantial effect on rate of hardening, or to achieve varying degrees of acceleration or retardation in rate of hardening of concrete; these admixtures may also include an air-entraining agent. Categories 1 and 2, especially calcium lignosulfonates, constitute the largest volume of use.

A new generation of water-reducing admixtures constitutes what are called *high-range water-reducing admixtures (HRWRAs)* or *superplasticizers* because of their improved effectiveness in portland cement concrete [Aignesberger 1971, Sasse 1975, Bonzel 1974b]. They usually belong to categories 5, 6 and 7 above.

6.5.2 Conventional Water-Reducing Admixtures

Water-reducing admixtures are used to obtain specified strength at lower cement content with no loss in slump, or to increase the strength without changes in slump and cement content, or to increase the slump of a given mixture without increase in water content [Popovics 1968f] (Fig. 6.10). It has been reported that properly used water-reducing and set-retarding admixtures also increase the strength of steam-cured concrete 135°F (57° C) over that of concrete without such admixtures [Mielenz 1960]. On the other hand, similar Soviet experiments provided numerous unfavorable results. Other effects resulting from the use of water-reducing admixtures include improvement in the freeze-thaw durability, primarily as a result of increased strength of concrete [Larson 1963]. They may also improve the impermeability of concrete, or improve the properties of concrete containing aggregates that are poorly graded, or may be used in concrete that must be placed under difficult conditions. They are useful when placing concrete by means of a pump, or when using a tremie because they improve the mixture by reducing water for equal plasticity and reducing tendency toward segregation. Such admixtures may increase the shrinkage and creep of concrete [Morgan 1974, Ivey 1967].

Water-reducing admixtures have been found to be effective with all types of portland cement, with portland blast-furnace slag cement, with portland pozzolan cement but not necessarily with high-alumina cement. However, the brand and kind of cement may influence the results more than differences in performance of a number of properly formulated admixtures of the same type [Kalousek 1974]. For instance, water reducers appear to be more effective with cements that have lower alkali and C_3A contents [Polivka 1960]. Thus, concretes made with Type II and Type V cements require lower admixture dosages than concretes containing Type I or Type III cements for the same slump. Also, the amount of water reduction is influenced by admixture

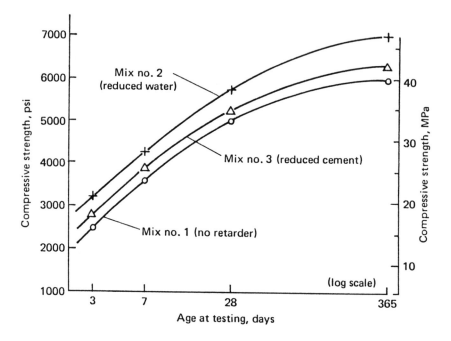

Figure 6.10: Average influence of various retarders on the compressive strength of concrete at various ages [Grieb 1962].

dosage [Gaynor 1962], mixing sequence [Bruere 1966], cement content, grading and type of aggregate, and presence of other admixtures, such as air-entraining admixtures, or pozzolans.

The principal role in the mechanism of water reduction and set retardation of the admixtures considered in the classification given above belongs to so-called surface-active agents [Duriez 1961, Ramachandran 1971, Mielenz 1969]. These substances are usually composed of long-chain organic molecules that are hydrophobic (nonwetting) at one end and hydrophilic (readily wet) at the other. Such molecules tend to become concentrated and form a film at the interface between two immiscible phases, such as cement and water, and alter the physicochemical forces acting at this interface. The water-reducing admixtures are absorbed on the cement particles, giving them a negative charge that then orients the water dipoles around each particle. Thus, the presence of such an admixture in a fresh mixture results in (1) a reduction of the interfacial tension; (2) an increase in the electrokinetic potential; and (3) a protective sheet of oriented water dipoles around each particle. That is, the mobility of the fresh mix becomes greater partly because of a reduction in the interparticle forces and partly because water is freed from the restraining influence of the highly flocculated system which is now available to lubricate the mixture. Less water is therefore required to achieve given consistency.

There are surfactants, such as hexadecyl trimethyl ammonium bromide, that increase the apparent viscosity for concentrations up to a certain value, beyond which further additions decrease viscosity. The elongated molecules of this surfactant are absorbed with their nonpolar ends oriented toward water, thus making the cement particles hydrophobic and increasing interparticle attraction. However, at higher concentrations a second layer begins to form on the first, and the molecules of the second layer being oppositely oriented. Thus, as the second layer builds up, the particles again become hydrophilic and interparticle attraction is accordingly reduced, with corresponding effects on apparent viscosity [Powers 1968].

6.5.3   High-Range Water-Reducing Admixtures

These admixtures. also called *superplasticizers* or *HRWRAs* have been in use intensively only during the last 15 years or so. [ACI 1979, 1981a, Transp. Res. B. 1979, Mielenz 1984]

ASTM C 494-86 specifies that high-range water-reducing admixtures (Types F and G) should reduce the quantity of mixing water required to produce concrete of a given consistency by at least 12 percent. This specification is quite conservative since a HRWRA may produce as much as 30 percent water reduction. In contrast, the range of water reduction of conventional plasticizers is around 5 to 10 percent.

One of the frequently used superplasticizers is *Melment L10* the composition of which is sulphonated melamine-formaldehyde condensate. It is usually available as a 20% aqueous solution. Another popular HRWRA is *Mighty 150* which is sulphonated naphtalene-formaldehyde condensate. It is usually available as a 42% aqueous solution. Both of these are sodium salts of high molecular weight condensates made from organic sulphonates of the type RSO_3^- where R is a complex organic group frequently of high molecular weight. (Fig. 6.11).[Malhotra 1979, 1981]

Superplasticizers are commonly batched in quantities such that the solids in the admixture are equal to 0.5 percent of the weight of the cement, but on occasion the dosage rate may be twice that. This figure is several times the dosage rate of conventional water-reducing admixtures. It has long been known that batching conventional admixtures in ever-increasing quantities would lead to large water reductions, comparable to those now attained with HRWRA. However, such massive dosages always led to undesirable side effects, chief of which were greatly prolonged setting time and excessively high air contents.

The plasticizing action by many high-range water-reducing admixtures, usually anionic in

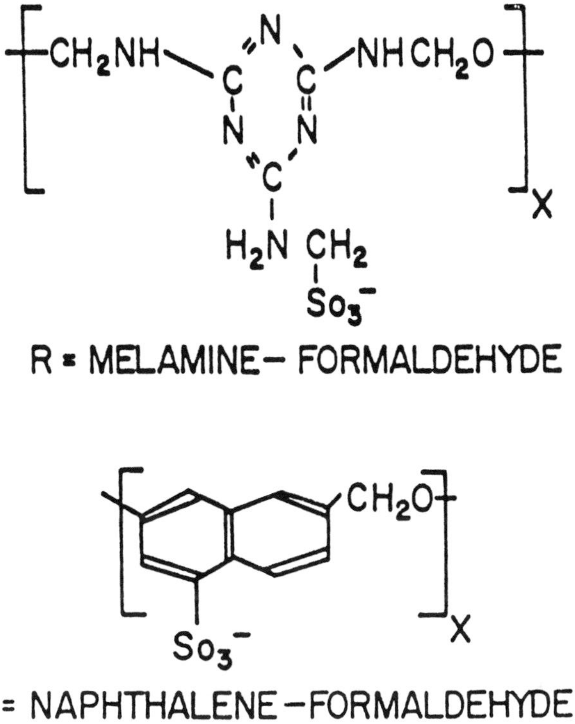

Figure 6.11: R–Organic group for naphthalene formaldehyde and melamine formaldehyde [Malhotra 1979, 1981].

nature, is similar to that of conventional
plasticizers. However, the superplasticizers produce
a much higher degree of dispersion [Cement Admix.
Assoc. 1976, Rauen 1976, Philleo 1986]. Adsorption,
viscosity and zeta potential values, obtained at
different concentrations of a HRWRA, show a definite
deviation at about the same concentration of the
superplasticizer, indicating the existence of an
electrostatic repulsion of the cement particles.
[Hattori 1979] However, HRWRA admixtures do not
affect the surface tension of water significantly;
therefore, they can be used at higher dosages
without obtaining excessive air entrainment as a
side effect. Note also that high-range water
reducers do influence air entrainment.[Tynes 1977]
Certain air-entraining admixtures have been reported
to be more effective with certain superplasticizers
in the production of adequate air entrainment. [ACI
1989]

    High-range water-reducing admixtures have been
found to be effective with all types of portland
cement and many other hydraulic cements similarly to
the conventional plasticizers, although exceptions
have also been reported.[Johnston 1987] It has been
found in some cases that higher SO_3 content may be
desirable when using HRWRA. Superplasticizers have
not been effective in concretes made with
high-alumina cement.[Quon 1981]

6.5.4   Set-Retarding Admixtures

Set-retarding admixtures are used primarily to
offset the accelerating effects of high temperature,
and to keep concrete workable during the entire
placing period. This method is of particular value
in preventing cracking resulting from form deflec-
tion of concrete beams, bridge decks, or composite
construction work. Set retarders are also used to
keep concrete plastic for a sufficiently long period
so that succeeding lifts can be placed without
development of discontinuities in the structural
unit. Note, however, that the use of a set retarding
or water-reducing admixture usually *does not* slow
down significantly the *slump loss* of a fresh

concrete. [Meyer 1979a]

The amount of retardation obtained is dependent on the specific admixture used, its dosage, the brand and type of cement, the temperature, mixing sequence [Bruere] (Fig. 6.12), and other job conditions. For instance, set retarders again appear to be more effective with cements that have lower alkali and C_3A contents. Also, in certain cases higher SO_3 content may be desirable in the cement to meet the competing sulfate demands of the C_3A and the admixture [Meyer 1979b, Newlon 1979]. The quantity of admixture added must be accurately determined and measured because a heavy overdosage can seriously slow down the setting and hardening of the concrete. This is particularly harmful when excess air is entrained in the concrete by the overdosage. Set-retarding effects of several organic compounds are summarized in Table 6.2.

There are water-reducing admixtures that also have set-retarding effects, and set-retarding admixtures with secondary water-reducing effects. For instance, refined white granulated sugar in the quantity of 0.1% by weight of the cement not only retards the setting very effectively but also provides a more workable mixture, especially with delayed addition, and increases the strength at the age of 7 days and later [Kuhl 1961b, Van Wallendael 1973].

The mechanism of set retardation is again based on adsorption [Hansen 1960, Diamond 1973]. The large admixture anions and molecules are absorbed on the surface of the cement particles, which hinders further reactions between cement and water, that is, retards the setting. Later, as a result of reaction between the organic salts and tricalcium aluminate from the cement, the former are removed from the liquid phase of the system, thus eliminating further retardation. This explanation is in qualitative agreement with observations that retarding admixtures are often particularly effective with low-tricalcium aluminate cements and that the initial retardation period is followed by rapid hydration and hardening of the paste [Foster 1960].

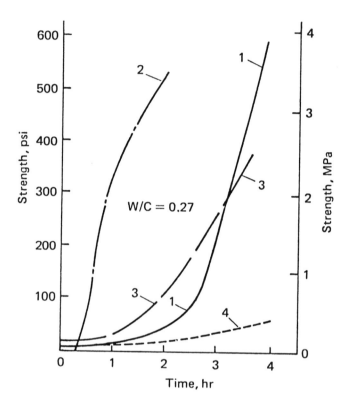

Figure 6.12: Effects of calcium lignosulfonate with various mixing sequences on the early strength developments of cement pastes, after Segalova et al. The portland cement is gypsum-free and contains 10.3% C_3A. Curve 1, no admixture, stirred for 5 min. Curve 2, 0.3% admixture added with the mixing water, stirred for 1 min. Curve 3, 0.3% admixture added belatedly, stirred for 20 min. Curve 4, 0.3% admixture added even later, after preliminary hydration [Eitel 1966].

Table 6.2: The Effect of Various Compounds as Admixtures on the Setting Time of Cement[a]

Admixture	Percent by wt. of cement	Tempera-ture, °F	Pumpability time, min		Thickening time, min	
			Without admixture	With admixture	Without admixture	With admixture
Arrowroot starch	0.10		127	336		
			100	236		
Dextrin	0.05		127	276		
			100	171		
Casein	0.4	140	342	533		
	0.4	200	185	535		
Sodium bicarbonate	0.14		100	133		
	0.14		58	85		
Tartaric acid	0.25		100	202		
			58	397		
Cream of Tartar	0.25		100	425		
			58	480		
Tartaric acid + sodium bicarbonate	0.20 +0.10		100	184		
			58	348		
Starch up to 10% soluble in cold water	0.15 −0.20	200			120	360
		100			380	405
		220			105	210
Starch 40–65% soluble in cold water	0.08 −0.10	100			380	960
		200			120	315
		220			105	120
Sodium salt of carboxymethyl cellulose	0.16	140			226	213
		220			71	639
Maleic acid	0.10	100				526
		220				127
	0.30	100				715
		220				292
Oxidized cellulose	0.09	140			204	786
		200			93	407
		220			72	121

Note: Reproduced from Orchard, *Concrete Technology,* Volume 2, 3rd edition by permission of the author and Applied Science Publishers.

[a]Abstracted by Orchard from information given by W.C. Hansen [Orchard 1973b].

6.6   POLYMERS

6.6.1   General

Polymers can be combined with portland cement concrete in several different ways. [Popovics 1974c] The one that is pertinent here is the form of polymer portland cement concrete (PPCC), or polymer modified concrete. This is a portland cement concrete to which one or more organic monomers or polymers are added before or during mixing. The polymers are used here in a way similar to the use of chemical admixtures with two exceptions: (a) the amount of chemicals in PPCC is much greater than the usual amount of any chemical admixture; and (b) the polymer in the concrete may supplement the cement in biding the mineral aggregate particles through the adhesive and cohesive properties of the polymer. [Popovics 1974b]

Polymer is a substance that is composed of thousands of simple molecules combined into giant molecules. The simple molecules are called *monomers*, and the reaction that combines them into polymer is called *polymerization*. The polymer can be in a liquid or a powdery form. When the polymer is in dispersion, it is called *latex*.

The latex is already in polymer form when it is added to the concrete, therefore the process is called *pre-mix polymerization*. Another process is the *post-mix polymerization* where the polymer components are added to the concrete in monomer form, and the polymerization takes place subsequently in the concrete along with the cement hydration.

Polymer modification, when made properly, can improve quite a few properties of the fresh and hardened concrete. Such properties are:

1. workability and finishability
2. cohesion, bleeding, and segregation
3. strengths, especially flexural and tensile strengths
4. bond

5. abrasion resistance
6. impermeability
7. physical and chemical durability

over similar but unmodified concretes and mortars.[Popovics 1984b] Note, however, that some of these properties are measurably lowered when tested in wet state.[Popovics 1986a, 1987b]

Modifying polymers may be compatible with other chemical or mineral admixtures.[Popovics 1984a, 1985a]

Polymer modified concretes and mortars have been used successfully in various applications, such as construction of bridge decks, industrial and garage floors, and pavements. They are also suitable for overlays and other repair and rehabilitation of concrete, floors, pavements and other structures.

There are no standard methods in this country for testing the suitability of a polymer system for modification of portland cement concrete, except for trial mixes. The general properties of a polymer, such as total solids content, pH value, coagulum, viscosity, stability, weight per gallon, particle size, surface tension, and minimum film-forming temperature, can be measured by various textbook methods. Similar methods have been specified in several foreign standards, for instance in Japan.[Schutz 1982]

There are no American standard methods for testing polymer modified mortars or concretes either, although there are foreign standards. For instance, there are Japanese standards for preparation of laboratory specimens; and testing the slump, unit weight, air content and strength of such specimens. Most of these are the modifications of the existing procedures for conventional cement mortar and concrete.

6.6.2 Latexes

Latexes are usually made by emulsion polymerization, and contain about 45 to 50 percent solids by weight. Most latexes are copolymer systems, that is, they consist of at least two monomers.(Walters 1987)

The first attempt for *latex modification of portland cement concrete* goes back to 1924 using natural rubber latex. Later synthetic rubber latexes of high molecular weight were tried along with synthetic resin latexes, such as polyvinyl acetate, polyvinyl propionate, and other latexes.[Hosek 1966, Kreijger 1968] In these compounds the carbon chain usually contains the $-CH_2-CH-$ group to which other atoms or radicals are attached. Most of the latexes used with portland cement are stabilized with surfactants that are nonionic. Today the most popular modifying polymer is probably the *styrene-butadiane (S-B)* latex but *ethylene-vinyl acetate (EVA)* and *polyacrylic ester (PAE)* emulsions are also used.

Addition of cement to latex leads to a breakdown of the latter with separation of the polymer particles. Rapid coagulation is induced by the multivalent ions (chiefly Ca^{++}) obtained from the cement, and by the incompatibility of the particles of the cement and polymer, respectively, which are oppositely charged. Addition of stabilizers keeps emulsions from separating out before the concrete has started to set.[Solomatov 1967] Maximum strength of latex modified mortars and concretes is obtained typically with 15 to 20 percent polymer by weight of cement.

As the cement hydrates, the latex particles become concentrated in the pores. The latex becomes effective only when the emulsion is broken through drying out. This takes place with continuing water removal by the cement hydration and evaporation. It follows then that, after an initial 24 hr of moist curing to eliminate cracking, additional moist curing of a latex modified concrete is undesirable since the latex will not have an opportunity to dry and develop enough strength.

The drying latex particles coalesce into a net which is interwoven in the hardened cement matrix. This acts as a comatrix that also coats the aggregate particles and the pores, thus improving the bond between the various phases in the concrete. This latex net also fills micropores and bridges microcracks that develop during the shrinking associated with curing. The combination of these two actions is the source of the strength contributed by the latex modification. Surfactants present in latex can entrap air and may require that a foam-suppressing agent be used. Dosage rates for air-entraining admixtures will be affected.

Strength results of several test series with an S-B latex, along with comparable results with epoxies, are presented in Figures 6.13 and 6.14. These figures illustrate not only the influence of polymer modification on the compressive strength but also the strength increasing effect of drying, and reducing effect of rewetting of concrete. [Popovics 1987b]

6.6.3   Epoxies

Several experiments have been reported in the literature for adding various polymerizable components (parts or intermediates) into fresh mortar or concrete where the polymerization was produced post mix either by irradiation or heat treatment of the specimens. [Steinberg 1968, Dikeou 1971, Gebauer 1971] Disappointing results as well as the complexity of the technique precluded further investigation in this direction.

A much simpler approach is to produce post-mix polymerization of the intermediates in the concrete solely by chemical means. Quite a few intermediates have been tried out but only epoxy systems have shown promise for improving the quality of concrete. [Lezy 1967, Valenta 1970, Sun 1975, Popovics 1976b, Nawy 1978]

Epoxy is a multi-component organic polymeric material. The epoxy components are added to the fresh concrete before or during mixing. The polymerization takes place subsequently in the concrete along with the cement hydration resulting in an *epoxy modified concrete*. Epoxy modified concretes have the same advantages and disadvantages as latex modified concretes, that is, improved workability, higher tensile and flexural strengths, etc.

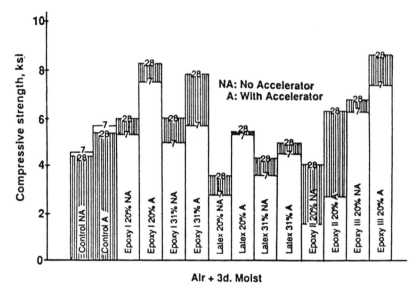

Figure 6.13: Effect of various polymer modifications as well as accelerator on compressive strength, I. Curing: the last three days in fog room, until then continuously in laboratory air, 1 ksi = 6.90 MPa [Popovics 1987b].

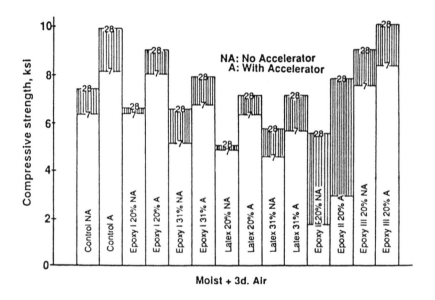

Figure 6.14: Effect of various polymer modifications as well as accelerator on compressive strength, II. Curing: the last three days in laboratory air, until then continuously in fog room, 1 ksi = 6.90 MPa [Popovics 1987b].

Epoxy compounds for concrete modification are usually formulated in two parts. Part A is most often the portion containing the *epoxy resin* and Part B is its *hardener* system. When mixed properly, the components polymerize and harden. This process is also called *curing*. Epoxy systems for concrete modification contain surfactant for emulsification.

The uncured components of an *epoxy resin* are either honey-colored liquids or brittle amber solids which become liquid when heated. Enlarged 10 million times, the molecules of resin might resemble short pieces of thread, varying in length from half an inch in some of the liquids to several inches in the solids. When polymerized, these threads are joined together at the ends and along the sides to form large cross-linked structures.[Skeist 1962]

Two types of reactive groups take part in the cross-linking: epoxy and hydroxyl. The epoxy group is a three-membered ring:

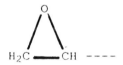

Such rings are under great strains, and will open under slight provocation. A variety of compounds with active hydrogen atoms, including amines, acids, phenols, alcohols and thiols, are capable of opening the ring and forming a larger molecule with a secondary hydroxyl group. The hydroxyl usually contributes less to the polymerization than the epoxy, especially at early ages.

Most epoxy resins are made from bisphenol A and epichlorohydrin. The bisphenol A, derived from acetone and phenol, is a diphenol--, it has two phenolic hydroxyl groups capable of reacting. Ignoring the unreactive ether linkages, this molecule may be symbolized as two connected triangles indicating the epoxide groups which are capable of further combination. This intermediate is called "diepoxide O," the "O" indicating absence of hydroxyls.   Three such molecules are shown in the upper part of Figure 6.15.[Schutz 1982]

The liquid resins in greatest use average 1.8 to 1.9 epoxides per molecule, as compared to perhaps 0.3 hydroxyls. It is the epoxide group, therefore, which is of greatest   utility in converting the liquid resin to a solid end product. Via the epoxide, one molecule of intermediate is linked to others.

For thorough curing a reactive *hardener* is required which ties several epoxy molecules together. Such hardeners are usually primary and secondary amines that function by reacting with the epoxides. Each hydrogen atom attached to a nitrogen atom is "active," that is, capable of opening the epoxide ring.   Thus a primary amine reacts to form a secondary amine, which in turn can affiliate with another epoxide.

During the *polymerization process* two types of linkage are possible: (1) directly to other epoxy intermediate molecules, with the aid of a catalyst; and (2) to a reactive hardener which combines with one or more additional molecules of resin.   In combination with concrete reactive hardeners are usually used.

The products of these reactions are secondary alcohols but these groups do not combine with other epoxides. They do, however, stimulate the remaining amine to react more quickly with epoxide. Therefore, the reaction is autocatalytic, it tends to build up at a continually accelerating rate, until most of the amino hydrogen or epoxide has been utilized. Some hardening agents, for instance aliphatic amines among the reactive hardeners, will bring about curing starting at room temperature, without application of heat. The curing of epoxy resin may be pictured as shown in the lower part of Figure 6.15.

When resin and hardener are combined in optimum proportions, the cured material will have the highest softening temperature and best combination of strength properties from these starting materials. For reactive hardeners, best results are

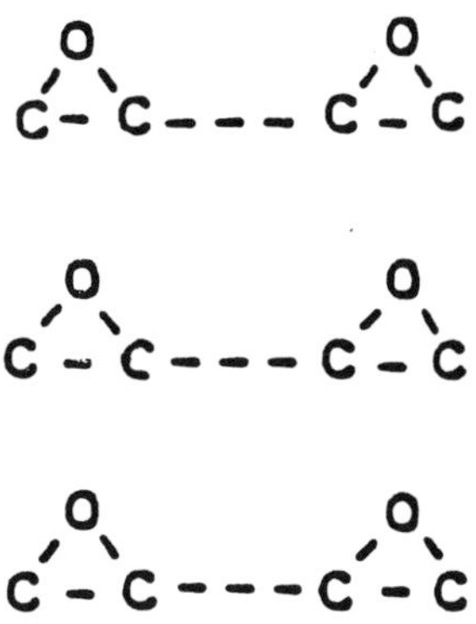

A typical epoxy resin before cure; the molecules are not crosslinked and are free to move. The resin is a liquid.

By addition of a curing agent, in this example a primary amine, the molecules are crosslinked into a dense solid.

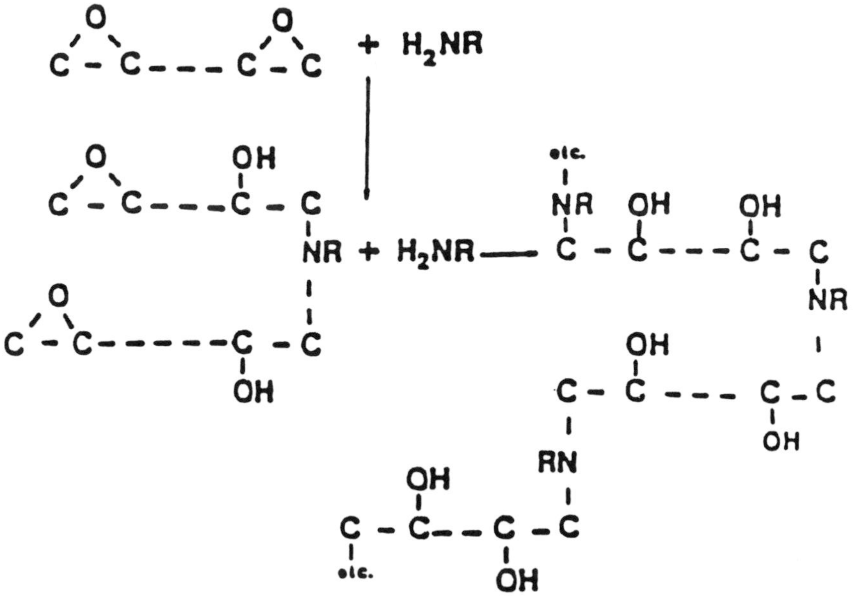

Figure 6.15: Curing of epoxy resin [Schutz 1982].

obtained when hardener and epoxide are combined in nearly stoichiometric amounts, i.e., when there is one active hydrogen atom for each epoxide group.

The polymerized products are thermoset; i. e., they are solids that may become softer when heated, but will never again liquefy. The polymerization of epoxies is irreversible. The physical properties of a polymerized epoxy depend on the extent of cure. The greater the *functionality*, that is, the *density of cross-links*, the better will be the mechanical characteristics and resistance to heat-softening, or attack by chemicals and water.

In addition to the regular requirements (strength, durability, etc.), an epoxy system which is to be used for modification of a portland cement concrete should have certain special properties, as follows:

1. the components should be soluble or dispersible in water
2. the polymerization of the components should be possible solely through chemical means even at room temperature, that is, without heating or radiation
3. neither the high liquid content nor the high pH of the cement paste should interfere with the strength or durability of the polymerized epoxy
4. the epoxy components, including the emulsifier, should not interfere excessively with the hydration of the portland cement and with technically important properties of the fresh and hardened concrete, even when other chemical admixtures are also present.

Epoxies are considered as generally resistant to chemical attacks.

6.7 OTHER CHEMICAL ADMIXTURES

6.7.1 Admixtures for Extreme Consistencies

Special admixtures are needed for *flowing concrete* since the production of such concrete by addition of water only would result in concrete of extremely low quality.[ACI 1989] ASTM C 1017-85 defines flowing concrete as concrete that has a slump greater than 7 1/2 in (190 mm) while maintaining a cohesive nature.

The plasticizing admixture for flowing concrete is either normal (Type 1) or retarding (Type 2). These materials generally are identical to those used as high-range water-reducing admixtures, and conform to ASTM C 949, Types F and G. (See Section 6.5.3.)

The dosage required to increase the slump to flowing consistency varies depending on the cement, initial slump, w/c, temperature, time of addition, and concrete mix proportions.

Flowing concrete is used in areas requiring maximum volume placement (slabs, mats, pavements) in congested locations. Flowing concrete is also useful for pumping.

Admixtures are used in *no-slump concrete* for essentially the same reasons as in other concretes. Consequently, they comprise most of the types and classes normally associated with other concrete mixtures. For instance, admixtures may be used to promote plasticity and cohesiveness in no-slump concrete. The admixtures most frequently used for these purposes are air-entraining admixtures and fly ash.[ACI 1981b]

No-slump concrete is defined as a concrete with a slump of 1 in. (25.4 mm) or less immediately after mixing.

6.7.2   Miscellaneous Admixtures [ACI 1989]

The gas-void content of a cement paste, mortar or concrete can be increased by the use of *gas-forming admixture*. These generate or liberate gas bubbles in the fresh mixture during and immediately following placement and prior to setting of the cement paste. Such materials are used to reduce the unit weight of the mixture, counteract settlement and bleeding, thus causing the concrete to retain more nearly the volume at which it was cast. They are *not* used for producing resistance to freezing and thawing; any such effort is incidental.

Admixtures that can produce these effects are hydrogen peroxide which generates oxygen; metallic aluminum which generates hydrogen; and certain forms of activated carbon from which adsorbed air is liberated. Present practice uses mostly aluminum powder.

Grouting admixtures are used to impart special properties to the grout. For instance, oil-well cementing grouts encounter high temperatures and pressures with considerable pumping distances involved. Grout for preplaced-aggregate concrete requires extreme fluidity and nonsettling of the heavier particles. Nonshrink grout requires a material that will not exhibit a reduction from its volume at placement. A wide variety of admixtures, such as retarders, plasticizers, gas-forming or expansion-producing or air-entraing admixtures, are used to produce the special grout property required.

Expansion-producing admixtures are used to minimize the magnitude of drying shrinkage. These chemicals expand in the concrete or react with other constituents of the concrete to cause the required expansion. They are used in restrained and unrestrained concrete placement. For unrestrained concrete, the expansion must not take place before the concrete gains sufficient tensile strength, or else the concrete will be disrupted. For restrained concrete, the concrete must be strong enough to withstand the compressive stresses developed.

The most common expansion-producing admixture is a combination of finely divided or granulated iron and chemicals to promote oxidation of the iron. Expansion is greatest when the mixture is exposed alternately to wetting and drying. This method is used in lieu of expansive cements. (Section 4.3)

Bonding admixtures are specifically formulated for use in portland cement mixtures to enhance bonding properties. These chemicals are usually organic polymers, either latexes or epoxies. The properties of these polymers are discussed in Section 6.6.

Pumping aids for concrete are admixtures with the sole function of improving concrete pumpability. That is, they are used to overcome difficulties that cannot be overcome, or not economically, by changes in the concrete composition. Many pumping aides are thickeners increasing the viscosity of mixing water.[Valore 1978, Fletcher 1971c] Thickeners can also cause air entrainment and/or set retardation.

Although concrete furnishes ample protection against corrosion to the steel embedded in it, there are cases where this protection is broken down, for instance in the presence of chloride ions. One approach to improve the steel protection is the use of *corrosion-inhibiting admixtures*. Numerous chemicals have been evaluated as potential corrosion-inhibiting admixtures. These include chromates, phosphates, hypophosphorites, alkalies, fluorides, sodium nitrite, sodium benzoate, etc. Recently calcium nitrite has been reported as an effective corrosion inhibitor.[Rosenberg 1977, Alonso 1990] Experience with corrosion-inhibiting admixtures has not been uniformly good.

Other concrete admixtures used in the industry are:

coloring; flocculating; fungicidal, germicidal, and insecticidal; dampproofing; permeability reducing; and, alkali-aggregate expansion reducing

admixtures.

6.8 FINELY DIVIDED MINERAL ADMIXTURES

These admixtures may be naturally occurring materials or artificially formed mineral substances. They may be classified into three types:

1. those that are relatively *inert* chemically, such as ground quartz, ground limestone, bentonite, hydrated lime, and talc;
2. those that are *pozzolanic*, such as fly ash, volcanic glass, diatomaceous earths, and some shales or clays; and
3. those that are *cementitious*, such as natural cements, hydraulic limes, slag cements (mixtures of blast-furnace slag and

lime),and granulated iron blast-furnace slag.

The most important of these three types is the group of pozzolanic materials or pozzolans. The pozzolanic and cementitious mineral admixtures have been discussed in Chap. 4.

The use of finely divided mineral admixtures can improve some of the properties of the fresh and hardened concrete. Such properties are:

1. workability and finishability of concrete that is too harsh
2. cohesion, bleeding and segregation
3. rate of heat of hydration development by replacement of a portion of portland cement
4. concrete strength at later ages
5. improvement of the chemical resistance of concrete
6. reduction of alakli-silica reaction.

6.9 STORAGE, SAMPLING, AND TESTING

Admixtures manufactured in liquid forms should be stored in watertight drums or tanks protected from freezing. The requirements for storage of powdered admixtures are the same as those for cementitious materials.

Samples for testing and inspection of admixtures should be obtained by procedures described for the respective types of materials in applicable speci- fications, such as ASTM C 260-86 for air-entraining agents.

Admixtures are tested for acceptance for one or more of three reasons: (1) to determine compliance with a purchase specification; (2) to evaluate the effect of the admixture on the properties of the concrete to be made with job materials under the anticipated ambient conditions and construction procedures ; and (3) to determine uniformity of the product.[ACI 1989]

ASTM C 494-86 specifies a laboratory method for testing the suitability of admixtures by making concrete mixtures with them and measuring certain properties of the fresh and hardened concrete. A summary of the pertinent physical requirements is presented in Table 6.3. Air-entraining admixtures can be tested according to the specifications of ASTM C 233-89. Standard methods are also available to test the effectiveness of mineral admixtures in preventing excessive expansion of concrete due to alkali-aggregate reaction (ASTM C 441-89), and to test fly ash or raw calcinated natural pozzolans for use in portland cement concrete (ASTM C 618-89). In addition, there are several other methods available in the literature [RILEM 1975, Halstead 1954, 1962, Fletcher 1971a, 1971b, Joisel 1968, Venuat 1967a, Walz 1975, Anon. 1976]. Although these tests afford valuable screening procedure for selection of admixture products, they do not eliminate the necessity of testing the admixture in trial mixtures of concrete prepared under the anticipated plant or field conditions.

There are no standardized methods for the determination of the quantity, or even the presence, of an admixture in fresh or hardened concrete, but special procedures are available in the literature [Andrews 1971, Kroone 1971].

6.10 FUTURE OF ADMIXTURES

The benefits from the use of admixtures will continue to accrue [ACI 1971b]. They will add further to the strength of concrete, reduce its energy demand and cost, and produce concrete that is more durable and easier to place. An example is an efficient water-reducing admixture for no-slump concretes.

"Inspector" admixtures will be developed. By changes of color, they will indicate if the water-cement ratio is too high, and if the temperature of the fresh concrete is too low or too high. Also, concretes with different properties on the same job will be "color coded" while in the plastic state. Such admixtures will also be helpful in maintaining uniformity. The economy of concrete will also be improved by developing admixtures that improve the

Table 6.3: Physical Requirements^A (ASTM C 494-86)

	Type A, Water Reducing	Type B, Retarding	Type C, Accelerating	Type D, Water Reducing and Retarding	Type E, Water Reducing and Accelerating	Type F, Water Reducing, High Range	Type G, Water Reducing, High Range and Retarding
Water content, max, % of control	95	...	...	95	95	88	88
Time of setting, allowable deviation from control, h:min:							
Initial: at least	...	1:00 later	1:00 earlier	1:00 later	1:00 earlier	...	1:00 later
not more than	1:00 earlier nor 1:30 later	3:30 later	3:30 earlier	3:30 later	3:30 earlier	1:00 earlier nor 1:30 later	3:30 later
Final: at least	...	...	1:00 earlier	...	1:00 earlier	...	...
not more than	1:00 earlier nor 1:30 later	3:30 later	...	3:30 later	...	1:00 earlier nor 1:30 later	3:30 later
Compressive strength, min, % of control:^B							
1 day	...	...	...	...	...	140	125
3 days	110	90	125	110	125	125	125
7 days	110	90	100	110	110	115	115
28 days	110	90	100	110	110	110	110
6 months	100	90	90	100	100	100	100
1 year	100	90	90	100	100	100	100
Flexural strength, min, % of control:^B							
3 days	100	90	110	100	110	110	110
7 days	100	90	100	100	100	100	100
28 days	100	90	90	100	100	100	100
Length change, max (alternative requirements):^C							
Percent of control	135	135	135	135	135	135	135
Increase over control	0.010	0.010	0.010	0.010	0.010	0.010	0.010
Relative durability factor, min^D	80	80	80	80	80	80	80

^A The values in the table include allowance for normal variation in test results. The object of the 90 % compressive strength requirement for a Type-B admixture is to require a level of performance comparable to that of the reference concrete.

^B The compressive and flexural strength of the concrete containing the admixture under test at any test age shall be not less than 90 % of that attained at any previous test age. The objective of this limit is to require that the compressive or flexural strength of the concrete containing the admixture under test shall not decrease with age.

^C Alternative requirements, see 17.1.4, % of control limit applies when length change of control is 0.030 % or greater, increase over control limit applies when length change of control is less than 0.030 %.

^D This requirement is applicable only when the admixture is to be used in air-entrained concrete which may be exposed to freezing and thawing while wet.

performance of blended cements and decrease the
energy content of concrete and concrete
construction. [Mielenz 1984]

Thermosetting and thermoplastic resin
admixtures will be added to concrete to control the
modulus of elasticity, and to increase the
resistance of the cement matrix to chemical attack.

7

Mineral Aggregates—General

SUMMARY

Concrete aggregate is a more or less inert, granular, usually inorganic material consisting normally of stone(s) or stonelike solid(s). Typical examples are sand, gravel, crushed stone, and crushed slag. The use of aggregate in concrete greatly reduces the needed amount of cement, which is important both from technical and economical standpoints.

Aggregates can be classified in several different ways: whether they are natural or manufactured; according to their petrography; according to their specific gravity; whether they are crushed or naturally processed; whether they are inert or reactive; and according to the sizes of their grains or particles. The last one is the most frequently used classification. On this basis one can distinguish between fine aggregates, consisting mostly of small particles, and coarse aggregates, consisting mostly of large particles. The dividing line between fine and coarse aggregates is arbitrary. In the countries of the Western Hemisphere, usually sieve No.4 separates the two size classes with a net opening of 3/16 in (4.75 mm).

The properties of aggregates are determined by appropriate testing of samples. Proper aggregate sampling is just as important as correctness of testing. Aggregate samples may be taken at various stages in the utilization process. The grading of the coarse aggregate is one of the properties that causes most concern. As a rule, a better estimate of

274

the actual grading is obtained by accumulating a sample randomly by means of a large number of small increments, such as scoopfuls, than by taking a few larger increments, such as shovelfuls. The number of field samples required depends on the criticality of, and variation in, the properties to be measured. The size of the field sample should be as large as practical.

7.1   INTRODUCTION

Mineral aggregate for concrete, or concrete aggregate, or simply *aggregate*, as it will be called in this book, is a more or less inert, granular, usually inorganic material consisting normally of stone(s) or stonelike solid(s). Typical examples are sand, gravel, crushed stone, and crushed slag.

In compacted hardened concrete the paste fills the voids between the aggregate particles and bonds them firmly together. That is, the paste holds the aggregate particles together as a matrix. Aggregate occupies roughly three-fourths of the volume of concrete, so its quality has a considerable importance on the concrete quality [Popovics 1968g]. Aggregate is used in concrete because it reduces greatly the needed amount of cement in the mixture, thus decreases the creep and shrinkage of the concrete and, at the same time, makes it cheaper.

7.2 CLASSIFICATION OF AGGREGATES

Aggregates can be classified in a number of different ways, as illustrated below. Note that some of the classifications overlap.

One can distinguish between naturally occurring or *natural*, and artificial or *manufactured* aggregates. The material of the particles in a so-called natural aggregate, such as crushed stone, is not changed artificially during aggregate production, although the aggregate itself may be submitted to manufacturing processes, such as crushing, washing, sieving, and so on [Legg 1974]. The material of particles of an artificial aggregate is produced, often as a by-product or waste, by certain manufacturing processes (heating, etc.) from naturally occurring materials. A typical example for this is crushed blast-furnace slag.

Natural *sands* and *gravels* are used most commonly as concrete aggregates because of their economical availability in most regions. Note however, that sand and gravel deposits are frequently quite variable in quality, especially in grading, there-fore, in order to obtain uniform aggregates from natural deposits, washing, screening, and separation into size fractions is recommended before use [Rockwood 1948, Blanks 1955].

The second most extensively used source of natural aggregates is quarried bedrock materials, that is, *crushed stone*. This is produced by breaking down the natural bedrocks into satisfactorily graded particles through a series of blasting, crushing, screening, and classifying operations.

Perhaps the most commonly used manufactured aggregate is crushed *blast-furnace slag*. This is a stonelike material that is a by-product of iron production. [Gutt 1978b] Blast furnace slag has been used as aggregate since the end of the last century in those areas where it was available, although not every slag is suitable for concrete making because of the potential unsoundness of the material.

Another group of manufactured aggregates in-cludes certain *lightweight concrete aggregates*. These aggregates are gaining popularity because of the rapidly increasing applications of lightweight concretes. Expanded shales, clays, slates, perlites, and slags are the most common manufactured light-weight aggregates. Examples of the utilization of *waste materials* as aggregates are crushed brick and crushed hardened concrete.

Manufactured aggregates have, as a rule, more uniform quality than unprocessed natural aggregates.

Another classification is based on the *petro-graphy* of the material of aggregate particles. Some of the more common or more important natural materials of which mineral aggregates are composed are silicate minerals, feldspars, ferromagnesian minerals, micaceous minerals, clay minerals, zeo-lites, carbonate minerals, sulfate minerals, iron sulfide minerals, and iron oxide minerals [Portland Cement Assoc. 1975]. These are described briefly, along with the various types of rocks, in ASTM C 294-86. Special aggregates, such as heavyweight aggregates, may contain different constituents (ASTM C 638-84).

There are three general classes of bedrocks, based on origin: *igneous, sedimentary,* and *metamorphic.* Igneous rocks are formed by cooling of molten lava (granite, basalt, diabase, gabbro, andesite, pumice, etc.). Sedimentary rocks are lime rocks made of calcium carbonate usually precipitated out of solution (limestone), or consolidated fragments of older rocks that have been transported and deposited by water, ice, wind, or gravity (sandstone, shale, conglomerate, etc.). Metamorphic rocks are either igneous or sedimentary rocks that have been changed by heat or pressure or both (slate, marble, quartzite, etc.). Generally, igneous rocks that are fine grained, well interlocked, and contain low percentages of feldspars have the best concrete-making properties. Among the sedimentary rocks, massive, hard limestones make usually the best aggregates. Deep-seated metamorphism can also produce excellent aggregate material. However, the degree of variation in the properties and composition of the various classes and subclasses of rocks is almost limitless. Any single type of rock may be porous or dense, strong or weak, hard or soft, decayed or unaltered, durable or unsound, fine or coarse grained; that is, the type of rock by itself does not determine whether or not it can provide an aggregate of adequate quality [Blanks 1955, US Bureau of Recl. 1966].

The mineralogic and petrographic composition of sands and gravels deposited by streams is usually very heterogeneous because many sources of parent rock are tapped by the stream system. Wind-blown materials are normally limited to finer sands of narrow size ranges. Their particles are predominately quartz.

Blast-furnace slags consist essentially of silicates and aluminum silicates of lime and other bases, although the details of composition vary from furnace to furnace. That is, slags differ in composition from rocks, but they are similar to certain igneous rocks in physical properties. Highly glassy slag tends to be less tough and more brittle than crystalline slags.

Accurate identification of rocks and minerals can, in many cases, be made only by a qualified geologist or petrographer. Test methods for such examinations are given in ASTM C 295, and in the literature [Mielenz 1966, Dolar-Mantuani 1966, Mather K. 1966, Soles 1980]

Aggregates can also be classified according to their (bulk) *specific gravity*, although the limits dividing the various groups are not definite. The usual, or *normal-weight* concrete aggregates have a bulk specific gravity around 2.6. Aggregates with a bulk specific gravity of less than about 2.4 are called *lightweight* aggregates, whereas those with a specific gravity greater than 2.8 are called *heavyweight* aggregates. The unit weight of a concrete can be controlled effectively through the specific gravity of the aggregate.

A distinction may be made between aggregate reduced to its present size by *natural processes* of weathering and abrasion, and so-called *crushed* aggregate, produced by manmade fragmentation of the parent rock. Crushed stone is the more harsh and angular.

There are many different types of *crushing machines* [Ujhelyi 1973, Waddell 1962, Hansen 1971]. Each crusher type provides a characteristic grading for the produced aggregate which depends primarily on the number of crushing effects to which a group of aggregate particles can be subjected during crushing [Lazar 1957]. Thus, the required grading of the crushed aggregate determines the type of crusher to be used. Typical curves for particle-size distributions are shown in Fig. 7.1. This figure demonstrates that a hammer crusher produces relatively much more smaller particles than a roll crusher would. It can also be seen that grading *b* can be approximated by a normal, and grading *a* by a logarithmic-normal distribution. Hammer crushers tend to produce better shaped particles, (See Chapter 10). Large crushing ratios in other types of crushers tend to produce poorer shaped material.

The majority of aggregates can be considered as *inert*, or inactive, in concrete, that is, such that it develops a bond with the hardening cement paste but otherwise does not influence significantly the hardening process or the soundness of the concrete. There are, however, aggregates that are *reactive*, that is, that develop noticeable reactions with the paste. Such reactions can be advantageous, such as

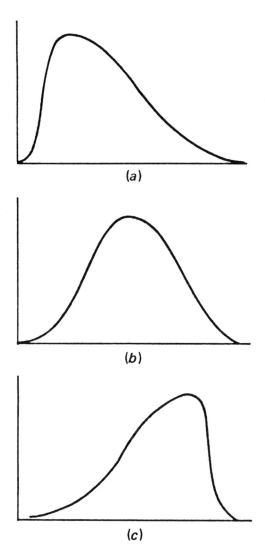

Figure 7.1: Typical grading types of crushed aggregates. (a) Numerous crushing effects (hammer crusher). (b) Several crushing effects (jaw crusher). (c) Single crushing effect (roll crusher). The horizontal axes are particle sizes, the vertical axes are frequencies [Lazar 1957].

the reaction of quartz aggregate in autoclaved concrete, or harmful, such as the alkali-aggregate reaction.

Perhaps the most frequently used aggregate classification is based on the *size* of the particle. An aggregate consisting of large particles is called *coarse* aggregate, whereas an aggregate consisting of small particles is called *fine* aggregate. The dividing line between fine and coarse aggregates is arbitrary and may vary from discipline to discipline, or from country to country. In concrete technology this dividing line is usually the No. 4 sieve in the United States (ASTM C 125-88) which, according to ASTM E 11-87, has a nominal net opening of 0.187 in (4.75 mm); this limit will be used in this book. In other words, particles passing through a No. 4 sieve are called fine aggregate particles, whereas particles retained on this sieve form coarse aggregate. Typical examples of fine aggregate are sand and crushed sand, respectively, and those of coarse aggregate are gravel, crushed stone, and crushed slag, respectively. Aggregate passing through a No. 16 sieve, with a nominal net opening of 0.0469 in (1.18 mm), may be called *fine sand*. Particles passing through a No. 200 sieve, which has a nominal net opening of 0.0029 in (0.075 mm), are called *silt* or, if they are smaller than 2 µm, are usually called *clay*. Aggregates consisting of both coarse and fine particles are called *combined* or mixed aggregates.

7.3 SAMPLING OF AGGREGATES

Proper sampling is very important for aggregates; it is perhaps even more important than for other materials because aggregates show greater fluctuations in quality than cements or other well-controlled products. Therefore, every precaution should be taken, for instance development of sampling plans, to obtain samples that will show the true nature and condition of the aggregates they represent.[Abdun-Nur 1980] Appropriate guidelines for sampling procedures help, but careful observation and the exercise of good judgment by the sampler are of primary importance in selecting a representative sample.

The methods specified for sampling coarse and fine aggregates in ASTM D 75-87 are suitable for the following purposes:

1. Preliminary investigation of the potential source of supply
2. Control of the product at the source of supply
3. Control of the operations at the site of use
4. Acceptance or rejection of the aggregates.

Aggregate samples may be taken from a conveyor belt (usually the best method for production control), from a flowing aggregate stream (bins or belt discharge), from stockpiles (usually the least desirable method), or from roadway (bases and subbases). They may be taken at the point of manufacturing, on route (not recommended), or at the job site [Legg 1974, Abdun-Nur 1966]. There are tools that make aggregate sampling easier [Proudley 1948]. Sampling procedures for the exploration of aggregate deposits are also available [US Bureau of Recl. 1966].

As a rule, the grading of coarse aggregate is the property whose variation causes most concern. There are several sources of this variation [Hudson 1969]. Figure 7.2 is scaled to show roughly the relative sizes of grading variations associated with stockpiling of coarse aggregates [Miller-W. 1965]. It can be seen that the primary factor in grading variation is segregation. Therefore, the variation of coarse aggregate grading is usually at minimum at the point of production, and tends to increase with each handling operation. Consequently, sampling at source or at intermediate points in the process stream does not appear to provide reliable information as to the variations in grading of the aggregate actually incorporated in the concrete.

In each case of sampling, a large number of sample increments, that is, small portions, are required to estimate the true average grading of a coarse aggregate with acceptable reliability. The greater the batch-to-batch segregation, the more test portions are required, whereas the greater the within-batch segregation, the more increments must be taken for each test portion to reduce sampling error [Miller-W. 1967a]. As a rule, a better

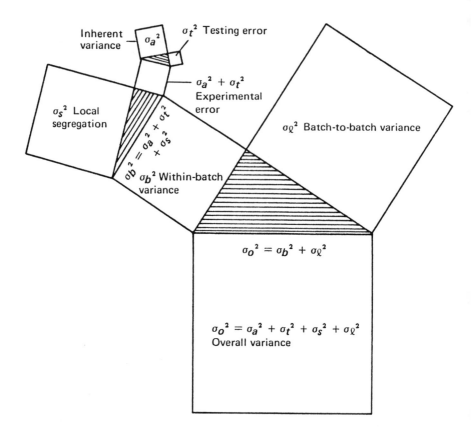

Figure 7.2: Sources and relative sizes of grading variances associated with stockpiling of coarse aggregates [Miller-Warden 1965].

estimate of actual grading is obtained by accumulating a test portion by a large number of small increments, such as scoopfuls, than by a few larger increments, such as shovelfuls. ASTM D 75-87 requires that at least three increments be selected at random from the unit being sampled and combined to form a field sample of adequate size. B.S. 812:1967 as well as ISO E 15 recommend the selection of at least 10 increments. Where information is required about the variability of the aggregate, several samples should be taken and packed separately for separate testing. Each sample should represent a particular part of the bulk quantity and may consist of a number of increments drawn from that particular part [Mills 1969].

The number of field samples required depends on the criticality of, and variation in, the properties to be measured, and should be sufficient to give the desired confidence in test results. Guidance for determining the number of samples required to obtain this desired level of confidence can be found in the following ASTM standards: D 2234, E 105, E 122, and E 141. The size of the field samples should be sufficiently large to provide for the proper execution of the planned tests. Beyond this, however, it is desirable to select as large samples as possible because the reliability of the test results increases with increase in sample size [Hudson 1969, Miller-W. 1964, ASTM E-1 1969, ASTM C-9 1989]. ASTM D 75-71 specifies that 25 lb (10 kg) should be the minimum size of a fine-aggregate sample for routine grading and quality analysis. For coarse aggregates the needed sample size increases with the increase in the maximum particle size, because grading variations in larger aggregates are greater than those in smaller aggregates [Gonsalves 1975]. For want of a pertinent specification, the following rules of thumb can be followed:

Needed size of coarse aggregate sample in lb should be at least 100 x nominal maximum particle size in inch.

or, according to ISO E 15
Needed size of coarse aggregate sample in kg should be at least 2 x nominal maximum particle size in mm.

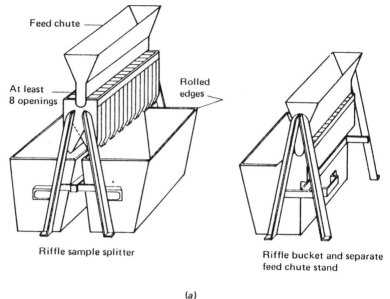

Feed chute

At least
8 openings

Rolled
edges

Riffle sample splitter

Riffle bucket and separate
feed chute stand

(a)

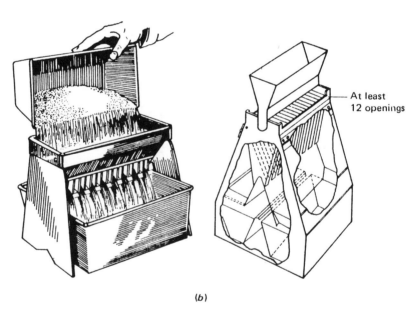

At least
12 openings

(b)

Figure 7.3: Riffle samplers for the reduction of the size of aggregate sample by splitting according to ASTM C 702-71T. (a) Large riffle samplers for coarse aggregate. (b) Small riffle sampler for fine aggregate. (Note: May be constructed as either closed or open type. Closed type is preferred.)

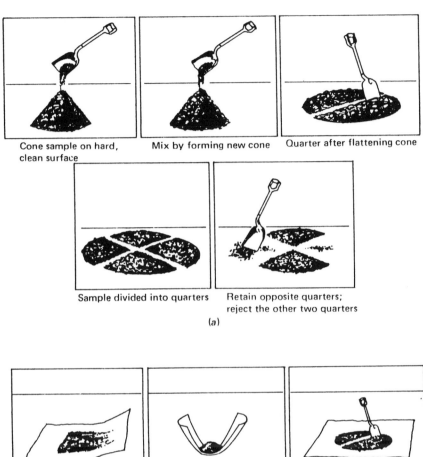

Cone sample on hard, Mix by forming new cone   Quarter after flattening cone
clean surface

Sample divided into quarters   Retain opposite quarters;
reject the other two quarters

(a)

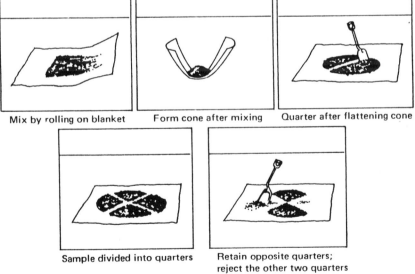

Mix by rolling on blanket   Form cone after mixing   Quarter after flattening cone

Sample divided into quarters   Retain opposite quarters;
reject the other two quarters

(b)

Figure 7.4: Reduction of the size of aggregate sample by quartering according to ASTM C 702-71T. (a) Quartering on a hard, clean, level surface. (b) Quartering on a canvas blanket.

Test proportions from the field sample may be extracted by splitting or by quartering (Figs. 7.3 and 7.4) as described, for instance, in ASTM C 702-87. Splitting is for fine-aggregate samples that are drier than the saturated surface-dry condition and for coarse aggregates, whereas quartering is for fine-aggregate samples having free moisture on the particle surface.

Aggregate samples should be transported in bags or containers so as to preclude loss or contamination of any part of the sample, or damage to the contents from mishandling during shipment, such as segregation or degradation.

8

Mineral Aggregates—Physical Properties

SUMMARY

Aggregate occupies roughly three-fourths of the volume of concrete. Therefore its properties have considerable importance to quality of the concrete.

The usual types of specific gravity of aggregate are essentially density concepts. They are used in the conversion from weight to solid volume, or vice versa, and in classification of aggregates. They control the unit weight of the concrete. Test methods specified for the determination of aggregate specific gravity are based on weight and volume measurements.

Water in an aggregate sample is either absorbed by the particles or may be on the surface of the particles as free water. The amount of water can be expressed on a weight or volume basis in different ways. Aggregate absorption may influence strongly many properties of the concrete, including durability. Standard methods measure the moisture content in aggregates by weight determinations in wet and dry states, although quicker methods are also available.

Unit weight is the weight of an aggregate sample filling up a container of unit volume. Voids content refers to the space between the particles. Free moisture may result in an increase in the voids content that is called bulking. The unit weight is used mostly in the conversion from weight to loose or compacted volume of aggregate, and vice versa. The voids content is used mostly to characterize the particle shape. The unit weight of aggregate can be determined by weight and volume measurements. The voids content can be measured directly, or may be calculated from the unit weight and bulk specific gravity.

Aggregate particles can be weak either because they are composed of inherently weak particles, or because the constituting grains are not well cemented together. Hard stones are not necessarily strong ones. Stronger rocks usually have higher moduli of elasticity. Weak aggregate cannot produce strong concrete. The modulus of elasticity of the rock determines the deformability of the concrete. There are many methods for the direct and indirect determination of aggregate strength, including the Los Angeles rattler test, but none of them is completely satisfactory. The same is true for the toughness and deformability tests for concrete aggregates.

The thermal properties of aggregates are important when the thermal propertie of the concrete are important (e.g., in mass concrete or insulating concrete).

Durability, that is, frost resistance, and soundness, that is, weather resistance, of an aggregate are controlled by its pore structure. The water in the pores freezes and the resulting volume increase of ice can generate internal hydraulic pressure high enough to cause the concrete to disintegrate. The best laboratory method is to test the durability of concrete made with the aggregate in question. The usual method for aggregate soundness is the sulfate soundness test.

Porosity refers to the volume inside the individual aggregate particles that is not occupied by solids. Many concrete properties are affected by the porosity of aggregates, the most critical being the durability. There are numerous methods to determine the various aspects of porosity (total porosity, pore-size distribution etc.).

8.1 INTRODUCTION

Aggregate occupies roughly three-fourths of the volume of concrete. Therefore its properties have considerable importance to the quality of the concrete. Some of the physical properties of a concrete aggregate, such as the specific gravity, determine the corresponding property (unit weight) of the concrete; others, such as strength control the corresponding concrete property (strength) only under certain conditions; whereas others again, such as absorption, may influence the corresponding concrete property (absorption) only to a limited extent.

Strict adherence to the requirements of the test methods is very important, because improper test procedures can result in inaccurate data and mistaken conclusions about the quality of the aggregate [ASTM 1976].

The physical properties of aggregates and aggregate materials, as well as the relationships between these properties, have been discussed extensively in the literature [Balazs 1975, Pilny 1974, El-Rawi 1973, Gaynor 1968a, Duriez 1961, Hansen 1971].

8.2 SPECIFIC GRAVITY AND SOLID VOLUME

ASTM C 127-88 for coarse aggregate and C 128-88 for fine aggregate distinguish three types of specific gravity, 73.4/73.4°F (23/23°C), for aggregates as follows:

Bulk specific gravity

$$G_b = \frac{A}{B - C} \qquad (8.1)$$

Bulk specific gravity (saturated surface dry)

$$G_{bssd} = \frac{B}{B - C} \qquad (8.2)$$

Apparent specific gravity

$$G_a = \frac{A}{A - C} \qquad (8.3)$$

where A = weight of oven-dry aggregate sample in air

   B = weight of saturated surface-dry sample in air

   C = weight of saturated sample in water.

The term *saturated* in Eq. 8.2 means actually *surface-saturated* according to ASTM C 127-88 because only the water quantity is considered that the aggregate can absorb during approximately 24-hr submersion in water. That is, only the outside part of the aggregate particle is saturated (see Fig. 8.2).

A fourth type is the "true" specific gravity, G_t, which is the ratio of A above to the volume of solids in the aggregate particles excluding all pores. A geometric interpretation of these specific gravity concepts is presented in Fig. 8.1.

In a porous material, such as stones,

$$G_b < G_{bssd} < G_a < G_t$$

The greater the porosity of the aggregate, the greater are the differences between the various specific gravities.

Equations 8.1 and 8.2 represent the corresponding definitions as given in ASTM E 12-70(1986), whereas

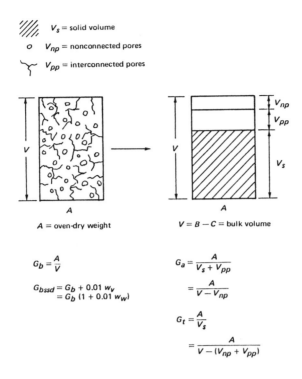

Figure 8.1: Geometric representation of the specific gravity concepts.

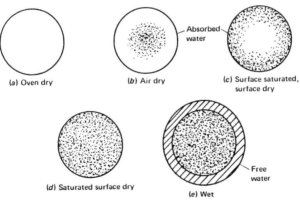

Figure 8.2: States of moisture in aggregate. Total moisture: (a) none; (b) less than the adsorption capacity, drying; (c) less than the absorption capacity, absorbing; (d) absorption capacity; (e) greater than absorption capacity. Surface moisture: (a), (b), and (c) negative, can absorb water; (d) none; (e) positive.

Eq. 8.3 corresponds to the definition given in ASTM C 125-88. For instance, the bulk specific gravity is defined as the ratio of the oven-dry weight in air of an aggregate particle at 73.4°F (23°C) to the weight of water displaced by the particle in its saturated surface-dry state at 73.4°F (23°C). Therefore, numerical values of densities are practically equal to the values of the corresponding types of specific gravity when the density is expressed in g/cm^3, or kg/dm^3. When the dimension of density is lb/ft^3, its numerical value is obtained by multiplying the specific gravity by 62.36 (which is the weight in pounds of one ft^3 of water).

The weight of an aggregate particle divided by its bulk density provides a "solid volume." This, as a rule, corresponds to the volume bounded by the visible surface of the particle, that is, to the volume which is occupied by aggregate in the concrete. When an oven-dry particle is added to a fresh concrete mixture, it displaces a quantity of the mixture equal to its solid volume. At the same time it may absorb some water from the mixture, thereby reducing the overall volume of the mixture by the volume of the water absorbed. When the aggregate is saturated but surface dry before batching by weight, the knowledge of G_{bssd} is useful. Most state highway specifications refer to this type of specific gravity. Note, however, that bringing a sample of aggregate to the saturated surface-dry state can be difficult, especially in the case of lightweight aggregates. The reciprocal value of the density may be called *specific solid volume*.

Example   8.1

You place 10 lb of saturated surface-dry aggregate into 20 lb of water in a container. G_{bssd} = 2.61.

What will be the total volume V_m of this mixture in ft^3?

Solution: Assuming that no materials are lost during the handling of the experiment due to absorption, evaporation, etc., the total volume of the mixture will be the sum of the water volume and the solid volume of the aggregate; that is

$$V_m = V_w + V_a = 20/62.36 + 10/(2.61 \times 62.36) =$$
$$= 0.321 + 0.061 = 0.382 \ ft^3$$

The aggregate may be a blend of several groups of aggregates having different specific gravities. This is the case, for instance, when lightweight coarse aggregate is combined with normal-weight sand, or even when both the fine and coarse aggregates are lightweight, since the bulk specific gravities of smaller lightweight aggregate particles are generally higher than those of the larger particles. In such cases it may be convenient to calculate the average specific gravity for the combined aggregate. Since the space occupied by the aggregate particles in the concrete is the decisive characteristic of the quantity of the aggregate, rather than their weight, this average specific gravity can be calculated either as the weighed arithmetic average of the specific gravities of the components where the weighing factors are the corresponding blending proportions by solid volume, or as the weighed harmonic average of the component specific gravities where the weighing factors are the corresponding blending proportions by weight. That is, for two components the average specific gravity G_{ave} is [Popovics 1964a]:

$$G_{ave} = \alpha G_1 + (1 - \alpha) G_2 \qquad (8.4)$$

$$\frac{1}{G_{ave}} = \frac{a}{G_1} + \frac{1 - a}{G_2} \qquad (8.5)$$

where α and a are the blending proportions for

aggregate 1 to the total aggregate by solid volume and by weight, respectively. This calculation is also discussed in ASTM C 127-88.

Example 8.2

A cubic foot of compacted concrete contains 60 lb of regular sand with specific gravity $G_1 = 2.60$, and 45 lb of lightweight coarse aggregate with $G_2 = 2.00$.

Calculate the volume of (cement + water + air) matrix (paste) V_m that is needed to fill up all the voids in this aggregate skeleton without any excess.

First solution: The needed volume of the matrix is the difference between 1 ft^3 and the solid volume of the combined aggregate V_s.

One can calculate the solid volume of the sand particles, that of the lightweight particles, and then sum them, as follows:

$$V_{sa} = \frac{60}{62.36 \times 2.60} = 0.37 \ ft^3$$

$$V_{1i} = \frac{45}{62.36 \times 2.00} = 0.36 \ ft^3$$

$$\overline{}$$

$$V_{so} = 0.73 \ ft^3$$

Therefore, the needed matrix volume is

$$V_m = 1 - 0.73 = 0.27 \ ft^3.$$

Second solution: Another method is to calculate the blending proportions, then the G_{ave} average specific gravity and the D_{ave} average density of the combined aggregate, and then calculate the solid volume. That is

$$a = \frac{60}{60 + 45} = 0.572 \text{ by weigth}$$

Then, from Eq. 8.5:

$$\frac{1}{G_{ave}} = \frac{0.572}{2.60} + \frac{0.428}{2.00} = 0.434$$

That is,

$$G_{ave} = 2.30$$

and

$$D_{ave} = 2.30 \times 62.36 = 143.5 \text{ lb/ft}^3$$

Therefore,

$$V_{so} = \frac{105}{143.5} = 0.73 \text{ ft}^3$$

So, again,

$$V_m = 1 - 0.73 = 0.27 \text{ ft}^3.$$

The significance of the specific gravity of aggregate lies in the fact that it is needed in the conversion from weight to solid volume or vice versa, which is a common task, for instance, in concrete proportioning. It is also used in the classification of aggregates as lightweight, normal weight, and heavyweight (see Chap. 14). It controls the density, or unit weight, of the concrete, and may influence other concrete properties, such as the tendency of the fresh concrete to segregate, or the heat conductivity.

Test methods for the determination of the specific gravities of aggregates are based on weight and volume measurements. Of course, the volume of a specimen for bulk specific gravity could be determined from dimensional measurements if the sample pieces were regular geometric shapes. Since this is not the case for aggregate particles, pertinent weights substituted into Eqs. 8.1 - 8.3 are used. Details of the test method for coarse aggregate are given in ASTM C 127-88, and those for fine aggregate in C 128-88. Note that both standards specify an immersion time of approximately 24 hr at room temperature. This means that in most cases only the outside part of the aggregate particle is actually saturated.

Repeated measurements on the same normal-weight aggregate sample usually do not deviate from the average of the values by more than ±0.01 in the case of coarse aggregate, and ±0.02 in the case of fine aggregate.

A coarse aggregate is considered to be surface dry when all visible water films have been removed from the surface of the particle with a cloth. The saturated surface-dry condition of a fine aggregate is that, by definition, at which a wet but drying sample just becomes free flowing. The necessity of using saturated surface-dry aggregate for specific gravity determination can be eliminated by using more complicated methods for the prevention of the liquid penetration, such as coating the particles with paraffin before the weight measurement in water, or by applying a nonpenetrating liquid, such as mercury. For lightweight aggregates, other test methods may be more suitable [Landgren 1964a, 1965, Jones 1957, Nelson 1958, Kunze 1974b].

The usual method for determing the true specific gravity of aggregate is to powder the sample to, say, 100 mesh (150 μm) and then to determine the specific gravity by some method, such as ASTM D 854 for soils, or ASTM C 188 for cements. The point to the powdering is not so much to destroy void spaces, since most aggregates have pores far smaller than any reasonable sieve opening, but rather to provide particles so small that their pores are readily and completely penetrated by the pycnometer fluid, for instance under vacuum and high-pressure saturation. In certain methods gas is used as the pycnometric fluid [Dolch

1966]. A method is specified in B.S. 812 for the determination of true specific gravity of aggregates, but other methods are also available [Van Keulen 1973].

Most countries, including the United States, do not use specific gravity as an acceptance criterion for normalweight aggregates except in special cases. Such cases are the assurance of durability of cherts or other frost-susceptible aggregates. The bulk specific gravities of rocks commonly used in normal-weight concrete are generally between 2.4 and 2.8. Special heavyweight aggregates may have specific gravities of 5 or even more [Brink 1966]. Specific gravities of lightweight coarse aggregates may be as low as 1.4.

8.3 ABSORPTION, MOISTURE CONTENT, AND PERMEABILITY

Since the materials of aggregates always contain more or less pores inside, practically all dry particles are capable of absorbing water. According to ASTM C 125-88, both this process and the amount of this water are called *absorption*. The amount of the absorbed water depends primarily on the abundance and continuity of the pores in the particle, whereas the rate of absorption depends on the size and continuity of these pores; secondary factors, such as the particle size, may also have an effect [Kucynski 1974]. *Absorption capacity* is the maximum amount of water the aggregate can absorb under the prevailing circumstances. Several states of moisture in aggregate are presented in Fig. 8.2.

The w_w amount of absorbed water can be expressed in several ways. According to ASTM C 127-88 and C 128-88, it is expressed in percent by weight in the saturated surface-dry state as

$$w_w = 100 \frac{B - A}{A} \qquad (8.6)$$

where *A* and *B* are as in Eqs. 8.1 - 8.3. Equation 8.6 corresponds to the definition of absorption given in ASTM C 125-88. Similar formulas have been adopted by many other countries, for instance, by the German DIN 52103.

Another possibility is to express the absorption in percentage of the volume of the dry aggregate particle, V

$$w_v = 100 \frac{V_w}{V} \qquad (8.7)$$

where V_w is the volume of the absorbed water at a given temperature. The relationship between the two absorption values is, to a good approximation,

$$w_v = G_b w_w \qquad (8.8)$$

where G_b is the bulk speciflc gravity of the aggregate. Absorption expressed in percent by volume is particularly suitable for lightweight aggregates.

A third possibility is to show the relative fillings of the accessible pores, that is, the percentage of the V_{pp} volume of the interconnected pores permeable to water that was filled up with absorbed water. This term is called the *degree of saturation* (d_s) and is, in percentage of volume, as follows:

$$d_s = 100 \frac{V_w}{V_{pp}} \qquad (8.9)$$

A practical form of this is the ratio of the actual absorption of the aggregate to the maximum absorption, that is, the absorption produced by vacuum treatment followed by high pressure. Saturated specimens have 100% degree of saturation. The moisture may be present not only inside the aggregate particles but also on their surface. This latter is called *free moisture*. The free moisture and the absorbed moisture together give the total moisture content (w_t) of aggregate. This is defined in percentage by weight as

$$w_t = 100 \frac{W_w - W_d}{W_d} \qquad (8.10)$$

where W_w and W_d are the weights of the aggregate sample in its wet and oven-dry state, respectively.

It also follows from the internal porosity of rocks that when an aggregate particle is subjected to a large enough fluid pressure on one side, it will be *permeable*. The extent of permeability again depends on the abundance, size, and interconnection of the internal pores and is usually characterized by the coefficient of permeability.

Absorption and permeability are important aggregate characteristics because, either per se or as characteristics of the internal porosity, they strongly influence the chemical stability, hardness, strength, deformability, and thermal properties of the aggregate. They also control soundness and durability, partly through their effect on the degree of saturation [Fagerlund 1971], on the ease with which water can penetrate or escape from the freezing zones, and partly through the deformations that wetting and drying cycles cause in an aggregate particle. When the aggregate is not fully saturated, it may absorb some of the mixing water in the concrete mixture. On the other hand, when free moisture is present, this adds to the mixing water. Thus, total moisture content, absorption, and free moisture in the aggregate used are useful data for the calculations of the effective water-cement ratio, and the adjustments in the proportions of concrete mixtures.

When the aggregate is a blend of several groups having different absorptions, the w_{wa} average absorption of the blend is the weighted average of the individual absorption values, weighted in proportion to the weight percentages of the groups in the blend (ASTM C 127-88). For two components [Helms 1962]

$$w_{sa} = aw_{w1} + (1 - a)w_{w2} \qquad (8.11)$$

where a is the blending proportion for aggregate 1 to the total aggregate by weight.

The absorption of an aggregate can be reduced by chemical surface treatments [Dutt 1971, Marek 1972], but economy of this is justified only in special cases.

A standard method for the determination of absorption, corresponding to Eq. 8.6, is given in ASTM C 127 for coarse aggregate, and in ASTM C 128 for fine aggregate by weight determinations in soaked and dry states. Repeated such absorption measurements on the same normal-weight sample usually do not deviate more than ±0.1 from the average percent absorption in the case of coarse aggregate, and ±0.3 in the case of fine aggregate. Note again that both standards specify an immersion time for the aggregate of approximately 24 hr at room temperature, therefore, the value determined is the 24-hr absorption. This relatively short immersion means that in most cases only the outside part of the aggregate particles is saturated (Table 8.1); that is, the term "saturated" in these two ASTM standards, is used in the sense of "surface saturated."

More complete saturation can be achieved by longer immersion, by boiling the water while the aggregate is in it, or by vacuum treatment [Reilly 1972]. In the latter, the sample is dried, evacuated, the water is admitted, and either atmospheric pressure is resumed or higher air pressure is applied. The sample then soaks for 24 hr. Naturally, vacuum-saturated

Table 8.1: Water Absorption for Some Aggregates of 5 to 10 mm Size[a]

Aggregate	Water absorption, % by weight			
	1 min	1 hr	1 day	max.
Thames Valley gravel	0.35	1.95	2.60	3.44
Mountsorrel granite	0.31	0.34	0.35	0.80
Bridport gravel	0.35	0.60	0.67	1.17
Leca lightweight	12.0	12.6	17.7	70.0
Lytag lightweight	10.0	11.0	13.5	24.0
Solite lightweight	3.0	6.2	7.5	20.0

[a]From [Lyndon 1975].

values of absorption are larger than values obtained
by boiling, which are again larger than values
obtained by simple immersion. Eqs. 8.6 - 8.9 are
valid for any immersion time or any kind of
absorption determination.

The determination of the absorbed water or the
total moisture content by drying is described in ASTM
C 566-84. Another method utilizes saturated air for
the determination of the absorption in the saturated
surface-dry state [Hughes 1970]. The amount of free
moisture can be determined as the difference between
the total moisture and the absorption, but there also
are other methods, based on displacement (ASTM C 70-
79(1985)), gas development, electrical resistance or
capacitance, nuclear radiation, microwave absorption,
and so on [Neville 1981, Orchard 1976, Novgorodsky
1973].

There are no standard methods in the United States
for measuring degree of saturation or aggregate
permeability, although pertinent methods are
described in the literature [Graf 1957]. Neither are
acceptable values specified for these aggregate
characteristics. Absorption by various aggregates,
including lightweight materials, range from virtually
zero to more than 30% by weight [Landgren 1964a].
Even rocks bearing the same name comprise a wide
range of absorption. The wide range of the rate of
absorption is illustrated in Table 8.1 for several
British aggregates. A method for measuring the rock
deformations caused by wetting and drying is
specified in DIN 52450 E (withdrawn). With this,
linear wetting coefficients as high as 50×10^{-6} to
100×10^{-6} were measured which, when repeated, are
large enough to damage certain rocks.

The finer the aggregate, the more free moisture it
can carry. The free moisture in a gravel or crushed
stone can hardly be greater than 2%, whereas that in
a fine sand can be 10% or even more. Much greater
variability exists in permeability. The values of
coefficient of permeability, presented by Powers
[Powers 1968], vary from 3.45×10^{-13} cm/sec for a
dense trap rock to 2.18×10^{-9} cm/sec for a porous

granite.

The above aggregate properties are characteristics mostly of single particles. The properties of aggregate samples, consisting of many particles, are discussed in the following sections.

8.4 UNIT WEIGHT, VOIDS CONTENT, AND BULKING

The unit weight U of an aggregate is defined as the weight of an aggregate sample filling up a container of unit volume. That is, the aggregate volume in question includes not only the solid volume of the particles but also the voids between aggregate particles. Mathematically,

$$U = \frac{W}{V} \qquad (8.12)$$

where W is the weight of the aggregate sample and where V is its total volume. Since both the weight and the volume depend on methods of compaction and on the moisture content of the aggregate [Fowkes 1974], the testing conditions under which the unit weight was determined should be known.

With respect to an aggregate sample, the term *voids* refers to the space between the aggregate particles. The total amount of voids is the difference between the total volume of the aggregate sample, including the voids, and the space occupied by the solid volume of the particles. It is usually expressed in percentage as

$$\text{Percentage of voids} = \upsilon = 100\frac{V_v}{V} \qquad (8.13)$$

where V_v is the total volume of voids in the sample and V is the total volume of the sample, including the voids.

Free or surface moisture holds the fine-aggregate particles apart; hence, there may result a marked decrease in the unit weight of the aggregate, that is, an increase in the percentage of voids. This phenomenon is known as *bulking*. Up to a certain limit in the free water content, the more water, the greater bulking. Also, the finer the aggregate, the greater the bulking. Sands that are completely submerged, or *inundated*, show no bulking. The bulking of coarse aggregate is also negligible [Troxell 1968]. The magnitude of bulking can be characterized by the *bulking factor*. This is the number of cubic feet of damp, loose aggregate resulting from one cubic foot of dry rodded aggregate.

The unit weight is used mostly in the conversion from weight to loose or compacted volume, and vice versa, in classification of aggregates as lightweight, normal-weight, or heavyweight [Brink 1966], and in the calculation of the voids content (percent of voids) in an aggregate sample. The voids content is used in certain concrete-proportioning methods, in certain methods of grading evaluation, and in certain methods of characterization of the shape of the aggregate particle [Powers 1968, Shergold 1953]. It should also be considered when aggregate is purchased, or used, on a volume basis.

The unit weight for any given set of conditions may be determined by weighing the aggregate sample required to fill an appropriate container of known volume. For comparison of different aggregates a standard method should be used, such as that described in ASTM C 29-89, where oven-dry aggregate is prescribed. Such results by an operator using the same sample and procedure should check within 1%.

The total amount of voids in an aggregate sample V_v can be determined experimentally by measuring the quantity of water required to inundate the sample that has been compacted into a container. The percentage of voids can also be calculated from the unit weight and the bulk specific gravity of the aggregate according to ASTM C 29-89. When the unit weight is expressed in kg/dm^3,   then

$$\upsilon = 100 \, \frac{G_b - U}{G_b}$$ (8.14)

When the unit weight is expressed in lb/ft^3,   then

$$\upsilon = 100 \, \frac{62.36 \, G_b - U}{62.36 \, G_b}$$ (8.15)

The experimental method may give lower values than Eq. 8.14 or 8.15 because the inundation may not drive out all the air from the voids.

There is no standard method in the United States for the determination of bulking.

There are no general standard specifications for the acceptable values of unit weight, voids content, or bulking except that air-cooled blast-furnace slag should have a minimum standard compacted unit weight of 70 lb/ft^3 (1120 kg/m^3). ASTM C 330-87 specifies for lightweight aggregates for structural concrete a maximum standard dry loose unit weight of 55 lb/ft^3 (880 kg/m^3) for coarse aggregate, 70 lb/ft^3 (1120 kg/m^3) for fine aggregate, and 65 lb/ft^3 (1040 kg/m^3) for combined aggregate.

The unit weight of sand in a damp loose state is around 90 lb/ft^3 (1450 kg/m^3), in a dry, compact state is around 105 lb/ft^3 (1680 kg/m^3), whereas that of a usual combined aggregate is around 115 lb/ft^3 (1850 kg/m^3). The unit weight of lightweight aggregate can be as low as 30 lb/ft^3 (500 kg/m^3) but this is not for structural purposes.

The maximum voids content of an aggregate consisting of one-size spherical, or ellipsoid [Tons 1968] particles is 48%, but particles with irregular shape provide higher, or with more compact particle arrangement, lower [Herdan 1960] voids contents. As Fig. 8.3 shows, well-graded sand and well-graded gravel can have voids contents as low as 30% in a dry, compacted state. Combinations of fine and coarse aggregates can have voids contents less than 25% [Walker 1930, Schwanda 1956, Albrecht 1965, Foth 1966, Joisel 1952, De Larrard 1987]. Specially graded aggregates can be even denser [Furnas 1931, Andreasen 1929]. Further effects of the grading and particle shape on the voids content can be seen in Figure 12.8

A comparison of Fig. 8.3 to Fig. 10.5 illustrates the fact, also demonstrated experimentally, that different gradings of identical fineness modulus (or specific surface) do not have necessarily identical voids content even when the maximum particle sizes are the same.

Powers pointed out [Powers 1968] that geometrically similar size distributions of the same type of particles have identical voids contents. This means, for instance, that if one aggregate is composed of particles ranging from 0.001 to 0.1 in, and another is composed of particles ranging from 0.01 to 1.0 in, and if the two have grading curves of the same shape in the semilog system of coordinates of Fig. 10.3, then they have equal voids contents. The simplest case of geometrically similar gradings is the group of one-size aggregates. Consequently, the voids contents of one-size aggregates having indentical particle shape are independent of the particle size. Geometric dissimilarity does not necessarily indicate unequal voids content.

The bulking capacity of a usual concrete sand is around 25%. Very fine sand may show bulking as much as 40% (Fig.8.4).

8.5 STRENGTH, TOUGHNESS, HARDNESS, AND DEFORMABILTTY

The strength and other related characteristics of rocks and aggregates can vary again within wide limits. Aggregate particles can be weak either

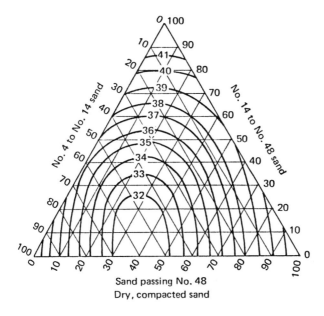

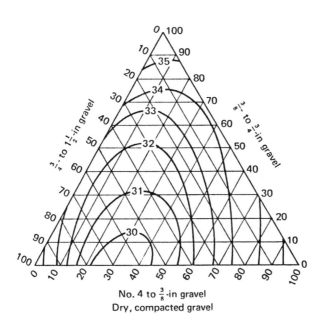

Figure 8.3: Percent of voids in a sand and in a gravel as a function of grading [Walker 1930]. (The use of triangular diagrams is discussed in Chapter 10.)

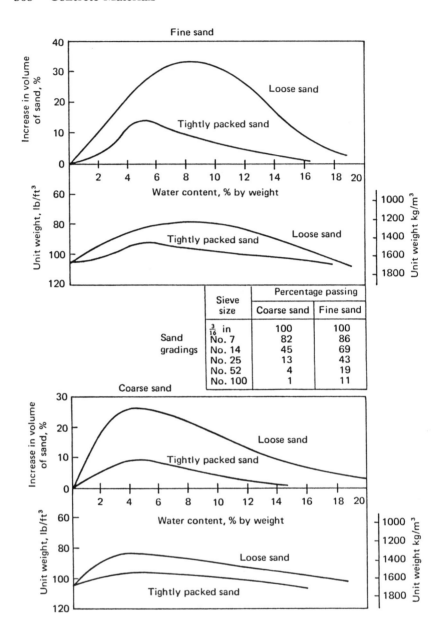

Figure 8.4: Examples for the bulking of fine aggregate. Reproduced from Orchard, *Concrete Technology,* Volume 1, 2nd edition, by permission of the author and Applied Science Publishers [Orchard 1973a].

because they are composed of inherently weak substances, or because the constituent grains are not well cemented together. For instance, granites composed of hard and strong crystals of quartz and feldspar typically posses lower strengths and elastic moduli than gabbros an diabeses because the grains in granite are poorly interlocked [Blanks 1955]. The porosity, or density, also has a significant influence in the strength or rock (Fig. 8.5). The same is true for toughness, which, with respect to aggregate, characterizes resistance to impact.

Hardness may be defined as the ability of one body to resist penetration by another. [McMahon 1966] Thus, it is a relative property of a material that depends on the properties of both the penetrating body and the penetrator. All hardness tests measure some combination of three material properties, namely resistance to elastic deformation (modulus of elasticity), plasticity (yield stress), and work-hardening capacity. Since each hardness test measures a different combination of these properties, hardness itself is not an absolute quantity, therefore, to be meaningful, any statement of the hardness of a body must include the method used for measurement. The three most frequently used general test methods are: scratch testing, indentation, and dynamic rebound testing.

Hardness testing is used mainly to categorize materials, such as the various minerals, and to aid in the indirect determination of mechanical properties of materials as affected by changes in chemical composition and/or microstructure. Such measurements find extensive applications in metallurgy but they are also used in testing nonmetallic materials, such as wood, rubber, plastics, and others, including stones and concretes [Popovics 1986b].

Hardness of aggregate particles depends on the hardness of the constituents, not on the firmness with which the constituents are cemented together. Thus, hard stones are not necessarily strong ones.

The deformability of rocks, which under uniaxial

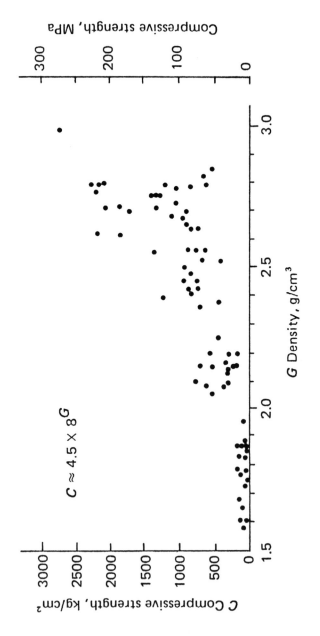

Figure 8.5: As a rule of thumb, heavier rocks are more likely to be stronger [Popovics 1955b].

loading is characterized by the modulus of elasticity, is similar to that of other brittle materials [Nadai 1950]. Stronger rocks usually have higher moduli of elasticity.

Weak aggregate particles, that is, particles significantly weaker than hardened cement paste, cannot produce concrete of reasonable strength. However, moderate strength or, more precisely, moderate compressibility of particles may be advantageous, because this may decrease stress concentrations in the cement paste of the concrete during loading, or wetting and drying, or heating and cooling, or freezing and thawing; thus it can reduce the danger of the cracking of concrete. Experimental evidence seems to suggest, for instance, that within certain limits, the higher the modulus of elasticity of the aggregate used, the lower the flexural strength of the concrete. Both weak and soft particles should be kept at a minimum when high abrasion resistance is required from the concrete. Also, the deformability of the aggregate is perhaps the most important single factor that influences the deformability of the concrete [Popovics 1969c].

The strength of a rock can be determined by the test methods accepted for other brittle materials. For instance, a cube test is specified for the compressive strength in the German DIN 52105, and prism for the flexural strength in DIN 52112. Since, however, such tests require relatively large rock pieces, they are usually unsuitable for testing the strength of aggregate pieces. The usual heterogeneity of aggregate also suggests a test that can be performed on the bulk aggregate instead of on single pieces. Apart from ASTM C 142-78(1984), which specifies the attempt to break up aggregate particles with the fingers in order to find friable particles, there is no standard method in the United States for the direct strength test of aggregate particles. A test of this type, however, is described in B.S 812. This is a crushing test in which a coarse aggregate sample is put in a steel cylinder, a steel plunger is placed on the aggregate, and a specified load is applied to the plunger. The amount of material broken

to pass a No. 7 sieve (2.41 mm net opening) is the
"crushing value", which is used to characterize the
strength of the aggregate sample. A similar test is
specified by the same British standard for testing
the toughness of the coarse aggregate, but in this
case the load is applied by an impact test machine.
The static crushing value usually agrees closely with
the dynamic impact value, but the impact test
requires only a small, portable apparatus [Road Res.
Lab. 1955].

Both of these tests are somewhat insensitive to
weak particles [Sweet 1948a]. One of the reasons for
this can be arching of the aggregate particles during
the test as well as the high friction between the
particles and the side wall of the steel cylinder.
These can reduce the load on the weak particles,
resulting in a better crushing value than the one
related to the actual strength of the particle.

American methods for testing the strength of
coarse aggregate indirectly are the Deval test (ASTM
D 2 and D 289, both discontinued), and the Los
Angeles rattler test (ASTM C 131-89 and C 535-89). In
these tests the aggregate sample is placed into a
metal cylinder usually with an abrasive charge of
steel spheres and then the cylinder is rotated,
subjecting the particles to impact and abrasion (Fig.
8.6). The percentage of aggregate worn off during the
test is characteristic of the combined toughness,
hardness, and wear resistance of the aggregate, thus
indirectly, and supposedly, of its strength. Of these
two types of tests, the Los Angeles method is
superior for normal-weight and heavyweight
aggregates, mainly because the correlation of its
results with service behavior is better than that of
the Deval test, and because it provides comparable
results for gravel and crushed aggregates [Sweet
1948, Woolf 1937, 1966, McLaughlin 1960]. The Los
Angeles method can be modified to test the
wearability of the aggregate under wet conditions
[Larson 1971]. Note also that (1) if the wear in the
Los Angeles test after 100 revolutions is considerab-

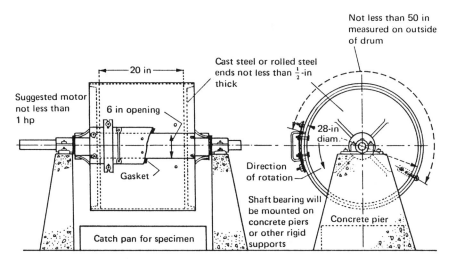

Figure 8.6: Los Angeles abrasion-testing machine. From ASTM C 131-89.

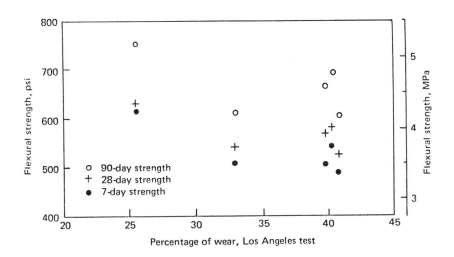

Figure 8.7: Relation between flexural strength of concrete and Los Angeles abrasion loss of coarse aggregate [Popovics 1961].

ly greater than 20% of the wear after 500
revolutions, this is an indication of the presence of
soft particles in the aggregate; (2) the wear
indicates the potential increase in the amount of
fines in the concrete when the fresh concrete is
subjected to prolonged mixing [Gaynor 1963a].
Unfortunately, the result of the Los Angeles test
does not indicate adequately the contribution of the
coarse aggregate to the concrete strength (Fig.8.7).

It is also possible that the pulse velocity in
coarse aggregate particles, or the dynamic modulus of
elasticity calculated from the pulse velocity, can be
used for the prediction of the concrete strength,
especially in the case of lightweight aggregates
[Wesche 1974, Krause 1975].   Another
indirect method for checking the strength of the
aggregate is to prepare concrete with it and judge
the adequacy of the aggregate strength from the
measured strength of the concrete.

The results of most of these strength tests for
aggregates show linear correlation with each other
[Kohler 1972].

A simple method for determining the quantity of
soft particles in a coarse aggregate sample on the
basis of scratch hardness is given in ASTM C 235
(withdrawn). Otherwise no direct methods for testing
the hardness are mentioned in present-day
specifications for the quality of aggregate. However,
there are many methods described in the literature
for other materials, as indicated below.

Perhaps to first method of measuring scratch
hardness still in wide use today by mineralogists was
published by Friedrich Mohs in 1822. This method
gives a relative ranking (scale) of minerals based
simply on their ability to scatch one another. His
scale is given in Table 8.2. The Mohs method is not
suitable for general use with materials of hardness
greater than 4 because in this range the intervals
are rather closely and unevenly spaced.[McMahon 1966]

The most widely used method of hardness testing
today is the *indentation method* in which a hard
indenter, such as, diamond, sapphire, quartz, or
hardened steel, of a particular geometry is forced

Table 8.2: The Mohs Scale of Hardness[a]

Mohs scale number	Standard mineral	Other materials
1	talc	
2	gypsum	human fingernail
3	calcite	copper
4	fluorite	iron
5	apatite	cobalt,hard portion of teeth
6	orthoclase	rhodium, tungsten, silicon
7	quartz	
8	topaz	chromium, hardened steel
9	corundum	sapphire
10	diamond	

a From [McMahon 1966]

into a test specimen. Measurement is made either of the size of indentation resulting from a fixed load on the indenter, or of the load necessary to force the indenter down to a predetermined depth. Many combinations of indenter, load, loading procedure, and means of indentation measurement are used among the various tests to accommodate the shapes, sizes and hardnesses of specimens. The most commonly used indenters for metals are shown in Table 8.3. When ball indenters are applied, the obtained hardness number is independent of load only when the ratio of load to indenter diameter is held constant. For cone and pyramid indenters, the hardness numbers are independent of load for all loads above a certain minimum. Variations of some of these methods have been tried for concrete, thus they may have limited applications for stones and aggregates. Some of these methods are: the Frank spring hammer, the Einbeck pendulum hammer and the Poldi-Waitzman hammer, the Williams testing pistol, and the Windsor probe. [Malhotra 1980]

A typical example for the *dynamic rebound testing* of metals is the Shore scleroscope. In this test a diamond-tipped hammer is dropped from a fixed height, and the height to which the hammer rebounds is read on a scale. This is taken to be the measure of hardness. The rationale of this is that the kinetic energy of the hammer is used up partly in rebound and partly in plastically deforming the specimen surface by making a small plastic deformation, that is, impression. The softer the material, the higher percentage of the kinetic energy is expended in making the impression, thus the rebound is small. The same principle is used in the various types of Schmidt hammer (ASTM C 805 and B.S. 4408) which is probably the most popular nondestructive test method for hardened concrete and may also be applicable for testing stones.

The abrasion resistance of rock pieces, which is related to the hardness, is discussed later in Section 8.9.1.

The elastic constants, characteristics of the deformability, can be determined directly again on regular rock specimens either by the static method,

Table 8.3: Indenter Geometry[a]

Name	Brinell	Rockwell		Vickers	Knoop	Monotron
Material of which indenter is made	Hardened steel or tungsten carbide	Diamond	Hardened steel	Diamond	Diamond	Diamond
Shape of indenter	Sphere	Cone	Sphere	Square-based pyramid	Rhomb-based pyramid	Hemisphere
Dimensions of indenter	D=10mm	$\theta = 120°$	$D = \frac{1}{16}$ in. $\frac{1}{8}$ $\frac{1}{4}$ $\frac{1}{2}$	$\theta = 136°$	$\alpha = 130°$ $\theta = 172° 30'$	D=0.75 mm

a From [McMahon 1966]

that is, measuring loads and related deformations in
a similar way as described for concrete in ASTM C496-
86; or by the dynamic method, that is, mea- suring
the resonance frequency (ASTM C 215-85) of, or the
pulse velocity (ASTM C 597-83) in the rock. Pulse
velocity measurements can also be performed on single
pieces of coarse aggregate. The static elastic
constants determined on duplicate cylinders from
different batches should not depart more than 5% from
the average of the two. The precision of the dynamic
methods is better.

The maximum permissible crushing value of
aggregates for concrete wearing surfaces is 30%, for
other concretes 45%, as described in B.S. 882. ASTM C
33-86 specifies that coarse aggregate for portland
cement concrete shall not have a Los Angeles loss
more than 50% unless the aggregate produces
satisfactory strengths in concrete. Several highway
departments are willing to accept aggregates with
losses up to 65%.

Most countries do not specify requested values for
direct strength, toughness, hardness, or deformability for concrete aggregates. All these
properties, of course, can vary within very wide
limits. For instance, the values of compressive
strength in Fig. 8.5 vary approximately between 1400
psi (100 MPa) and 40,000 psi (2800 MPa). The flexural
strength is usually 7-12% of the compressive
strength, where the higher percentages normally
belong to weaker rocks. The standard crushing value
for good igneous rock is 15% or less [Neville 1981].
Gravel, limestone, and slags usually have higher
values; slags may have crushing values of 30% or even
more [Road Res.Lab. 1955]. Los Angeles abrasion
values of aggregates used for highway concrete are
usually within 15 and 60% [Woolf 1966]. Values of
static modulus of elasticity of similar aggregates
are 7×10^6 to 15×10^6 psi (approximately 0.5×10^5
to 10^5 MPa) for good igneous rocks, and 3.5×10^6 to
10×10^6 psi (approximately 0.25×10^5 to 0.75×10^5
MPa) for good sedimentary rocks [Graf 1960]. The
modulus of elasticity of siliceous gravel is around
9.0×10^6 psi (0.6×10^5 MPa). The moduli of

elasticity of lightweight aggregates are, of course, less. The compression modulus for igneous rocks is 4.2×10^6 to 12.5×10^6 psi (approximately 0.3×10^5 to 0.9×10^5 MPa), and for sedimentary rocks is 5.5×10^6 to 16.5×10^6 psi (approximately 0.4×10^5 to 1.2×10^5 MPa).

Values of the dynamic Poisson's ratio of normal-weight aggregates are usually between 0.15 and 0.35 [Kaplan 1959]. Data concerning toughness and hardness of aggregate materials can be found in the literature [Pilny 1974, Blanks 1955, Woolf 1966].

8.6 THERMAL PROPERTIES

The three most important thermal properties of mineral aggregates are (1) the *coefficient of expansion*, which characterizes the length or volume changes of the material in response of changes in temperature; (2) the *thermal conductivity*, which indicates the ease with which heat can move in the material; and (3) the *specific heat*, which is the amount of heat needed to raise the temperature of the material 1°C. *Diffusivity* is conductivity divided by the product of specific heat and density, and indicates the rate at which temperature changes can take place within a mass. The first property is important for the designer, the other two for calculation of heat transfer.

The coefficient of expansion is dependent primarily on the composition and internal structure of the rock; thus it can vary greatly even within one rock type. Thermal conductivity is influenced primarily by porosity, thus its value varies within narrower limits for normal-weight aggregates than does the coefficient of expansion. The specific heat values of various rocks are practically the same if the calculation is based on the apparent volume of the specimen rather than on weight (except for materials of high porosity, such as lightweight aggregates). It should be noted that the values of these thermal properties of aggregate change considerably with the absorption of relatively small amounts of water [Lewis 1966]. This significant influence is due to the relatively large differences in thermal properties between water and dry aggregates [Powers 1968].

In most cases one need not worry about the thermal properties of mineral aggregates in everyday practice, but there are several exceptions. The most important of these is mass concrete, that is, a concrete structure having a minimum dimension of at least several feet, because these aggregate properties influence the temperature difference between the surface and interior regions and can produce tensile stresses of damaging magnitude in the hardening concrete. The coefficient of expansion of the aggregate may be a matter of concern when the thermal expansion of the concrete is restricted, particularly in long concrete elements, or where the concrete is subjected to frequent and sizable warming-cooling cycles [Cook 1966]. Thermal conductivity is important when the insulating capability of the concrete is important. The specific heat of the aggregate is used mostly for calculations related to heating or cooling the aggregate or concrete.

The generally used aggregate specifications do not prescribe any standard method for the determination of the thermal properties of aggregates: neither are acceptable values specified. Pertinent test methods are described in the literature [U.S. Dept. of Int. 1940, Verbek 1951].

Powers quotes [Powers 1968] numerous measured values for thermal properties of aggregates. Accordingly, the coefficient of expansion of the usual rocks varies between 0.5×10^{-6} and $8.9 \times 10^{-6}/^{\circ}F$ (0.9×10^{-6} and $16.0 \times 10^{-6}/^{\circ}C$) [Rhoades 1946]. The thermal expansion of a rock in saturated state is approximately 10% less than that in the dry state [Pilny 1974]. Powers also points out that the apparent coefficient of expansion of mature, partially saturated hardened cement paste is much greater than these aggregate values; it can be as high as $18 \times 10^{-6}/^{\circ}F$ ($33 \times 10^{-6}/^{\circ}C$). The thermal conductivity of usual rocks at normal temperatures varies between 1.1 and 2.7 Btu/f·hr·°F, whereas the specific heat is around 0.18 Btu/lb·°F or cal/g·°C [U.S. Dept. of Int. 1940].

8.7 DURABILITY AND SOUNDNESS

The *durability* of a concrete aggregate is inter-preted as the measure of how successfully a properly made concrete containing this aggregate can with-stand the damaging effects of repeated cycles of freezing and thawing [Popovics 1976c] .

The *soundness* of the aggregate is used as a more general term that includes not only frost resistance but also the ability to withstand the aggressive actions to which the concrete containing it may be exposed, particularly those due to weather [ACI 1985d, Woods 1968]. In this sense, soundness includes the ability of aggregate to resist excessive volume changes caused not only by freezing and thawing but also by other temperature changes and alternating wetting and drying [ACI 1961].

An excellent review of the history and development of concepts related to aggregate durability has been prepared by Larson and co-workers [Larson 1964b].

8.7.1 Mechanism of Frost Action in Aggregate

Powers [Powers 1955a] was the first to propose a hypothetical mechanism for frost action in concrete aggregates [Cady 1969]. This mechanism attributes the damaging effect of frost to the generation of hydraulic pressure by an advancing ice front in critically saturated aggregate pores. More speci-fically, the water in the pores of the rock freezes at low temperatures; the resulting approximately 9% volume increase in the ice generates internal ice pressure or, by forcing the remaining water to move into the pores, internal hydraulic pressure in places where the interconnected pore system in the aggregate particles had been filled with water at more than the critical value of approximately 91%. This internal pressure can be high enough to cause the concrete to disintegrate, crack, or suffer other damage. The average critical value of saturation in a batch of aggregate particles can be far below the theoretical limit of 91% [Schulze 1962, Buth 1970].

A different hypothetical mechanism was proposed by Dunn and Hudec [Dunn 1965]. They determined by differential thermal analysis the quantities of water that froze in clay-bearing dolomitic limestones of widely varying frost susceptibilities. The findings indicated that smaller percentages of the contained water froze in unsound rocks than in sound rocks. In several instances no freezing at all was detected in deteriorating unsound rocks, and when the freezing did occur in those rocks, it was very gradual. On the other hand, sound rocks invariably displayed some freezing, and it occurred as a single pulse. They hypothesized, therefore, that the major factor in frost destruction of rock is not the ice but a temperature-dependent volume change of adsorbed, or "ordered", water contained in certain pores, so-called force spaces. Thus, this hypothesis is in direct contradiction to the Powers hypothesis mentioned above regarding the state of water during the period in which the coarse aggregate undergoes destructive volume change.

Further investigation by Cady has shown that the hydraulic pressure produced by icing is the primary contributor to destructive volume changes in the majority of concrete aggregates. However, the simultaneous existence of the mechanism resulting from adsorbed water in these aggregates is also consistently in evidence. The latter mechanism, generally, is a minor contributor to destructive volume changes, but it can be the major contributor for certain aggregate types, such as clay-bearing dolomitic limestones.

Thus, it appears that at least two basic mechanisms are involved in the frost-damaging process in concrete aggregates.

8.7.2 Aggregate and the Frost Resistance of Concrete

The frost resistance of properly air-entrained

concretes made with different aggregates is determined by the internal structure, especially the *pore structure* of the aggregate, by its composition, and by the particle size.[Popovics 1976c, 1984c] Another way of expressing this is that the frost resistance of concrete, as influenced by the aggregate, depends on the composition, rate and amount of water absorption, permeability, modulus of elasticity, and tensile strength of the coarse aggregate particles [Lewis 1953, Iyer 1975]. For instance, rocks having absorption of about 0.5% by weight or less, which is perhaps typical of some quartzites and traprocks, are usually frost resistant under any practical circumstances because the amount of absorbed water is so little even at critical saturation that the internal pressure developed upon freezing is elastically accommodated within the aggregate particles without any damage. On the other hand, critically saturated aggregates of higher absorption but of low permeability, that is, of fine pore structure, perhaps typical of cherts, can cause freezing failure [Walker 1956, Schuster 1961, Wuerpel 1940, Sweet 1942] because of high internal hydraulic pressure developed in the aggregate. Several investigators have found specifically that the amount of small, micron-size pores is a significantly separating parameter of concrete aggregates with good and bad field histories [Dolch 1966, Rhoades 1946, Walker 1968, Sweet 1948b, Hiltrop 1960, Shakoor 1985].

The larger the aggregate particle, the more difficult it is for the water to escape from the particle, therefore the higher can be the developed internal pressure. Indeed, fine aggregate particles have shown little effect on the resistance to repeated freezing and thawing of concretes even when the same rock as a coarse aggregate produced a nondurable concrete [Wray 1940, Stark 1973, 1976]. Therefore, this latter failure is called the *critical-size* effect [Powers 1955a] . Aggregates of higher absorption but of relatively high permeability, perhaps typical of many limestones, dolomites, and sandstones with a relatively coarse pore structure, do not develop high internal hydraulic pressures, yet they can also cause freezing failure of concretes because the water forced out of the aggregate pores by the volume increase can develop an excessively high external pressure in the surrounding mortar [Verbeck 1960]. This failure is called the *expanditure* effect, and can be reduced by air entrainment.

The general deterioration of concrete by freezing and thawing starts with cracking. Because of the typical shape of such cracks, this phenomenon is called *D-cracking*. [Bukovatz 1974, Philleo 1986, Schwartz 1987] It should be noted that some individual aggregate pieces may be critically saturated in a concrete even though the average degree of saturation of the total aggregate particles may be considerably lower. In this case, freezing produces, instead of a general disintegration, the phenomenon known as *popouts*, in which coarse aggregate particles near the surface push off the surface layer of mortar when they expand, leaving holes in the surface [Bache 1968]. Chert particles of low specific gravity, limestone containing clay, and shaly material are well known for this behavior.

A potentially vulnerable aggregate, for which significant summertime drying is presumed, can give satisfactory performance in concrete if it is subjected to periodic drying in service so that it never becomes critically saturated. This can happen, for instance, if the length of time required for the aggregate to become critically saturated is longer than its winter exposure to wetting and freezing, even though the aggregate would cause disruption if frozen while saturated. Therefore, important factors are the porosity and pore-size distribution of the aggregate, and the permeability and thickness of the mortar cover protecting the aggregates from water. Verbeck and Landgren [Verbeck 1960] calculated that an aggregate with a fine pore structure would reach a high degree of saturation much more rapidly than an aggregate with a coarse pore structure, even if the aggregates had the same total porosity. For aggregates with similar pore-size distributions, the one with high porosity should require more time to attain any particular degree of saturation than the one with low porosity. A long saturation time is beneficial; therefore aggregates some intermediate porosity in concrete may actually show the poorest performance. Also, as a rule of thumb, stronger rocks may be expected to have better frost resistance (Fig. 8.8).

In brief, if an aggregate of doubtful durability should be used in concrete, the potential frost damage can be reduced in several different ways: (1) selection of an aggregate in which the amount of the micron-size pores is small; (2) drying of the aggregate to a meaningful extent either naturally or artificially before using it in the concrete; (3) use of the smallest acceptable maximum particle size of aggregate; (4) elimination of the particles of low specific gravity, for instance by jigs or by heavy-media separation; (5) coating of the coarse particles with an impermeable material; and (6) reduction of the amount of flaky or elongated particles. [Dutt 1971, Marek 1972, Lin 1976]. Of course, other aspects of the production of quality concrete are also necessary to avoid frost damage in concrete, primarily proper air entrainment, low water-cement ratio, and adequate compaction, as well as proper drainage [Stark 1970] of the concrete in service.

8.7.3 Aggregate and the Soundness of Concrete

The influence of aggregate on the soundness of concrete subjected to wetting and drying is also controlled by the pore structure of aggregate. Although certain aggregates produce sizable expansion during wetting and drying cycles, as was discussed in connection with absorption, this problem is not nearly as serious as the problem of freezing and thawing. Neither is the potential damage in concrete produced by repeated heating and cooling. However, large differences between the values of hardened cement paste and aggregate of the thermal coefficient of expansion or thermal diffusivity could produce excessive stress in concrete [Callan 1952].

The durability and, to a certain extent, the soundness of the aggregate are extremely important whenever the durability and soundness of the concrete

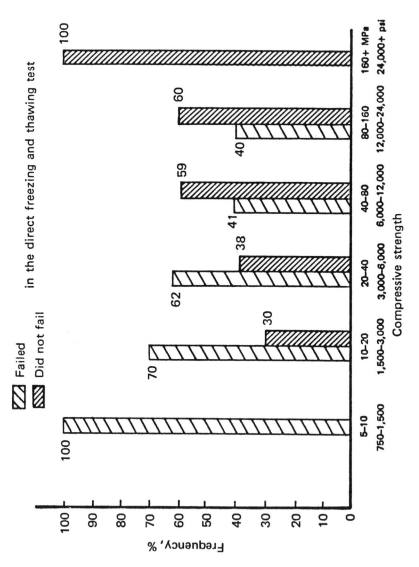

Figure 8.8: More often than not, stronger stones have better frost resistance [Popovics 1955b].

are important. This is, of course, the situation in any outdoor and some indoor concretes, that is, in the majority of concrete applications. Location of the weathering regions is presented in Fig. 9.2.

8.7.4 Methods of Testing Frost Resistance

Historically, the first method for testing the durability of an aggregate was to test directly its frost resistance; that is, unconfined particles, or the parent rock of the particles, were subjected to a direct freezing and thawing test. Generally, in these tests either the damage that the sample suffered after having been soaked in water and then submitted to repeated cycles of freezing and thawing up to a specified number of cycles was determined, or the freezing and thawing was continued until the specified degree of deterioration had taken place in the material. Such a method is specified, for instance, in AASHTO T 103 and in the German standard, DIN B1.3 [Bonzel 1976b]. The testing time can be reduced considerably by special techniques, such as intensified water saturation [MacInnis 1971], or by the use of a 0.5% alcohol-water solution instead of water in the above test [Brink 1958]. However, the destructive effects of freezing on the aggregate in concrete and those produced by the particles in freezing concrete usually are not simulated adequately by freezing the aggregate in an unconfined state. For instance, rocks susceptible to critical-size failure will show a large overmagnified loss in unconfined freezing and thawing, whereas rocks subjected to expanditure difficulty will not. Therefore the results obtained by the unconfined freezing of aggregate do not correlate closely with the service records of the aggregate in concrete. The same is true for measurements of other properties of the aggregate that are related to the frost resistance [Sweet 1942, 1948b, Higginson 1953].

Better correlations have been obtained between field performance of concrete aggregates and resistance to freezing and thawing in the laboratory of concretes made with them [Jackson 1955, Cordon 1966]. Presently two such test methods are specified

in ASTM C 666-84: procedure A, which is rapid
freezing and thawing in water (formerly ASTM C 290);
and procedure B, which is rapid freezing in air and
thawing in water (formerly ASTM C 291). Two other
tentative methods, namely, slow freezing and thawing
in water or brine (C 292), and slow freezing in air
and thawing in water (C 310), were dropped from the
ASTM specifications because of their large labor cost
and the long time required for results [Arni 1966].
Unfortunately, the reproducibility of all these
results is poor [Highway Res. B. 1959a]. The method
in procedure A is the more severe. Slower freezing
rates usually produce faster deterioration in the
concrete [Lin 1975].

Deterioration of the concrete specimens is shown
by any of the following: (1) loss in weight; (2) loss
in strength; (3) expansion of the specimen; (4) the
slope of the first-cycle temperature versus length-
change curve, or the slope of the length-change
versus time curve [Walker 1966]; (5) the condition of
the specimen by visual examination or by microscopy;
(6) increase in absorption; (7) reduction in pulse
velocity; and (8) reduction in sonic modulus of
elasticity. Among these the last one appears to be
the best. By using the resonant-frequency method
[ASTM C 215-85], the reduction in the sonic modulus
can be characterized numerically by the durability
factor DF as follows:

$$DF = \frac{PN}{M} \qquad (8.16)$$

where P = $100n_1^2/n^2$ = relative dynamic modulus of
elasticity after N cycles of freezing and
thawing, %

n = fundamental transverse frequency at zero
cycles of freezing and thawing

n_1 = fundamental transverse frequency after N
cycles of freezing and thawing

N = number of cycles at which P reaches the
specified minimum value for discontinuing

the test, or the specified number of cycles
at which the exposure is to be terminated,
whichever is less

M = specified number of cycles at which the
exposure is to be terminated.

The value of DF can vary between 0 and 100. The
higher DF values represent more frost-resistant concretes.
An ASTM recommended practice is also available in C
682-87 for the evaluation of frost resistance of
coarse aggregates in concrete by critical dilation
procedures. This utilizes a systematic approach taking advantage of the services of a trained petrographer and a battery of tests to provide needed
information about the frost resistance of coarse aggregate in concrete. Such an approach is represented
schematically in Fig. 8.9 as adopted by ASTM from
Larson and Cady [Larson 1969].
Despite the use of concrete, the ASTM methods
above do not simulate the conditions under which
concrete is subjected to freezing and thawing under
natural weathering. For instance, the important
effects of seasonal drying are not taken into
consideration. Therefore, Powers modified these test
procedures by letting the hardened concrete dry and
then resoaking it before a slow freezing in air
[Powers 1955]. Results obtained with the Powers test
were supported better by field experience than the
ASTM tests [Tremper 1961], although a correlation was
found between the Powers results and those obtained
by the method of procedure B [Wills 1963]. The Powers
test method is quite lengthy.
Considering everything, at present time the
service record is the engineer's most reliable
criterion for predicting the aggregate performance
concerning durability of concrete. Existing methods
for testing the frost resistance of aggragates can
show, at best, the relative frost resistance of
aggregates, but they cannot predict the performance
of a concrete made with an aggregate under specific
field conditions.
Comparison and evaluation of various methods for
testing the frost susceptibility of concrete

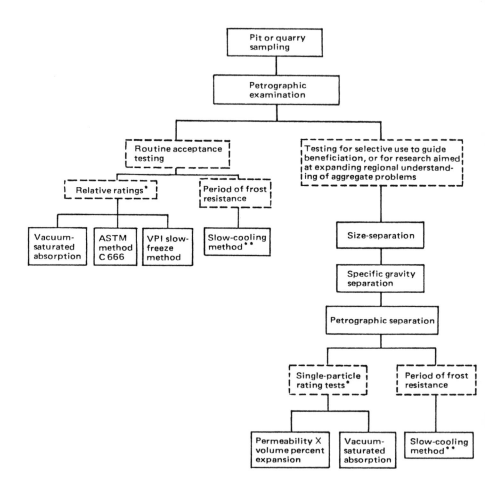

Figure 8.9: Procedural approaches to frost susceptibility tests of coarse aggregate in concrete as recommended in ASTM C 682-87. *Interpretation depends on previous experience relating test results to field performance. This information generally is not available. **As described in ASTM C 682-87.

aggregates can be found in the literature [Larson 1965, Gjorv 1971]. These also indicate that petrographic analysis is a useful supplement to any such test method.

8.7.5 Methods of Testing Soundness

The usual method for estimating overall aggregate soundness is the *sulfate soundness test*, described in ASTM C 88-83. The method has also been standardized in many other countries, for instance, in DIN 52111.

The sulfate soundness test consists of alternate immersion of a carefully graded, and weighed aggregate sample in a solution of sodium or magnesium sulfate and oven drying it under specified conditions. The accumulation and growth of salt crystals in the pores of the particles is thought to produce disruptive internal forces similar to the action of freezing water [Bloem 1966]. The exact mechanism of this process is not known, but it is probable that the effects of heating and cooling, wetting and drying, and pressure developed by migration of solution are also involved to a certain extent. A simplified analysis suggests that the magnesium sulfate test should be more severe than the sodium sulfate test, because the hydrated salt of magnesium sulfate can fill all available pore volume during the third cycle rather than the fourth [Garrity 1935]. The damage is measured after a specified number of cycles, usually either 5 or 10, in terms of the amount of sample that will pass a sieve smaller than the size on which it was originally retained.

The advantage of the sulfate soundness test is that it is quick. However, since the mechanism of disruption in the sulfate soundness test is different from that in freezing concrete, one can say quite reasonably that the sulfate test provides nothing, only a rough empirical correlation with concrete performance [Verbeck 1960].

Sulfate soundness test loss depends more on the quantity of pores than on their more critical characteristics of size and distribution. Based on this and also on experimental evidence [Harman 1970, Gaynor 1968b], Bloem feels that it would be more logical to use the simple absorption test rather than the more complicated sulfate test for the rough estimation of aggregate soundness (or durability) [Bloem 1966, Dolch 1959]. It has also been suggested that the sulfate soundness test might be used to accept aggregate if they withstand the test, but not to reject on the basis of great loss. Unfortunately, the test is not reliable even to that extent. An additional disadvantage is that the reproducibility of the sulfate soundness test, especially the laboratory-to-laboratory reproducibility, is poor. Here again, past experience, that is, the service record is the engineer's most reliable criterion for predicting aggregate performance concerning soundness of the concrete.

ASTM C 33-86 specifies that acceptable coarse aggregate subjected to five cycles of the soundness test (ASTM C 88-83) should show a weight loss not greater than 18% when magnesium sulfate is used, or 12% when sodium sulfate is used Coarse aggregate failing to meet this requirement may be acceptable if concrete of comparable properties, made from similar aggregate from the same source, has given satisfactory service when exposed to weathering similar to that to be encountered; or, in the absence of such a service record, if the aggregate in question gives satisfactory results in concrete subjected to freezing and thawing tests, and produces concrete of adequate strength. Somewhat stricter loss values in the soundness test are specified for fine aggregates.

There are no standard limit values in the United States for the acceptance of concrete aggregate based on the results of the standard unconfined freezing and thawing tests. DIN 4226 Bl.1 and 2 specify 4% as the upper limit for the permissible loss in weight as a result of unconfined freezing test in Bl.3 of the same standard.

Actual values of frost resistance and sulfate soundness of various aggregates can vary within very wide limits. For instance, air-entrained concretes with certain aggregates showed hardly any measurable damage after 200 cycles of freezing and thawing; on the other hand, concrete specimens made with saturated chert gravel showed 70-75% loss in dynamic modulus in only 5 cycles [Verbeck 1960, Highway Res. B. 1959a].

8.8 POROSITY IN AGGREGATES

The term *porosity* refers to the volume inside the individual aggregate pieces that is not occupied by solids [Dolch 1966] . It does not refer, therefore, to the interparticle voids in a packing of pieces. The porosity p is defined numerically as

$$p = \frac{V_v}{V} \qquad (8.17)$$

or

$$p\% = 100 \frac{V_v}{V} \qquad (8.18)$$

where V_v = volume of the pores inside the particles, that is, $V_{np} + V_{pp}$ in Fig. 8.1
    V = total bulk volume of the porous aggregate particles.

A distinction is frequently made between the total porosity and that part that is interconnected (V_{pp}), which is usually called the *effective porosity*. For many aggregates the assumption is reasonable that total and effective porosities are equal.

8.8.1 Pore Size

Anotner important aspect of porosity is the size of the pores. Since the pores twist and turn, that is, have highly irregular shapes, there is no exact definition for the pore size. The solution of this dilemma can be approached by using a model that has a pore system of such simplicity that it can be described analytically, yet the model presents ade-

quately the property in question of the real material. The usual model for porous solids is one in
which the pore space is made up of a bundle of
capillary tubes, perhaps of different sizes. More
complicated models are also described in the
literature [Scheidegger 1957].

Another practical way to obtain a working number
for the pore size is to calculate the *hydraulic
radius*, which is defined as the cross-sectional area
of a pore space devided by its wetted perimeter when
a fluid flows through the porous solid. But when the
pore space is saturated, the wetted perimeter becomes
the total perimeter and the the hydraulic radius,
r_h, is

$$r_h = \frac{p}{s_b} \qquad (8.19)$$

where p = porosity
s_b = specific surface inside the aggregate.

The amounts of various pore sizes can be presented
by the *pore-size distribution*. The usual cumulative
distribution curve provides the volume of pore spaces
smaller than r, whereas its differential curve
provides the volume of pores with sizes between r_1
and r_2.

A final parameter of interest is the *tortuosity*,
which is characteristic of the shapes of the pores
and is defined as the square of the ratio of the
real, or tortuous, length of the flow path of a fluid
permeating a porous solid to the bulk length of the
flow path.

8.8.2 Significance of Porosity

Many properties of the aggregate, and therefore of
the concrete, are significantly or even critically
affected by its porosity. This is so for two basic
reasons:

1. Any pore space, that is, the total porosity, decreases the volume of solids in bulk volume of the aggregates; this reduces its density, probably proportionally with the pore content, although a nonlinear relationship has also been reported [Lange 1969]; it also reduces its strength, modulus of elasticity, and abrasion resistance, which reductions are approximately exponential functions of the pore content [Verbeck 1969]

2. The "effective" pore space permits the penetration and retention of water or other liquids into the aggregate.

Except for extreme cases, the decrease in the volume of solids in the aggregate that is, the reduction in aggregate properties listed above under reason 1, does not seriously affect concrete strength. However, it has significant effects on the unit weight, mix proportion by weight, wear resistance, modulus of elasticity, and shrinkage of the concrete.

Of even greater significance is category 2 above, that is, the penetration and retention of water or solutions by the aggregate particle and the control of the movement of these liquids through the material [Larson 1964b]. The most important consequence of this is the great influence the aggregate pore system has on the durability of concrete [Larson 1965, Gjorv 1971].

8.8.3 Methods of Testing Porosity

There are no standardized methods in the United States for the measurement of porosity or pore size in aggregates. Some of the more frequently used methods are described below [Lewis 1953].

A common test method for the determination of *total porosity* is to measure the bulk specific gravity, G_b, and the true specific gravity, G_t, as described before, and to calculate the porosity p by

$$p = 1 - \frac{G_b}{G_t} \qquad (8.20)$$

The disadvantage of this method is that an adequately accurate determination of G_t is not easy.

Another method for the determination of total porosity is to test visually a polished or thin section of the aggregate. The area of void spaces can be measured by various camera lucida or photomicrographic methods [Sweet 1948b, Verbeck 1947], and image analyzers connected to electron microscopes. Extensions of this idea are the linear travers and the point-count methods, described in ASTM C 457-82a for use on the air-void system of hardened concrete. These methods also require a polished or thin section of material.

The main objection to the use of these microscopic methods to determine aggregate porosity is that in many aggregates considerable portions of their pore spaces are smaller than the limit of optical microscopes, which is usually a few microns. But even for the visible pores the method is tedious and not particularly precise.

Probably the most common method used in casual investigations of the *effective porosity* of aggregate is to measure the absorption and assume the volume of water to equal the pore volume, that is, to assume complete saturation. However, one hardly ever gets complete saturation. Even the evacuation procedure results typically in around 90% average degrees of saturation with usual normal-weight aggregates. The volume of absorbed liquid is, therefore, an imperfect measure of the effective porosity of aggregate. Another method for the measurement of effective porosity is based on the so-called *McLeod gauge* porosimeter. This measures the volume of air expelled from the pores of a dry sample while it is immersed in mercury [Dolch 1959].

An advanced but still relatively simple form of the absorption test for the determination of the volme of micropores is the *Iowa pore index test*. The apparatus is a modified version of the pressure-type air meter (Popovics 1982a). 9000 g of ovendried aggregate, graded between 1/2 in (12.7 mm) and 1-1/2 in (38.1 mm) is placed in the apparatus, covered with water, and 35-psi (241-kPa) constant air pressure is introduced. The amount of water forced into the aggregate during the first minute is taken to represent the macropores, thus omitted from further consideration. The amount of water forced into the aggregate during the next 14 minutes is taken to represent the volume of micropores and called the *pore index*. Pore index higher than 50 ml usually indicates frost-susceptible aggregate, especially if silt and clay is present in the rock in sizable quantity. [Shakoor 1985]

Determinations of *pore size* are subject to the difficulties of definition discussed previously. Therefore, all such determinations are approximate in nature. Pore sizes of the total porosity can be estimated by the various microscopic methods previously discussed or by using photomicrography. By assuming a shape for the pores, usually spheres, the specific surface can also be calculated.

The most important method of measuring the *specific surface* of a solid is the sorption method [Scheidegger 1957, Young 1962, Sneck 1970]. Only the effective pore space participates in such measurements. The principle is to determine the magnitude of uptake of a vapor, usually nitrogen or water vapor, by the solid as a function of vapor pressure and at a constant temperature, that is, the adsorption isotherm [Landgren 1964b] . These data are then interpreted by one of several theories and the result is a value for the surface area of the solids. The theory that has been most useful for such calculationsis the BET method [Brunauer 1963] (see Section 3.5.4). The sorption method is applicable for aggregate [Blaine 1953, Feldman 1961], but since the usual aggregates have relatively small internal specific surface, relatively large samples should be used to minimize experimental error.

A second major way of determinig the *hydraulic radii* and specific surfaces of the aggregate porosity is by means of permeability measurements. The principle involved is that a fluid, either a gas or a liquid, in viscous flow through a porous solid is retarded by a drag on the pore walls that is proportional to the magnitude of the solid surfaces contacted. In such measurements it is quicker to use gases because of their lower viscosity, and the method is just as good for rocks as liquids. The quantitative relationship between permeability and solid surface area involves, among other things, the actual path the flowing fluid should traverse, and good estimation of this is the weakest aspect of this method. In any case, the permeability method must be expected to give a lower limit of the effective specific surface because any pore space containing stagnant fluid does not contribute to the viscous retardation. On the other hand, if one is interested in the pore space that supports fluid flow, as for instance in studies of the critical size of aggregates in freezing, then the permeability method may give a more useful answer. A third group of ways to determine the size of effective pores in solids is based on the *meniscus phenomenon*. The basic idea of this method is to measure the capillary pressure, or some property associated with it, of the liquid in the pore from which the radius of the tube containing the liquid can be calculated.

The meniscus phenomenon can also be used for the determination of *pore-size distribution*. All that is needed here is to determine the capillary pressure as a function of the degree of saturation and then construct a cumulative pore-volume versus pore-size curve.

Another major technique for the determination of the size distribution of effective pores utilizes the so-called *mercury porosity meter* [Washburn 1921]. In this apparatus mercury is forced into the evacuated sample, and the amount that goes into the pores is measured as a function of pressure causing ingress, that is, capillary pressure [Hiltrop 1960, Sneck 1970]. Several commercial instruments using this principle are available.

Although the results of the mercury porosimetry have shown good correlation with concrete frost resistence in the field, the method has weaknesses. The main weakness is that it uses very small samples. In addition it is expensive and complicated to run the test.

A third method combines the meniscus effect with permeability. Here water is allowed to enter the pore system of the dry solid by capillarity. The progress of the liquid is retarded by viscous forces, so the forces causing both advancement and retardation are dependent on pore size. The quantitative measure of the effect is called *absorptivity* [Powers 1946]. The advantage of this method is that it is easy to perform because it requires only a balance and a stopwatch.

The size distribution of total porosity can be determined by microscopic methods, and image analyzers. Other methods for specific surface and pore size exist. Capillary condensation, small-angle x-ray scattering, heat of immersion, rate of dissolution, as well as radioactive and electrical methods, can be used. These methods, however, have been little used on aggregates.

A collection of recent European studies on porosity and porosity measurements was published by RILEM [RILEM 1973]. Similar measurements on American aggregates as well as analyses concerning the relationships between various porosity characteristics have also been performed [Meininger 1964].

There are no specifications concerning the permissible limits of aggregate porosity or pore sizes. The porosity of usual aggregate materials ranges from nearly zero to as much as 20%, most commonly from about 1 to 5%. For lightweight aggregates the porosity may be in the range of 30 to 50%. The sizes of the pores also vary greatly among different aggregates. In dense aggregates, such as granite or basalt, they are small, whereas certain sandstones and lightweight aggregate have very coarse pore systents [Verbeck 1966].

8.9   WEAR AND SKID RESISTANCE

Concretes in pavements, floors, and similar applications should have adequate wear resistance, or skid resistance, or both [Popovics 1968g]. The nature and extent to which these concrete properties are influenced by the aggregate used are discussed below.

8.9.1 Wear   Resistance

Removal of surface material and severe transformation of surface appearance may be called *wear*. Wear may occur because of chemical, physical or mechanical attacks, or any combination of these. Mechanical wear takes place through asperity contacts and junction breakages. A model based on asperity contact has been developed that relates load, distance slid, and wear volume produced. This suggests a fatiguelike phenomenon in wear particle production. [Burton 1980]

The abrasive wear on concrete road surfaces caused by rubber-tired vehicles (without chains or studs) is usually negligible, but other kinds of heavy vehicles or abrasive effects may cause considerable wear.

Harder rocks usually have better abrasion resistance. However, the influence of the *type of mineral aggregate* on the abrasion resistance of concrete, above and beyond its contribution to the strength of the concrete, is less clear. Smith, for instance, drew the following conclusions from the results of his tests on 60 concrete mixtures tested by three different methods of abrasion [Smith 1958]:

1. The abrasion resistance of a concrete varies directly with its compressive strength, otherwise regardless of the aggregate quality.
2. No significant correlation was found to exist between the quality of the coarse aggregate, as determined either by the sodium sulfate soundness test or by the Los Angeles abrasion test, and the abrasion resistance of concretes containing these aggregates.

However, these conclusions hold only for the aggregates that were used in this test series, that is, for aggregates having wears less than 40% as measured by the Los Angeles abrasion test, and a maximum particle size of 0.75 in (19 mm). Use of softer aggregates may reduce the abrasion resistance of the concrete. It was found, for instance, that if the compressive strength of the concrete is 6000 psi

(42 MPa) or higher, the type of aggregate has little effect on the abrasion resistance; for concretes of lower strengths, however, the effect of the aggregate type can be considerable [Road Res. Lab. 1955, Graf 1960]. Graf also found that the effect of aggregate is even more pronounced if the abrasion test is performed with the addition of water. Under such circumstances the standard abrasion (DIN 52 108) of a hard aggregate, such as quartz, basalt, or porphir, is about half the standard abrasion of the concrete made with the aggregate.

The *grading* of aggregate can also influence abrasion. A decrease in its coarse aggregate content, if it is of good quality, decreases the abrasion resistance os the concrete [Smith 1958, Price 1947]. Experimental data also show that a considerable increase in concrete abrasion is caused by the substitution of an overly fine sand for a well-graded sand [Graf 1960, a'Court 1954].

The effect of *particle shape* on concrete abrasion has not been fully investigated yet. It appears that angular but not flaky particle shape is desirable for high abrasion resistance [Li 1959].

A special form of abrasion is *erosion*. This is defined, according to ASTM G 40 as progressive loss of original material from a solid surface due to mechanical interaction between the surface and a fluid, a multicomponent fluid, impinging liquid, or solid particles. The action of the abrasive particles in the water or air is controlled largely by the velocity of the water or air, the abrasive material, the general surrounding conditions, and the angle of impact. When the relative motion of the solid particles is nearly normal to the solid surface, the wear is called impact erosion or impingement erosion. When this relative motion is nearly parallel to the solid surface, the wear is called abrasive erosion. [Popovics 1989]

These two different actions produce different types of erosion. In the case of an ideally pure impingement erosion, the matrix, that is hardened cement paste, is gradually displaced by the repeated impacts causing localized, deep *cavitation* in the concrete. This mechanism indicates that the resistance to this type of erosion is controlled by the softer component of the concrete, which is usually the matrix. In contrast, a pure abrasive erosion is less localized, that is, the wear is distributed more or less uniformly producing a relatively smooth concrete surface. This type of erosion is more sensitive to the hardness of the coarse aggregate. In most cases, however, the two types of mechanisms act simultaneously, so the character of such erosion will be a combination of the impingement and abrasion erosions. Such combined erosion produces typically the so-called *differential* wear where the worn surface is uneven, with the harder component protruding from the softer material.

The relative abrasion resistance of rock pieces can be tested by the method specified in ASTM C 418-89 for concrete. This procedure simulates the action of waterborne abrasives and abrasives under traffic on concrete surfaces. It performs a cutting action which tends to abrade severely the less resistant components of the specimen.

The stone specimen should be immersed in water for 24 hours and then surface dry with a damp cloth to obtain a surface dry condition at the time of test. The the selected surface of the specimen is submitted to a blast of air-driven silica sand for one minute. It is repeated at least eight different spots. The abraded volume is determined, for instance by filling the abrasion cavities with an oil base modeling clay. From this the abrasion coefficient can be calculated, as follows:

$$A_c = V/A \qquad\qquad (8.21)$$

where

A_c = abrasion coefficient, cm^3/cm^2

V = abraded volume, cm^3

A = area of surface abraded, cm^2.

Adjustments in the pressure used and the type of abrasive permit a variation in the severity of abrasion which may be used to simulate other types of wear.

The British method for abrasion resistance of aggregates in B.S. 812 specifies two aggregate pieces between 1/2 and 3/8 in. (12.5 and 9.5 mm) in size set in pitch in a shallow tray. The aggregate pieces held in this way are pressed with a total force of 2 kg at two points diametrically opposite on a metal disk which can revolve for 500 revolutions in a horizontal plane at 28 to 30 revolutions per minute. Standard quatz sand is fed onto the disk just in front of the specimen being tested. The percentage loss in weight of the two samples is determined and the average of the two results is taken as the aggregate abrasion value.

8.9.2 Skid Resistance
Skid resistance is the force developed when a tire that is prevented from rotating slides along the pavement surface. [Highway Res. B. 1972c] It originates from the *friction* between the tire and the pavement surface. In general terms friction is the resistance that is encountered when two contacting surfaces slide or tend to slide over each other. In case of solids, the surfaces may be dry causing *dry* friction or lubricated causing *wet* friction.

There is no generally accepted theory at present for the phenomenon of friction. The first theory was offered by Coulomb in 1785 in which he attributed friction primarily to grippage as well as to the force required to lift one asperity over another. Despite its apparent logic, this theory can explain only some of the observations on friction, not all, thus it has been largely discarded. [Burton 1980] So it appears that surface roughness is only a limited cause of friction. A more recent theory states that friction may be ascribed to the molecular forces at the sliding surfaces; that is, when two molecules come into contact, i. e., come into each other's repulsion field, and then separate, a loss of energy occurs which is manifest as friction and cause heat. This theory seems plausible but lacks complete experimental justification. A third theory of friction is the quite popular "stick - slip" theory.

According to this, when one solid slides over another, the surfaces make contact only at the tips of their asperities, that is, the actual area of contact is very small. Thus high pressure and heat develop that are concentrated in small discrete areas rather than dispersed over the apparent contact area. The resultant is high local loading and hot spots which weld and prevent further sliding. As the surfaces are made to slide over one another, the just welded junctions are sheared, and so one surface jerks over the other quickly. Exact proof of this theory is also missing. [Batchelor 1966, Szeri 1980] Despite the lack of a satisfactory theory, it is generally accepted presently that the origin of friction is a combination of cohesion, abrasion and adhesion with the latter the dominant factor.

The magnitude of friction between any two materials can be measured and represented by a number known as the *coefficient of friction*   the usual symbol of which is μ. It is defined as

$$\mu = \frac{\text{moving force parallel to surface}}{\text{pressing force perpendicular to surface}} \qquad (8.22)$$

The *dynamic* coefficient of friction is the ratio of {the minimum force parallel to the contacting surfaces required to slide steadily one surface over the other} to {the normal force pressing the two surfaces together}. The *static* coefficient is the ratio of {the minimum parallel force that can start to move one surface over the other} to {the normal force}. Static frictions are usually slightly higher than the corresponding dynamic coefficients. A brake with $\mu = 0.5$ is twice as effective as a brake with $\mu = 0.25$ in retarding the moving member.

It is apparent that the coefficient of friction is not a definite, invariable quantity even for two given materials, as commonly tabulated in textbooks. Rather, it may be highly sensitive to the test method applied and the prevailing testing conditions.

Since friction comes always from two material

components, it is expected that properties of the rubber tire have a role, and as experience has clearly demonstrated, a decisive one, in vehicel skidding along with the finish and wetness of the pavement surface. [Moyer 1963, Highway Res. B. 1972b] Nevertheless, it is also agreed that certain properties of the mineral aggregate in the concrete influence the skid resistance of the pavement significantly. This effect of the *coarse aggregate* particles is due to the fact that they form a large portion of the riding surface of a pavement after the mortar portion of concrete is worn away. The *fine aggregate* particles can impart microstructure of the mortar portion of concrete that is necessary for the required friction. [Rose 1971]

Test results show that the antiskid properties of a well-designed and well-constructed pavement surface are dependent to a large degree on the *polishing characteristics* of the coarse particles of the mineral aggregate used [Shupe 1960]. This is not surprising because the surface of a pavement is continuously polished by traffic, thus, the greater the polishing tendency of the aggregate used, the smoother the surface will become and, consequently, the greater will be the reduction in skid resistance of the surface. The relationship between the extent of the polish of stone in laboratory (B.S. 812:1960) and the skid resistance of pavement made with the chippings of the stone is presented in Fig. 8.10.

Experience has also shown that surfaces containig *crushed* aggregate exhibit better inital skid resistance than surfaces made from comparable but rounded aggreate. After a certain perod of wear, however, the skid resistance of the two surfaces will become practically identical.

As far as the *composition* of the stone is concerned, limestones, as a group, have acquired a poor reputation with regard to susceptibility to polishing, although not all limestone-concrete surfaces become slippery, even when wet. Those limestones that readily polish possess either very fine-grained crystalline structures, or consist of rounded oolitic grains supported in a calcite matrix of similar hardness. The limestone that exhibit the

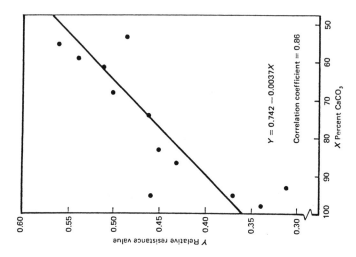

Figure 8.11: Correlation between CaCO₃ content and skidding susceptibility of a limestone [Shupe 1959].

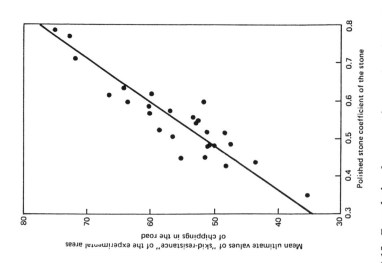

Figure 8.10: Correlation between the results of the standard British laboratory polishing test and the ultimate state of polish of chippings on a straight, heavily trafficked road [MacLean 1959].

best antiskid characteristics have crystalline granular internal structures composed of angular interlocking grains and are relatively low in calcium carbonate (Fig. 8.11). Sandstones, in a variety of forms, have consistently exhibited exellent antiskid characteristics. Other stones having polishing characteristics similar to those of sandstones, such as coarse-grained granites and traprocks, also possess excellent antiskid properties. Conversely, stones having polishing characteristics similar to those of a pure limestone, such as chert and rhyolite, have poor antiskid properties [Shupe 1960, Shelburn 1963].

The good skid resistance of the sandstone group is attributed to the internal structure of the stone. In general, sandstones are composed of small, angular quartz grains supported by a weaker cementing matrix. The individual quartz grains are highly resistant to polishing, since quartz is No. 7 on Mohs's relative hardness scale as shown in Table 8.2. (Calcite, the major constituent of limestone, is No. 3.) In addition, there is a so-called particle-by-particle wear on the sandstone, which means that before an exposed quartz grain has the opportunity to become highly polished, it is dislodged from the weaker cementing matrix and a fresh, harsh particle appears at the surface.

Concrete pavements, as a rule, do not have significant particle-by-particle wear. However, it is possible to produce a "differential wear." If the aggregate has less resistance to wear than the supporting cement paste or mortar, which is usual with limestone aggregates, then the particles are worn slightly below the mortar surface, thus increasing the roughness and the skid resistance of the concrete surface. On the other hand, if the aggregate is more resistant (e.g. quartz gravel), the particles remain slightly above the surface of mortar, and again increase the skid resistance. Consequently, if an aggregate with unfavorable polishing characteristics, for instance limestone, is used in a concrete pavement, which may be necessary for economic reasons, then natural quartz sand is recommended as a fine aggregate to provide for differential wear at the pavement surface [Woods 1963].

There is no ASTM method for the measurement of the skid resistance of an aggregate. The polishing characteristic on an aggregate can be represented by the *polished-stone coefficient* specified in B.S. 812. This is determined by setting pieces of the aggregate in the periphery of a 16-in. (400-mm) diameter wheel which is rotated in a vertical plane. An 8-in. (200-mm) diameter rubber-tyred wheel is rub on top of this, sand and finally emery flour being fed in as an abrasive. After abrasion, the coefficient of friction on the aggregate surface is measured in a standard skid resistance tester and recorded as the polished stone coefficient of the material. There are many other test methods both in this country and abroad for the measurement of skid resistance of concrete and aggregates.[Highway Res. B. 1958c, 1959b, 1970, 1972c]

9

Chemical Properties of Aggregates

SUMMARY

Although mineral aggregates are considered essentially inert, that is, chemically inactive in concrete, they may frequently contain reactive substances that are harmful to concrete if present in excess quantities. Such substances are called *deleterious materials*.

One type of deleterious material is the size fraction of particles that are *finer than a No. 200 (75-μm) sieve*. The permissible upper limit for these very fine particles depends on their composition, the composition of the concrete, and whether these particles are present in the aggregate in dispersed form, or as lumps, or as coatings. Of these three possibilities, the dispersed form is most common. Excessive *lightweight and/or soft particles* reduce the strength, durability, and/or wear resistance of concrete. These effects are mostly physical in nature. *Organic impurities* in aggregates, such as tannic acid, can interfere chemically with the strength development of concrete. The standard test method utilizes the darkening of a NaOH solution for the presence of organic impurities, but a positive result with this test is not conclusive. *Other harmful impurities* include certain unstable minerals, such as pyrite and CaO, because they can produce popouts and/or staining of the surface of the concrete.

Various kinds of chemical reactions take place between aggregate particles of specific natures and the alkalies of portland cement in the presence of moisture. Such reactions may cause cracking in the concrete through excessive expansion. The most common case is when the cement reacts with certain siliceous particles (opal or other low-crystalline silicas). This is called the *alkali-silica reaction*. The resulting expansion of concrete may exceed 0.5%, and the cracks may be as wide as 1 in (25 mm). These cracks are caused by the development and swelling of a gel composed of sodium and potassium silica. The expansion can be prevented, or at least reduced, in several ways, such as by using cement of low alkali content, low cement content, replacing a portion of the cement with a pozzolan, and so on.

Another harmful reaction is the *alkali-carbonate rock* reaction. In this case the dolomite constituent of argillaceous dolomitic limestones reacts with the alkalies of the cement to produce large expansion and cracking of the concrete. This expansion cannot be reduced by pozzolan addition. There are also other kinds of harmful cement-aggregate reactions, with or without the presence of alkalies. When the reaction between aggregate and cement paste is not accompanied by excessive volume changes, the effect is usually beneficial.

Presently there is no sure laboratory method to determine whether or not a given aggregate will cause excessive expansion. Even the standard test methods can provide estimates only of the potential reactivity of the aggregate. Some of these methods are quick, but they are usually less reliable than more time-consuming methods, such as the mortar bar test for the alkali-silica reaction.

In most cases the aggregate does not have a decisive role in the deterioration of concrete produced by chemical attack from outside sources.

9.1   INTRODUCTION

Although mineral aggregates are considered inert,

that is, chemically inactive, in concrete, they may frequently contain substances any of which, individually or in combination, are harmful to concrete if present in excess quantities. The types and quantities of such deleterious substances vary greatly from aggregate to aggregate.

9.2 DELETERIOUS MATERIALS

Certain reactive materials, present in the aggregate usually as various kinds of particles mixed with the inert aggregate particles, are deleterious, that is, harmful for concrete. Their effects are usually chemical, although physical effects may also be present.

9.2.1 Particles Finer than a No. 200 (75-μm) Sieve

Particles finer than a No. 200 (75-μm) sieve may be present in the aggregate in three different forms, as follws.

They may be dispersed in the aggregate in the form of *clay, silt,* or *stone dust.* Such materials reduce bleeding and may improve the workability of the fresh concrete when they improve the aggregate grading, that is, when they are present in small quantities and the grading of the sand is rather coarse. The presence of fines may also reduce the permeability of the concrete. However, an excess of fines

1. causes reduction in the workability, particularly in the case of granite dust [Road Res. Lab 1955, Hughes 1968], presumably because of the mica content [U.S. Bureau of Recl.1966];
2. increases the shrinkage of concrete;
3. may reduce the entrained air content; and
4. impairs its durability [Gruenwald 1939, Poijarvi 1966, Buth 1966, 1968b, Meyers 1963].

Prolonged mixing of the concrete (30 min) may

also increase the amount of fines by grinding [Gaynor 1963].

Perhaps because the composition of the fine material is also important [Hansen 1963a, Balazs 1957], it has not been clarified what is meant by the term "excessive." Fine clay had a noticeable effect on the workability in the quantity as low as 3% [Graf 1960]. On the other hand, there is an opinion that "dust" may be permissible up to 20% depending on the circumstances [Road Res.Lab. 1955, Ramirez 1987].

ASTM C 117-87 provides a measure for the quantity of fines by washing the aggregate sample and by passing the decanted wash water containing suspended and dissolved material through a No. 200 (75-μm) sieve. Its disadvantage is that it does not indicate the degree of harmfulness of the fines. The AASHTO T 176-86 *sand-equivalent test* was developed to overcome this difficulty. [Hveem 1953] Sand equivalent SE is defined as

$$SE = 100(\text{Sand Reading})/(\text{Clay Reading}) \qquad (9.1)$$

where Sand Reading and Clay Reading represent the volumes of the sand and clay, respectiveky, as measured by AASHTO T 176. The relationship between SE and the fine particle content of the sand is shown in Fig. 9.1.

The result of this test is a good indicator of the effect of the fine particles present on concrete properties [Buth 1966, 1968b]. For instance, Duriez concluded on the basis of 28-day mortar strength tests that when the result of a sand-equivalent test is (a) less than 60, approximately two out of three such sands provide a mortar of poor quality; and (b) more than 70, approximately six out of seven sands provide a mortar of good quality [Duriez 1961].

The sand-equivalent test evaluates the fines in terms of the extent to which a specified suspension prepared from the aggregate will settle in a given period. A somewhat similar sedimentation method is specified in B.S. 812:1967. DIN 4226 specifies a simple test for the determination of clay and silt in fine aggregate, as follows: About 500 g of

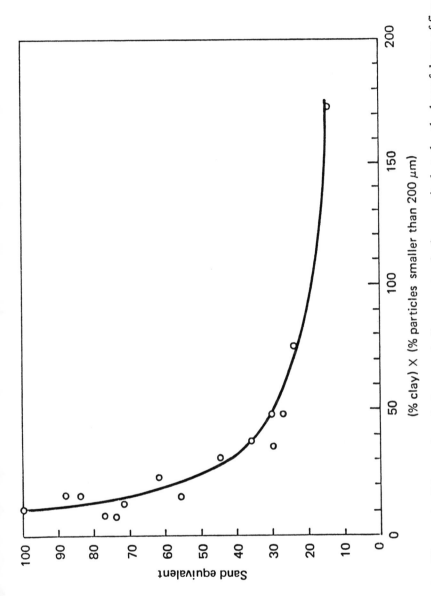

Figure 9.1: The result of the sand-equivalent test indicates not only the quantity but also the harmfulness of fine particles [Duriez 1961].

aggregate and 750 ml of water are placed in a large graduated glass cylinder. The cylinder is shaken vigorously, and then allowed to stand for at least an hour. The clay and silt will now settle in a well-defined layer above the sand, thus its amount can be determined from the scale of the cylinder. This method is less accurate than the ASTM method.

The second possibility of the occurance of fine particles in the aggregate is in the form of *lumps*. The harmfulness of such lumps depend on their survival as lumps in the concrete, where they may break down from freezing and thawing, or wetting and drying, to produce unsound concrete, or unsightly surface pits [Bloem 1966]. Those clay lumps that break down during mixing should be considered as minus sieve No.200 (75-μm) material. The quantity of the remaining lumps can be tested by the method given in ASTM C 142-78(1984) by soaking the aggregate sample and then pressing the particles with the fingers to try to break them up into fines.

The third possibility is that fine particles form a layer deposit on the larger particles. This is called *coating*. Coatings that are removed from aggregate particles during the mixing of the concrete should be considered again as minus sieve No. 200 (75-μm) material added to the mixture [Mather B. 1966]. Fine particles that remain on the particles after the concrete has been placed reduce the concrete strength, especially the flexural strength, considerably. Therefore, such coatings should be eliminated beforehand; this is done usually by washing the coarse aggregate. No standard method is available in the United States for testing the nature or size of coating, although ASTM C 295-85 recommends that pieces of coarse aggregate be examined petrographically to establish whether coatings are present and if they are, what kind.

9.2.2 Lightweight and Soft Particles

The relatively light coal and lignite in finely divided form may reduce the concrete strength. In the more common form of small quantities of larger particles these have no significant effect on concrete strength, but they may detract from appearance of the concrete by producing surface pits and stains [Bloem 1966]. There are other particles, too, for which harmfulness is associated with low specific gravity, for instance, chert or chalk. In general, particles of very low density and shale are considered to be unsound because of their porosity [Walker 1953]. ASTM C 123-83 describes a method that separates the particles of low density by flotation in a liquid of suitable specific gravity. Petrographic examination may be necessary for positive identification.

Soft particles must be present in rather large quantities in order to cause reduction in the strength of concrete, but they can be harmful in smaller quantities when the concrete is subjected to freezing or severe abrasion.

The rather arbitrary method prescribed in ASTM C 235 (withdrawn) for identification of soft particles is a simple determination of whether or not a brass needle of designated hardness will scratch the particle.

9.2.3 Organic Impurities

Certain types of organic matters may occur with natural aggregates. Among these, tannic acid and its compounds, derived from the decay of vegetable matter, interfere with the strength development of concrete, particularly at early ages [Hansen 1963a, Mather 1951].

The colorimetric test in ASTM C 40-84 uses the darkening of a 3% solution of NaOH as an indicator for the presence of organic impurities. This method has been accepted by many other countries, despite the unfortunate fact that this test reacts to other organics, such as bits of wood, which are not significantly harmful to the concrete strength. A negative test result, therefore, is conclusive evidence of freedom from organic matter, but a positive reaction, that is, dark color, may or may not indicate danger. Thus, In the case of positive reaction, ASTM C 87-83 can be used as a check. The usual laboratory method of checking compares the suspected sand with the suspected sand treated to remove the organic matters in mortars of fixed water-cement ratio.

9.2.4 Other Impurities

Unstable minerals are susceptible to oxidation, hydration, or carbonation if exposed to atmospheric agents. Concrete may become unsightly or even distressed by these actions. For instance, the minerals pyrite and marcasite (FeS_2) are oxidized and hydrated in the presence of water and oxygen to sulfuric acid and hydrated iron oxides with an accompanying increase in volume. This can cause popouts and surface staining of concrete, particularly in warm and humid areas [Hansen 1963, Blanks 1955, Midgley 1958, Mielenz 1963]. It should be noted, however, that not all pyrites are reactive. Continued oxidation and hydration of ferric and ferrous oxides as well as hydration and carbonization of magnesia (MgO) and lime (CaO) in sparse fragments in the aggregate can also cause popouts in the concrete. Aggregates with high mica content should be avoided because they may not remain stable in concrete. The presence of gypsum and other sulfates in the aggregate may also have harmful effects on concrete. Certain lead compounds, as well as boron, may greatly reduce the early strength of concrete [Midgley 1970]. Aggregates containg excessive amount of alkalies should also be avoided.

Also, fine aggregate from the seashore may contain large quantities of salt (chlorides) [Figg 1975] which may produce undesirable effects in certain applications.

9.2.5 Specifications for Deleterious Materials

The maximum allowable amounts for impurities in fine and coarse aggregates, as specified in ASTM C 33-86, are listed in Table 9.1 and 9.2, respectively. The location of the corresponding weathering regions is presented in Fig. 9.2. The same ASTM speci-

fication permits the use of a coarse aggregate
having test results that exceed the limits specified
in Table 9.2, provided that the concrete made with
similar aggregate from the same source has given
satisfactory field service or satisfactory
laboratory results. Other specifications, such as
B.S. 882 or DIN 4226 contain more or less similar
limiting values.

Table 9.1: Allowable Upper Limits for Deleterious Substances in Fine Aggregate for Concrete (from ASTM C 33-86)

Item	Weight percent of total sample max.
Clay lumps and friable particles	3.0
Material finer than No. 200 (75-μm) sieve:	
Concrete subject to abrasion	3 0[a]
All other concrete	5 0[a]
Coal and lignite:	
Where surface appearance of concrete is of importance	0.5
All other concrete	1.0

[a]In the case of manufactured sand, if the material
finer than the No. 200 (75,-μm) sieve consists of
the dust of fracture, essentially free from clay or
shale, these limits may be increased to 5 and 7%,
respectively.

Table 9.2: Limits for Deleterious Substances and Physical Property Requirements of Coarse Aggregate for Concrete (ASTM C 33-90)

Note: See Figure 9.2 for the location of the weathering regions. The weathering regions are defined as follows:

(S) Severe Weathering Region—A cold climate where concrete is exposed to deicing chemicals or other aggressive agents, or where concrete may become saturated by continued contact with moisture or free water prior to repeated freezing and thawing.
(M) Moderate Weathering Region—A climate where occasional freezing is expected, but where concrete in outdoor service will not be continually exposed to freezing and thawing in the presence of moisture or to deicing chemicals.
(N) Negligible Weathering Region—A climate where concrete is rarely exposed to freezing in the presence of moisture.

Class Designation	Type or Location of Concrete Construction	Maximum Allowable, %						
		Clay Lumps and Friable Particles	Chert^C (Less Than 2.40 sp gr SSD)	Sum of Clay Lumps, Friable Particles, and Chert (Less Than 2.40 sp gr SSD)^C	Material Finer Than No. 200 (75-μm) Sieve	Coal and Lignite	Abrasion^A	Magnesium Sulfate Soundness (5 cycles)^B
	Severe Weathering Regions							
1S	Footings, foundations, columns and beams not exposed to the weather, interior floor slabs to be given coverings	10.0	...	...	1.0^D	1.0	50	...
2S	Interior floors without coverings	5.0	...	...	1.0^D	0.5	50	...
3S	Foundation walls above grade, retaining walls, abutments, piers, girders, and beams exposed to the weather	5.0	5.0	7.0	1.0^D	0.5	50	18
4S	Pavements, bridge decks, driveways and curbs, walks, patios, garage floors, exposed floors and porches, or waterfront structures, subject to frequent wetting	3.0	5.0	5.0	1.0^D	0.5	50	18
5S	Exposed architectural concrete	2.0	3.0	3.0	1.0^D	0.5	50	18

(continued)

Class Designation	Type or Location of Concrete Construction	Maximum Allowable, %						
		Clay Lumps and Friable Particles	Chert[C] (Less Than 2.40 sp gr SSD)	Sum of Clay Lumps, Friable Particles, and Chert (Less Than 2.40 sp gr SSD)[C]	Material Finer Than No. 200 (75-μm) Sieve	Coal and Lignite	Abrasion[A]	Magnesium Sulfate Soundness (5 cycles)[B]
	Moderate Weathering Regions							
1M	Footings, foundations, columns, and beams not exposed to the weather, interior floor slabs to be given coverings	10.0	...	...	1.0[D]	1.0	50	...
2M	Interior floors without coverings	5.0	...	...	1.0[D]	0.5	50	...
3M	Foundation walls above grade, retaining walls, abutments, piers, girders, and beams exposed to the weather	5.0	8.0	10.0	1.0[D]	0.5	50	18
4M	Pavements, bridge decks, driveways and curbs, walks, patios, garage floors, exposed floors and porches, or waterfront structures subject to frequent wetting	5.0	5.0	7.0	1.0[D]	0.5	50	18
5M	Exposed architectural concrete	3.0	3.0	5.0	1.0[D]	0.5	50	18
	Negligible Weathering Regions							
1N	Slabs subject to traffic abrasion, bridge decks, floors, sidewalks, pavements	5.0	...	...	1.0[D]	0.5	50	...
2N	All other classes of concrete	10.0	...	...	1.0[D]	1.0	50	...

[A]Crushed air-cooled blast-furnace slag is excluded from the abrasion requirements. The rodded or jigged unit weight of crushed air-cooled blast-furnace slag shall be not less than 70 lb/ft³ (1120 kg/m³). The grading of slag used in the unit weight test shall conform to the grading to be used in the concrete. Abrasion loss of gravel, crushed gravel, or crushed stone shall be determined on the test size or sizes most nearly corresponding to the grading or gradings to be used in the concrete. When more than one grading is to be used, the limit on abrasion loss shall apply to each.

[B]The allowable limits for soundness shall be 12% if sodium sulfate is used.

[C]These limitations apply only to aggregates in which chert appears as an impurity. They are not applicable to gravels that are predominantly chert. Limitations on soundness of such aggregates must be based on service records in the environment in which they are used.

[D]This percentage may be increased under either of the following conditions: (1) if the material finer than the No. 200 (75-μm) sieve is essentially free of clay or shale the percentage may be increased to 1.5; or (2) if the source of the fine aggregate to be used in the concrete is known to contain less than the specified maximum amount passing the No. 200 (75-μm) sieve the percentage limit (L) on the amount in the coarse aggregate may be increased to L = 1 + [(P)/(100 - P)] (T - A), where P = percentage of sand in the concrete as a percent of total aggregate, T = the limit for the amount permitted in the fine aggregate, and A = the actual amount in the fine aggregate. (This provides a weighted calculation designed to limit the maximum mass of material passing the No. 200 (75-μm) sieve in the concrete to that which would be obtained if both the fine and coarse aggregate were supplied at the maximum tabulated percentage for each of these ingredients.

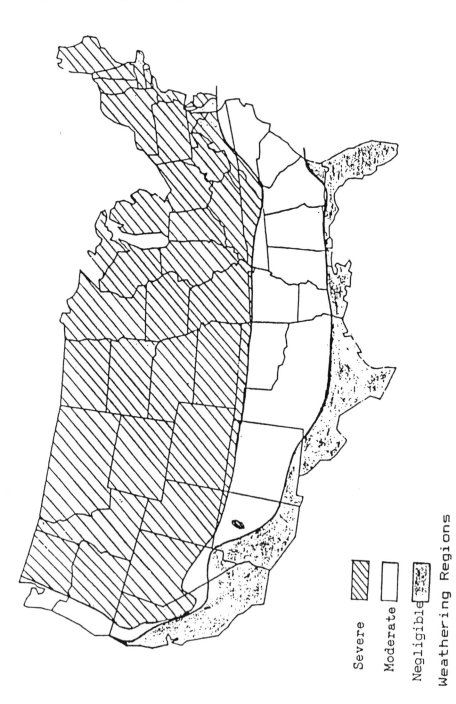

Figure 9.2: Location of weathering regions [from ASTM C 33-90].

9.3 REACTIVITY OF CONCRETE AGGREGATES

9.3.1   Alkali-Silica   Reaction

It has long been recognized that aggregates in
portland cement concrete (or mortar) may contain
particles other than the previously discussed
deleterious materials that are not inert, that is,
not inactive chemically. When such particles are
present in the aggregates in injurious amounts and
the aggregates are used in concrete that will be
subject to any kind of wetting, harmful reactions
usually take place with the alkalies in the cement.
Such reactions can cause cracking in the concrete
through excessive expansion, unless a suitable
method is applied to prevent or reduce the
expansion. [Grattan-Bellew 1987] Although the major
factor controlling these reactions is the cement
used, the aggregate also has a significant role, as
discussed below.

Among the several possible aggregate reactions,
the *alkali-silica reacition* is the most common and
was recognized first [Stanton 1940a, 1940b].
Expansion of concrete in structures as a result of
this reaction can exceed 0.5% linearly. Cracks
relating directly to such expansions develop an
irregular yet characteristic pattern, usually called
map cracking, or *pattern cracking*. These cracks are
widest at the surface, possibly as wide as 1 in (25
mm), or even more, but they invariably extend into
the concrete only for short distances, usually only
a few inches, being lost in a maze of ramifying
fractures of microscopic dimensions [Blanks 1955,
Dolar-Mantuani 1969, 1983, Bonzel 1974a]. Also, they
develop more where the concrete is less restrained.
In concrete pavements, for instance, the cracks run
down the road typically longitudinally because
abutting pavement slabs prevent expansion in this
direction. Thus the slab expands more in the lateral
direction causing longitudinal cracks.

Affected concretes in the presence of moisture

frequently develop gelatinous exudations and white amorphous deposits on their surfaces or in their pores. Chemical analyses show this gel to be composed essentially of sodium and potassium silica gels. Also, it is commonly apparent on fractured surfaces that many of the aggregate particles in affected concretes are surrounded by a darkened zone within the margin of the pebbles. This zone represents the depth to which the particle has been attacked by the alkalis.[Idorn 1967]

The literature [Gillott 1973, Mehta 1986] indicates general agreement that the excessive expansion concrete resulting from alkali-silica reaction is caused by the swelling of the sodium-potassium gel. This gel is produced by the reaction between the alkalies deriving from the cement and certain active siliceous minerals and rock types, such as opal, opaline cherts, rocks containing an amorphous or low-crystalline phase of silica (Figs. 9.3 and 9.4) [Powers 1955b, Hester 1958, Hansen 1966] or strained quartz [Buck 1983]. The rates of reactions of these various siliceous minerals may be different, according to the degree of instability of the SiO_2, but it seems certain that the basic reactions are alike, namely, rupture of O—Si—O bonds through reactions with protons and OH^- ions [Hansen 1967, Diamond 1975]. Swelling pressures are believed to be due to the imbibition of water by the gel, and this can exceed the tensile strength of the mortar or concrete [Verbeck 1955]. It appears also that swelling occurs in the outer shell of aggregate particles and that this is responsible for the ultimate breakdown of the concrete.

Expansion of concrete resulting from alkali-silica reaction is influenced by many factors [Locher 1974b, Pike 1960, Verein D. Z. 1973, Hobbs 1979]. For instance, expansion at a given age tends to increase with increase in the alkali content of cement and its fineness, with increase in the cement content (Fig. 9.5), with the reactivity of the aggregate, and with increase in temperature of curing, at least in the range 50-100°F (10-38°C) [Guomundsson 1975]. In addition, German experiments

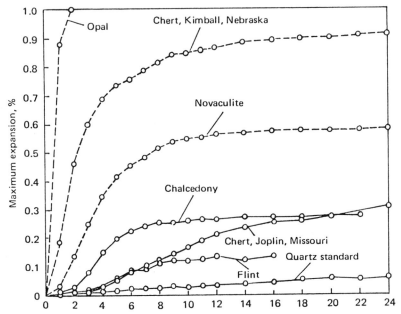

Figure 9.3: Effect of the type of rock on the expansion of portland cement mortar bars containing high-alkali cement, I: siliceous rocks and minerals. From Blanks and Kennedy [1955].

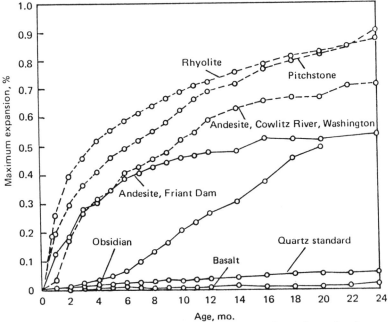

Figure 9.4: Effect of the type of rock on the expansion of portland cement mortar bars containing high-alkali cement, II: volcanic rocks. From Blanks and Kennedy [1955].

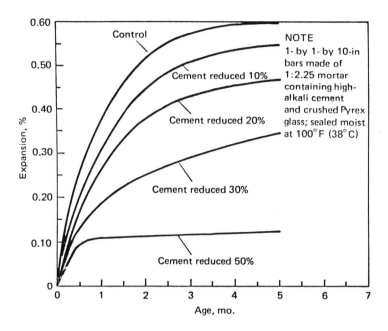

Figure 9.5: Expansion of reactive combinations in mortar bars is reduced by reduced cement content. From Blanks and Kennedy [1955].

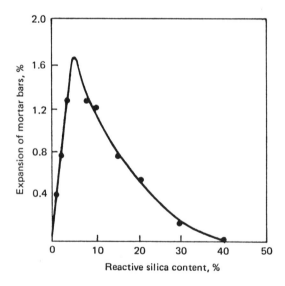

Figure 9.6: Relation between expansion after 224 days and reactive silica content in the aggregate [Vivian 1950].

have shown that higher water-cement ratios and
higher early strengths also increase the expansion
due to alkali-silica reaction.[Bonzel 1989] On the
other hand, many laboratory tests as well as field
experience have demonstrated that the replacement of
a portion of the portland cement with certain
cementitious or pozzolanic mineral admixtures
reduces or even eliminates the excessive
expansion.[Hobbs 1982, Grattan-Bellew 1987] Such
admixtures are, among others, fly ash, silica fume
(see Sect. 4.6), granulated blast-furnace slag,
pumice, calcined clay, etc. The expansion-reducing
effect of pozzolans is probably due to the
production of additional calcium silicate hydrates
that can retain additional alkali; thus, they reduce
the alkali available for reaction with the
aggregate.[Bhatti 1985] Generally, the alkali
content of fly ash is higher than a typical portland
cement. Although these alkalies are acid-insoluble,
thus probably they do not react with aggregate to a
large extent, it is advisable to use fly ash having
a low alkali content.[Carrasquillo 1987]

The rate of expansion increases with decreasing
particle sizes of the reactive aggregate down to 20
μm [Diamond 1974]. Maximum expansion of small mortar
specimens is obtained when the specimens are stored
in saturated air. Also, maximum expansion was
measured if the aggregate contained not less than
2.5 and not more than 5% opal (Fig. 9.6), because
above this level the greater the surface area of the
reactive material, the lower the quantity of the
alkali available per unit area, thus the less
alkali-silica gel could be formed [Verbeck 1955,
Diamond 1974]. It should be noted, however, that
this limit percentage is dependent on particle size
and type of reactive constituent [Blanks 1955,
Highway R.B. 1958b].

In brief, the expansion of concrete as a result
of alkali-silica reaction can be prevented, or at
least reduced, in several different ways:

1. Don't use reactive rock! Or, at least, blend it
   with nonreactive aggregate [ACI 1962].
2. Don't use high-early-strength cement.

3. Use cement containing less than 0.6% alkalies calculated as sodium oxide.
4. Don't use admixture or mixing water that contain alkalies.
5. Use low cement content and low water-cement ratio.
6. Replace a portion of the cement with an adequate amount of certain fine powdered materials containing reactive silica, such as a pozzolan. [Powers 1955b, Mehta 1976].
7. Increase the void space present in the concrete.
8. Eliminate the water supply to the hardened concrete, for instance, by placing a watertight membrane between slab and grade.
9. Do not seal the concrete surface until the concrete has had adequate time to dry out [Campbell 1974].
10. Coat the aggregate particles with an impermeable material.

At present the first six methods, or combinations of these, are the most practical and effective [Woods 1968]. No excessive volume changes have been observed in Europe when reactive aggregates were used with cements of high blast-furnace slag contents (Sect. 2.6) [Wischers 1976b].

Note from Fig. 9.6 that inadequate amounts of reactive silica would increase, rather than reduce, the expansion. More specifically, laboratory tests showed the following [Carrasquillo 1987]:

1.when the available alkali content of fly ash is less than 1.5 percent, its beneficial effect in preventing expansion due to alkali-silica re-action increases as the percentage of cement replaced increases, regardless of class of fly ash, alkali content of cement, alkali content of the fly ash, and aggregate reactivity;

2. when the available alkali content of fly ash is greater than 1.5 percent, there is a *minimum* pecentage of cement replaced, below which the fly ash causes expansions larger than those of a com-parable mixture but without fly ash, and above which the fly ash causes smaller expansions. This

minimum is called the *pessimum limit.*

9.3.2   Other Aggregate Reactions

Another, more recently observed aggregate reaction
capable of producing cracking in concrete is the
alkali-carbonate rock reaction [Swenson 1957,
Highway Res. B. 1958a 1964, Transportation Res. B.
1974, Newlon 1962]. In one case, relatively rare
argillaceous dolomitic limestones used as coarse
aggregate have been shown to react with excessive
cement alkalies in moist environment to produce
large expansion of concrete with or without a
preceding shrinkage period [Dolar-Mantuani 1971]. In
another case, a laboratory series this expansion has
reached 0.1% within 6 weeks [Swenson 1960]. In side
walks and floor slabs the cracks penetrated about
two-thirds of the depth of the slab and fracturing
penetrated the rock particles as well. The areas
surrounded by the cracks appeared to be relatively
sound and unaffected by other environmental factors
[Newlon 1962]. Nondolomitic limestones have also
been found to be reactive.[Buck 1966] Many of the
attacked carbonate rock particles exhibit rims: some
are narrow and dark, others are wider and less well
defined. As a rule, only a very small quantity of
alkali-silica gel can be detected.

Carbonate rocks capable of such reaction possess
a characteristic texture and composition. The
characteristic texture is that in which large
crystals of dolomite are scattered in a fine-grained
matrix of calcite and clay. The characteristic
composition is that in which the carbonate portion
consists of substantial amounts of both dolomite and
calcite, and the acid-insoluble residue contains a
significant amount of clay.

One of the marks of the alkali-carbonate
reaction is that the dolomite constituent of the
aggregate is chemicalty attacked by the alkali of
the cement. This so-called *dedolomitization* reaction
resutts in the formation of brucite, alkali
carbonate, and calcium carbonate [Hansen 1967,
Woods 1968, Hadley 1961]. Also, there is a moisture
uptake by the previousty "unwetted-clay" constituent

of the aggregate. It has been suggested that this latter is responsible for the excessive expansion in the alkali-carbonate rock reaction and that the dedolomitization is necessary only to provide access of moisture to the clay [Swenson 1967].

In brief, expansion of concrete resulting from alkali-carbonate rock reaction increases with increase in the alkali content of the cement, with increase in the maximum particle size, and with increase in the curing temperature. The reaction also requires the presence of moisture. Therefore, this expansion can be prevented or at least reduced in several ways [Swenson 1964]:

1. Don't use reactive rock! Or, at least blend it with a nonreactive aggregate
2. Use cement containing less than 0.4% alkalies calculated as sodium oxide
3. Use minimum cement content and minimum coarse aggregate content consistent with good-quality concrete
4. Use the smallest acceptable maximum size of aggregate
5. Elminate the water supply to the hardened concrete, for instance, by placing a water-tight membrane between slab and grade
6. Do not seal the concrete surface until the concrete has had adequate time to dry out
7. Cat the coarse particles with an impermeable material.

Note that inhibitors normally found effective for alkali-silica reaction cannot be relied upon to reduce expansion in the alkali-carbonate rock reaction.

A third kind of alkali-aggregate reaction has been reported in Nova Scotia by Canadian researchers [Gillot 1973, Duncan 1973a, 1973b, 1973c]. Also, there are harmful aggregate reactions in concrete that take place without the presence of alkalies. For instance, CaO, which can occur in blast-furnace slags, reacts with water; or, ferrous sulfides (pyrite, etc.) in aggregate particles may react with water in the presence of oxygen and calcium hydroxide. Such particles expand individually, producing uneven stress, and those that are close to an exposed surface produce popouts [Hansen 1966, Highway Res. B. 1963].

In contradistinction with the aggregate reactions discussed above. there are reactions between aggregate and cement paste that are *favorable* for the concrete. The characteristic of these is that the reaction is not accompanied by excessive volume changes. For instance, a limited chemical reaction is believed to be beneficial with respect to the bond between the aggregate and paste [Lerch 1956]. The lime-silica reaction at high temperatures is also a useful one that has been utilized for many years in the high-pressure steam curing of concrete.

9.3.3   Test Methods and Evaluation of the Results

Because of the complexity of aggregate reactivities in concrete, presently there is no sure laboratory method to determine whether or not a given aggregate will cause excessive expansion [Mather 1975, Heck 1983]. Perhaps the best criterion is to rely on the service record of the aggregate. If such a record is not available, certain laboratory methods may be used, but these provide only estimates of the *potential* reactivity of the aggregate; they cannot provide quantitative information with adequate reliability on the degree of reactivity to be expected or tolerated in service. Quite a few such laboratory tests have been recommended in the literature [Bredsdorff 1960, Anon. 1974]. Several of these have been standardized by ASTM and are described below.

ASTM C 227-87 and C 289-87 were developed for testing the potential reactivity of cement-aggregate combinations, primarily the *alkali-silica* reaction. C 227-87 specifies the measurements at various ages of the increases in the length of 1-in x 1-in x 11 1/4-in (25-mm x 25-mm x 285-mm) mortar bars containing the suspected cement-fine aggregate combination in 1:2.25 weight proportion, during storage over water at $100^{O}F$ ($30^{O}C$). The cement to be used in such tests should contain alkalis more than 0.6%. If coarse aggregate is to be tested, it should be crushed first to fine aggregate and a specified grading should be produced for the motar bar test.

The mortar-bar test provides information on the likelihood of harmful reactions occurring, and it has shown good correlation with field experience. On the other hand, usually 6 months are needed before judgment on the aggregate can be made, since lack of expansion in a short period is no guaranty against a delayed reaction.

ASTM C 289-87 covers chemical determination of the potential reactivity of an aggregate with alkalies in portland cement. The reactivity is indicated by the amount of reaction during 24 hr at $176^{O}F$ ($80^{O}C$) between 1 N NaOH solution and the pulverized aggregate in question. More specifically, the reduction in alkalinity and the amount of dissolved silica are measured. The main advantage of this test method is that it can be made quickly and although it is less reliable than the mortar-bar test, it provides helpful information when results of the more time-consuming tests are not available [Lerch 1956, Mielenz 1948, 1958, Chaiken 1960, Lerch 1950]. For an improvement of the reliability, it is recommended that interpretations based on this chemical method be correlated with ASTM C 295- 85.

ASTM C 295-85 is a recommended practice for petrographic examination of concrete aggregates with an optical microscope. Research laboratories often supplement optical examination by additional procedures, such as x-ray and electron diffraction, differential thermal analysis, infrared spectros-copy, and so on. All these methods, of course, are useful not only for testing the potential alkali-silica reaction but also any other potential reactivity of aggregate in concrete by identifying the mineral constituents.[Dolar-Mantuani 1983] However, again, they cannot establish definitely whether or not a given mineral will result in excessive expansion.

A fourth method, ASTM C 342-79(1985), was

designed for testing the potential volume change of cement-sand-gravel combinations. The essence of the method is to measure the linear expansion of mortar bars that are similar to the bars specified in C 227. These bars are subjected to variations of temperature and water saturation during storage under prescribed conditions.

Another pertinent standard method is ASTM C 586-69(1986), which tests the potential alkali reactivity of *carbonate rocks*. The method covers the periodic determinations of the expansive characteristics of cylinders of carbonate rocks with an overall length of 1.38 in (35 mm) and a diameter of 0.35 in (9 mm), while immersed in a solution of 1 *N* NaOH at room temperature. The observable length change occurring during such immersion indicates the general level of reactivity of rocks. This test was originally intended as a research screening method rather than for specification enforcement and to supplement data from field service records, petrographic examinations, and tests of aggregate in concrete. More recent experiments [Buck 1975], however, have demonstrated that such length changes of the rock greater than 0.1% by 84 days can be considered as a routine indicator for potentially harmful rocks from the standpoint of cement-carbonate aggregate reaction. It is also advantageous that the method is simple. The time for indication of the potential expansive reactivity of certain carbonate rocks may be as short as 2 weeks [Swenson 1960, Hadley 1961].

Finally, there is an ASTM method (C 441-89) that covers the determination of the effectiveness of mineral admixtures in preventing the excessive expansion caused by reaction between aggregates and alkalis in portland cement. The method is similar to the one described in C 227 except that here crushed Pyrex glass is used as a reactive aggregate with cement-admixture combinations in the mortar bars.

According to ASTM C 33-86, cement-aggregate combinations that, when tested by the method in C 227-87, show expansions greater than 0.1% at 6 months usually should be considered capable of harmful reactivity. When 6-month results are not available, combinations should be considered potentially capable of harmful alkali-silica reactivity if they show expansions greater tnan 0.05% at 3 months.

If the test method specified in ASTM C 289-87 provides an R_c reduction in alkalinity exceeding 70 millimoles/liter, the aggregate is considered potentially reactive if the S_c dissolved silica is greater than R_c; if the provided R_c is less than 70, the aggregate is considered potentially reactive if S_c is greater than 35 + 0.5R_c (Fig. 9.7).

As far as the results of petrographic examinations (C 295) are concerned, C 33-86 also states that reactive materials, such as several forms of silica, silica-rich volcanic glass, and certain zeolites, can render an aggregate harmfully reactive when present in quantities as little as 1.0% or even less.

If cement-aggregate combinations tested by the method specified in ASTM C 342-79(1985) show expansions that equal or exceed 0.20% at an age of 1 yr, they may be considered unsatisfactory for use in concrete that is exposed to wide variations of temperature and degree of saturation witn water.

There are no limit values specified for the results of the test method in ASTM C 586-69(1986).

9.4 DETERIORATION OF AGGREGATES BY CHEMICAL ATTACKS FROM OUTSIDE

Portland cement concrete is vulnerable to several types of chemical attack that are harmless as far as the aggregate is concemed. These are typically the reactions that attack the free calcium hydroxide in the concrete developed during the hydration of the portland cement. Examples are the deteriorating attacks of water solutions of sulfates, or those of sugars. There are other types of chemical reactions that can attack the aggregate, such as those produced by acids; however, the same reactions, as a rule, deteriorate simultaneously the cement paste of the concrete even more.[Popovics 1987d]

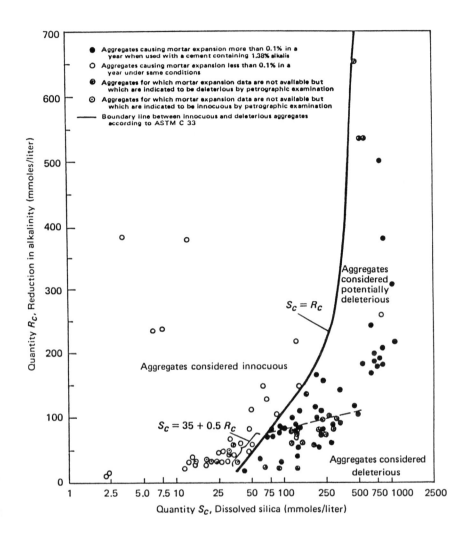

Figure 9.7: Illustration of division between innocuous and deleterious aggregates on the basis of Reduction in Alkalinity Test according to ASTM C 289-71. Reprinted by permission of the American Society for Testing and Materials, Copyright.

Thus, in the overwhelming majority of the practical cases, the aggregate does not have a decisive role in the deterioration of portland cement concrete produced by chemical attack from outside sources.

10

Geometric Properties of Aggregates

SUMMARY

The shape of an aggregate particle is irregular from a geometric standpoint, therefore it cannot be defined, or measured, adequately and its influence on the properties of concrete cannot be evaluated precisely. Particle shape is controlled by two properties: roundness (angular, round, etc.), and sphericity (elongated, flat, etc.). The surface texture can be polished or dull, smooth or rough, and so on. Particle shape and surface texture are the results of the interaction of the nature, structure, and texture of the rock. Equidimensional particles are preferred to flat or elongated particles in concrete aggregates. The water requirement of round and/or smooth particles is less than that of angular and/or rough particles, especially for fine aggregates. On the other hand, coarse aggregate roughness increases the tensile and flexural strengths of the concrete. No ASTM methods exist by which the roundness, sphericity, or surface texture can be measured adequately. There are, however, practical methods that measure a combination of these aggregate properties. One such method is the measurement of the percent of voids in a one-size aggregate fraction; another is the orifice flow test. Both of these tests can be used to estimate the effect of aggregate on the mixing-water requirement of the concrete.

The *particle size* of irregular shapes cannot be characterized either by a single number without ambiguity. This has led to an approximate but simple method, the *sieve test,* for the definition and determination of particle sizes. The sieve test can also provide the distribution of particle sizes, called *grading,* in an aggregate sample by separating the particles according to size with sieves of gradually decreasing openings. Results accurate enough for most aggregate testing can be obtained by sieve series where the ratio of the openings of consecutive sieves is 2. Aggregate grading can influence many properties of concrete. The grading expresses the percentages of the various particle sizes, or size fractions, either by absolute volume or, when the specific gravity of all size fractions is the same, by weight. In most cases these percentages are presented in a graphical form providing the so-called grading curve either on a total percentage-passing or total percentage-retained basis in a rectangular system of "percentage" versus "particle-size" coordinates. A simplified form of grading characterization is by a *grading point* in a triangular diagram.

Another kind of grading characterization is *numerical characterization.* Here the grading is represented essentially by a single number. This number can be either an average of the particle-size distribution, that is, an average particle size, or an empirically established number. These numbers characterize the grading to the extent that averages can characterize distributions. That is, a numerical characterization cannot provide information about the details of the particle size distribution, but rather characterizes the coarseness of the grading as a whole. Experimental results support the fineness-modulus method as the best for concrete aggregates among the available numerical characterizations. The various types of average particle size of a grading can be determined mathematically or graphically from the particle-size distribution.

10.1 SHAPE AND SURFACE TEXTURE OF PARTICLES

The shape and surface texture of particles are aggregate properties that have not been adequately defined, therefore they cannot be measured adequately and their influence on the properties of the concrete cannot be evaluated precisely. A relatively large literature dealing with these problems is listed and partially discussed in the literature by Mather [Mather 1966]. He states that *particle shape* is controlled by two relatively independent properties: *roundness* and *sphericity*.

Roundness, or angularity, is the property the measure of which depends on the relative sharpness of the edges and corners of the particle. Roundness can be defined numerically as the ratio of the average radius of curvature of the corners and edges of the particle to the radius of the maximum inscribed circle. The use of descriptive terms, however, is more common. For instance:

Angular: little evidence of wear on the particle
     surface
Subangular: evidence of some wear, faces untouched
Subrounded: considerable wear, faces reduced in area
Rounded: faces almost gone
Well rounded: no original faces

Sphericity is the property that measures, depends on, or varies with the ratio of the surface area of the particle to its volume, the relative lengths of its principal axes or those of the circumscribing rectangular prism, the relative settling velocity, and the ratio of the volume of the particle to that of the circumscribing sphere.[Harr 1977] If, for instance, two of the principal axes are much shorter than the third axis, the particle is called elongated; if two of these axes are much longer than the third one, the particle is called flat. Approximate methods have also been recommended in B.S. 812 and elsewhere to characterize the sphericity numerically either by the ratios of the three principal axes, or by combinations of these. One example is

$$\text{Sphericity} = \frac{d}{a} \qquad\qquad (10.1)$$

another is

$$\text{Sphericity} = \sqrt{\frac{bc}{a^2}} \qquad\qquad (10.2)$$

where a, b, and c are the longest, intermediate, and shortest axes of the particle, respectively, and d is the diameter of a sphere of the same volume as the particle. Note, however, that inherent ambiguities of how to measure the a, b, and c values make these characteristics questionable [Tons 1968, Lees 1964].

Descriptive terms are given to particle-shape classifications in B.S. 812, such as rounded, irregular, flaky, elongated, or angular; in this specification, however, roundness and sphericity are not treated separately.

Surface texture is the property the measure of which depends on the relative degree to which particle surfaces are polished or dull, smooth or rough, and the type of roughness. Usually the particles are simply described as rough, fairly rough, smooth or very smooth; or, in B.S. 812, as glassy, smooth, granular, rough, crystalline, or honeycombed; or by some other series of qualitative terms. Numerical characterization of the surface texture, such as the ones developed for metals, has not been applied to aggregates.

Mather points out that the degree of sphericity, roundness, and smoothness of a particle are the result of the interaction of the nature, structure, and texture of the rock of which the particle consists. For instance, rocks having closely spaced partings or cleavages in one or two directions tend to yield flat or elongated particles, that is, shapes of low sphericity. Also, the design and operation of the crushing equipment influence the sphericity of the crushed particles; generally, the greater the reduction ratio, the lower the sphericity. Roundness results primarily from wear to which the particle has been subjected. Surface texture depends on hardness, grain size, pore structure, and texture of the rock, and the degree to which forces acting on the particle surfaces have smoothed or roughened it. Hard, dense, fine-grained materials will generally have smooth fracture surfaces.

Equidimensional particles are generally preferred to flat or elongated particles for use as concrete aggregates because they produce tighter packing when consolidated and, therefore, require less water, cement paste, or mortar for a given degree of workability of concrete. Pertinent experimental evidence for this is discussed in Mather's paper [Mather 1966]. Another way to put this is to say that it is easier for packed spheres to move relative to each other in fresh concrete than it is for flat or elongated particles to move under similar circumstances. Excess amounts of elongated and flat particles can also adversely affect the strength and durability of concrete, as they tend to be oriented in one plane, with water and air voids forming underneath. Well-rounded particles may also be expected to require less water or cement paste for equal workability than angular particles of equal sphericity and similar surface texture. This is particularly true for fine aggregates [Bloem 1963, Wills 1974]. As far as the influence of the surface texture is concerned, the water requirement of smooth particles is less than that of rough particles. As shown in Fig. 10.1, the greater friction between the rough aggregate particles may have something to do with this [Jaeger 1971, Clanton 1952]. On the other hand, experimental evidence indicates that certain types of (coarse) aggregate roughness increase the tensile and flexural strengths of the concrete by improving the bond between the cement paste and the aggregate surface. The same bond is decreased, however, by the presence of coating formed, for instance, by very fine particles on the aggregate surface. (Section 9.2.1) This, in turn, reduces the concrete strength, especially the tensile and flexural strengths. Note that it is much easier to reduce the paste-aggregate

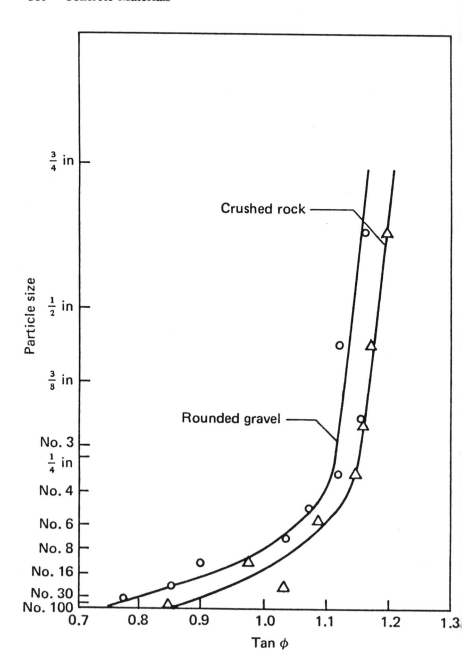

Figure 10.1: Variation of the coefficient of friction between aggregate particles with particle shape and particle size [Clanton 1952].

bond than to improve it. [Popovics 1987g]

Apart from extreme and special cases [Popovics 1976d], particle shape and surface texture (except for coating) are not critical properties affecting the quality of a concrete aggregate. They are important enough, however, for us to mention that whenever concrete of good tensile or flexural strength is needed, general practice is to use well-rounded concrete sand and crushed coarse aggregate of high sphericity.

No ASTM methods exist by which quantitative determinations of particle roundness, sphericity, or surface texture can be made. Published test methods for the determination of the roundness are based on measurements of radii of curvature; on particle outline; and so on. Methods for testing the sphericity of aggregate are based on the determinations of the three principal axes of particles with a caliper [Walz 1936], as specified by the Corps of Engineers, and many others, or with an optical method [Plagemann 1964a]; on thickness and length measurements with gauges (B.S. 812); on settling-velocity determinations and sieving; on sieving the aggregate through sieves of differing opening shapes [Schiel 1948]; on the number of particles of a given size that can be held in a container of a given volume [Stern 1937]; on the rate at which (fine) aggregates of a single size fraction run through a specified orifice [Rex 1956, King 1972]; on volume measurements of particles and determination of the actual specific surface or surface area of the aggregate by wax coating the coarse particles or by the permeability method for small particles [Loudon 1952-1953, Shacklock 1957]; and so on. Surface texture can be measured by certain methods developed for metals; with a microscope [Wright 1955] or electron microscope [Plagemann 1964b, Ozol 1972, Barnes 1979]; and so on [Orchard 1973b]. Also, there are methods that were developed for measuring, say, roundness (or angularity), but the test results are affected by the sphericity, or by the surface texture, or by both [Herden 1960]. One such method is the measurement of the denseness of packing of a one-sized fraction of the given aggregate under well-defined conditions [Hughes 1966, Shergold 1953]. The denser the packing, the more favorable is the particle shape. For instance, an *angularity number*, *AN*, defined by Shergold for coarse particles is

$$AN = v - 33 \qquad (10.3)$$

where *v* is the percent of voids in a coarse aggregate sample when the sample of a narrow size range is *compacted* in a prescribed manner in a specific container. The value of *AN* is 0 for an ideal aggregate, 5 for a typical gravel, 10 for an angular crushed stone, and 30 for crushed reef shell [Popovics 1973b]. The cut surfaces of concretes made with the last three aggregates are shown in Figs. 13.2 and 13.3. They provide not only two-dimensional pictures of the respective particle shapes, but also Fig. 13.2 reveals that highly elongated or flat aggregate particles have a preferred orientation in the concrete in which they have their minimum dimension in the direction of the compaction.

Another, similar particle shape characteristic is the *angularity factor*, *AF*, proposed by Powers [Powers 1968], which can be written in the following form:

$$AF = \frac{1}{\psi} = 1 + 0.0444(v_1 - 42) \qquad (10.4)$$

where ψ = sphericity factor = specific surface diameter/volume diameter (see Section 10.2)

v_1 = the percent of voids in a coarse or fine aggregate sample of narrow size range when the sample is in a *loose* state.

In the same sense, results of the British Standard compacting factor test (B.S. 1881:70) carried out on dry aggregate can also be used for measuring angularity [Kaplan 1958].

A test method based on the packing principles has been incorporated in B.S. 812 for coarse particles.

The percentage-of-voids concept appears applicable for particle-shape characterization of fine aggregates, too [Wills 1974].

Recently the internal structures of concretes were analyzed made with aggregates of differing gradings and differing particle shapes [Popovics 1973b]. The analysis is discussed in Chap. 13 in the section, "Aggregate Grading and the Internal Structure of Concrete." Nevertheless, it is fitting to point out here that the results of this analysis support the angularity-number and angularity-factor concepts for the characterization of particle shape. It should be noted, however, that good correlation was obtained between the results of the void-content test and those of the orifice-flow test, indicating that either could be used to predict the effects of fine and coarse aggregates on concrete mixing-water requirement [Bloem 1963, Gaynor 1968a, Wills 1967, Malhotra 1964]. The equipment required for the void-content test is less expensive.

None of the major specifications contains any specific requirements regarding roundness and surface texture. Specifications of the Corps of Engineers provide an upper limit of 25% on flat and elongated particles in coarse aggregate. B.S. 1984 permits not more than 40% flat particles for 1 1/2 in (38.1 mm) nominal-size aggregate, and 35% for 1-, 3/4-,and 1/2-in (25.4-, 19.1-, and 12.4-mm) nominal-size aggregates, as determined by B.S. 812.

10.2 PARTICLE SIZE

If particles were perfect spheres, the radius, or the diameter, would determine the size completely. For any other particle shape the term "size" cannot be characterized by a single number without ambiguity. For instance, in the simple case of cubes, the size can be, equally logically, the edge length, or the diagonal of the cube face, or the body diagonal. This situation is even more ambiguous in the case of irregular shapes that aggregate particles always have. In other words, the size of the particle is not independent of its shape, unless it is specified what size-dependent property (surface, volume, etc.) is to be assessed. A consequence of this is that the optimum particle size distribution, that is, the grading that is optimum for a given concrete, is also dependent on the particle shape, as will be shown later.

Some of the attempts to provide theoretical size characteristics for particles are as follows [Powers 1968]:

Volume diameter, d_v is the diameter of a sphere of the same volume V as the irregular particle; that is,

$$d_v = \left(\frac{6V}{\pi}\right)^{1/3} = 1.24 \ V^{1/3}$$

(10.5)

Surface diameter d_s is the diameter of a sphere of the same surface area S as the irregular particle; that is,

$$d_s = \sqrt{\frac{S}{\pi}} = 0.564\sqrt{S}$$

(10.6)

Specific surface diameter d_{sp} is the diameter of a sphere of the same specific surface, that is, surface area per unit solid volume, as the irregular particle; that is,

$$d_{sp} = \frac{6V}{S}$$

(10.7)

Stokes's diameter is the diameter of a sphere, in a given fluid, of the same free-falling terminal velocity as the irregular particle.

A different concept is the *particle-packing volume,* which is the volume that a piece of aggregate occupies in a mass of other particles. This is a function of the solid volume, internal pores, and surface roughness of the particle [Tons 1968]. Other definitions for particle size can be found in the literature [Cadle 1965, Hatch 1957].

The lack of a single practical definition of size led technical people to an approximate but simple method of defining and determining the particle sizes of aggregates. This method is based on a test carried out by means of sieves of conveniently selected openings. The size of a particle, regardless of its shape, is designated as d_i if that particle just can be passed through a sieve with an opening of d_i size. It is more practical as well as rational to use a series of differing sieves for particle size determination: If a particle passes the sieve with d_i opening but is retained on the next smaller sieve with a net opening of d_{i-1}, then it is said that the particle size is within d_{i-1} and d_i, or that the particle belongs to the d_{i-1} - d_i size fraction. For instance, the 1/2 - 3/4-in (12.5 - 19.1-mm) fraction contains essentially particles that passed the 3/4-in sieve but were retained on the 1/2-in sieve, that is, particles the nominal "sizes" of which are between 1/2 and 3/4 in, respectively. It is in this sense that the term "particle size" is used in this book, unless it is indicated otherwise.

Theoretically, the *maximum aggregate size*, or *maximum particle size*, present in a d_{i-1} - d_i fraction is d_i. Sometimes this d_i is called the nominal maximum size. In practice, however, there are particles in the aggregate fraction larger than the nominal maximum size; this is called the oversize. Therefore, the maximum size of aggregate D is defined as the "largest size aggregate particles present in sufficient quantity to affect the physical properties of concrete; generally designated by the sieve size on which the maximum amount permitted to be retained is 5 to 10 percent by weight" [ACI 1985d]. Similar consideration is applied to the term "minimum particle size" d_{min} except that the amount permitted to pass through the d_{i-1} sieve (the undersize) is usually different.

Sometimes it is convenient to use the relative

particle size d_r. This is defined as $d_r = d/D$. For instance, if $d_r = 0.25$, this means that this d size is 25% of the maximum particle size D.

From the maximum particle sizes of an aggregate fraction one can also calculate an average particle size d_{ave} by using pertinent simple formulas. This average may be the arithmetic average of the maximum and minimum sizes $0.5(d_i + d_{i-1})$, the geometric average $(d_i d_{i-1})^{1/2}$, the harmonic average $2d_i$-$d_{i-1}/(d_i + d_{i-1})$, or the logarithmic average $0.5 \log (d_i d_{i-1})$. When the ratio d_i/d_{i-1} is 2 or less, the first three of these averages do not differ much.

Another method for the estimate of, say, the average volume diameter of a fraction (for instance, for the calculation of sphericity) is to count a known weight of particles and compute the equivalent-sphere diameter by substituting V, the average solid-volume per particle, into the definition of d_v given above as Eq. (10.5). Similar calculations can be performed for the average surface diameter or the average specific surface diameter.

The average particle size d_{ave} gives more accurate numerical characteristic of the coarseness, or fineness, of the aggregate fraction than either D or d_{min}, especially when the fraction limits are wide. This is important since many properties of fresh concrete, and to a certain extent those of hardened concrete, are influenced directly or indirectly by the coarseness of the aggregate.

There are no standard methods for the determination of the theoretical sizes, such as the volume or the surface diameters, or the average sizes of particles.

10.3 SIEVES AND SCREENS

The terms *sieve* and *screen* can be defined as a plate or sheet or a woven cloth or other device with regularly spaced apertures of uniform size mounted in a suitable frame or holder for use in separating

material according to size. The term sieve usually
apples to an apparatus in which the apertures are
square, and the term screen to an apparatus in which
the apertures are circular.

As was mentioned before, test sieves (screens)
are the accepted means for determining the particle
size or the upper and lower limits of a size fraction
of an aggregate within wide ranges. The U.S. standard
sieve series is specified in ASTM E 11-87, and the
perforated-plate sieve series in ASTM E 323, so that
the openings of two consecutive sieves in the series
form, in most cases, the fixed ratio of $(2)^{0.25}$ =
1.19. Other sieve series are specified in B.S.410 and
in the international standards ISO R 20/3 and R 40/3.

For aggregate testing, square-hole sieves are
used except when the aggregate specification is
based on round-hole apertures. Results with a given
square aperture and with the same diameter round
aperture are not compatible [ASTM Comm E-1 1969]
because of the interference of the different
sphericities of the aggregate particles. A rule of
thumb is that , for equivalency, d_{sieve} = $0.8d_{screen}$
[Rothfuchs 1934]. More refined approximate
procedures have also been recommended [Luhr 1971].
It has also been found that results accurate enough
for most aggregate testing can be obtained by sieve
series where the ratio of the apertures of the
consecutive sieves is 2. However, it is not
compulsory to use such a mathematically regular
sieve series: In many cases certain sieves can be
omitted without hurting the accuracy of the result,
or supplementary sieves can be added to the series
to obtain the required accuracy, depending on the
nature of the test and the application of the
aggregate. A sieve series suitable for the majority
of aggregate testing is presented in Table 10.1.

Test sieves for aggregates used to be described
by the size of the apertures in inches for larger
sizes, and by the number of openings per linear inch
for sieves smaller than about 1/4 in (about 6 mm).
Thus, a No. 100 sieve has 100 openings per inch,
that is, 100 X 100 openings in each square inch.
More recently, sieve sizes are designated by the
nominal aperture size expressed in millimeters (mm)

Table 10.1: Test Sieves for Aggregates[a,b]

| | Sieve designation | | ISO designation[c] | |
| | | | ISO R 565 1967 R 20/3 | ISO R 40/3 |
Standard	Alternative	Aperture, in		
50.8 mm	2 in[d]	2.00	50 mm[f]	
38.1	$1\frac{1}{2}$	1.50		37.5 mm
25.4	1[d]	1.00	25[f]	
19.0	$\frac{3}{4}$	0.750		19
12.7	$\frac{1}{2}$[d]	0.500	12.5[f]	
9.51	$\frac{3}{8}$	0.375		9.5
4.76	No. 4[e]	0.187		4.75
2.38	No. 8	0.0937		2.36
1.19	No. 16	0.0469		1.18
595 μm	No. 30	0.0234		600 μm
297	No. 50	0.0117		300
149	No. 100	0.0059		150
74	No. 200	0.0029		75

[a]From ASTM Committee E-1, *Manual on Test Sieving Methods*, STP 447, 1969.   Reprinted by permission of the American Society for Testing and Materials, Copyright.

[b]For complete specifications, including permissible variations from nominal apertures and wire diameters and method of checking and calibration, see the most recent ASTM Designation E 11.

[c]Issued by the International Standards Organization, Geneva, Switzerland.

[d]These sieves are not in the standard fixed ratio series but have been included because they are in common usage.

[e]These numbers (4–200) are the approximate numbers of openings per linear inch, but it is preferred that the sieve be identified by the standard designation size of opening in millimeters or micrometers.

[f]These sieves are supplementary sizes included because of being in common usage.

or in micrometers (μm).

10.4 GRADING

Any batch of aggregate contains particles of different sizes. The distribution of the particle sizes d within the batch is called *grading* or *gradation*. It is usually expressed as the quantities of the different size fractions of the particles as the percentages of the total quantity of the batch, or, even more frequently, as a cumulative percentage of the particles that are smaller (or larger) than each or a series of sieve openings [ACI 1985d]. If the aggregate contains every size fraction between the minimum and maximum particle size, the grading is called *continuous* grading; when one or more fractions are missing, the grading is called *gap* grading. The particle-size distribution have averages, that is, average particle sizes, that also characterize the grading to extent that the averages can characterize distributions.

The significance of the aggregate grading is that it influences directly many technically important properties of fresh concrete, such as consistency and segregation, and to a certain extent the properties of hardened concrete as well [Baker 1973]. For instance, an aggregate with 1-1/2-in. (38.1-mm) maximum particle size that contains twice as much sand as the optimum quantity (therefore being heavily oversanded) would require about 50-75% more water to obtain the required workability than the optimum grading; this, in turn, would require either a corresponding increase in the cement content of the concrete, or would result in a strong reduction of the concrete strength. The consequences of using a seriously undersanded grading can be even more harmful because such gradings cannot produce workable and dense concrete [Kennedy 1940].

Several direct and indirect methods have been recommended for the testing of the aggregate grading. The indirect methods intend to estimate certain effects of the grading on the properties of concrete by testing some grading-affected properties of the aggregate. Such methods are, for instance, the determination of the water retention of, or the percent of voids in an aggregate [Wise 1952, Hummel 1959, Walker 1930, Schwanda 1956, King 1962, Fowkes 1974, Furnas 1931, Anderegg 1931, McGeary 1961, Tynes 1968, Rothfuchs 1935, Campen 1940, Joisel 1952]. Since the indirect methods for grading testing and evaluations are not used in practice, and rarely in laboratories, for concrete aggregates, these methods will no be discussed here.

10.5 SIEVE TEST

The direct methods are intended to determine the particle-size distribution by dry or wet sieving, sedimentation, optical methods, etc. Of all the methods, sieving is by far the most commonly used testing procedure for the determination of the grading of an aggregate. Here a weighed sample of dry aggregate is separated into size fractions by shaking it through a series of sieves of progressively smaller openings. The weights of each size fraction are determined and their quantities are calculated as percentages of the total quantity of the sample. Dry sieving alone is usually satisfactory for routine testing of aggregates. However, when accurate determination of the total amount passing the no.200 (75-μm) sieve is desired, first the sample should be tested by wet sieving, according to ASTM C 117-87. Details of the performance of the sieve analysis have been standardized in many specifications, including ASTM C 136-84a and B.S.812. The standard deviation for the repeatability of the sieve test on coarse aggregate, under routine conditions, is approximately 0.4%.

The percentages of the size fractions should express the relative solid volumes occupied by the fraction. Therefore, the percentages should be by absolute volume, not only theoretically but also in practice, whenever the specific gravities of all the size fractions are not the same. Examples of this are the combination of normal-weight sand with either a lightweight or a heavyweight coarse aggregate, or, in certain cases, an aggregate consisting of lightweight fine and coarse particles, since smaller lightweight particles tend to have higher bulk specific gravities than larger particles of the same origin. When the specific gravities of the fractions do not differ significantly, the percentage by weight is the same as, or at least close enough to, the percentage by absolute volume. Since the calculation of the former is very simple, the percentages of grading are expressed in most cases on a weight basis. Percentages that do not express the relative solid volume of the aggregate fraction, such as percentage by loose volume, or by the number of particles, are not used in concrete technology.

There are three usual forms to express the grading by the percentages of the size fractions. These are the "total (or cumulative) percentage passing" (y_{TPP} or simply TPP), the "total (or cumulative) percentage retained" (y_{TPR} or TPR), and the "percentage passing-retained" (Δy or PPR). Note that TPP = 100 - TPR at any point; that is, the two types of percentages are not independent of each other. Thus, it is not worthwhile to calculate both of them. Example 10.1 illustrates these three forms.

Example 10.1 The grading of a 5000-g aggregate sample consisting of particles of the same specific gravity was tested by sieve analysis. The sieves used are shown in the first column of Table 10.2. The weights of the size fractions retained on each sieve, obtained by the analysis, are presented in the third column of the table.
Determine the grading by calculating the percentage passing-retained for each fraction, the total percentage retained, and the total percentage passing. The solution is shown in Table 10.2.

10.6 GRADING CURVES

Any of the three sets of percentages calculated in Example 10.1 provides the grading of the aggregate. Nevertheless, for the sake of an easier evaluation,

Table 10.2: Data for Example 10.1

Sieve	Particle size (d), mm	Weight retained, g	Percentage passing-retained (PPR), by weight	Total percentage retained (TPR), by weight	Total percentage passing (TPP), by weight
2-in	50.00	0		0.00	100.00
			6.20		
1½ in	37.50	310		6.20	93.80
			18.46		
1-in	25.00	923		24.66	75.34
			14.84		
½-in	12.50	742		39.50	60.50
			12.96		
⅜-in	9.50	648		52.46	47.54
			11.86		
No. 4	4.75	593		64.32	35.68
			12.26		
No. 16	1.18	613		76.58	23.42
			11.72		
No. 30	0.60	586		88.30	11.70
			7.40		
No. 100	0.15	370		95.70	4.30
			4.30		
Pan	0.00	215		100.00	0.00
Total		5000	100.00		

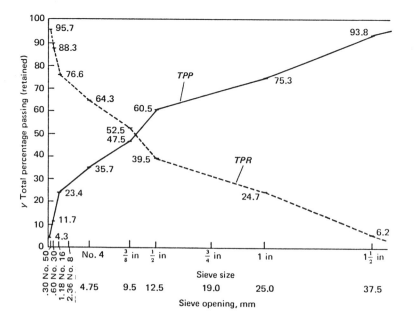

Figure 10.2: Example of grading curves on total percentage passing (TPP) and total percentage retained (TPR) basis, respectively, in the linear system of coordinates. (See Example 10.1.)

the percentages are customarily presented in graphical form. The most frequently used form for this presentation is the so-called *grading curve* or *sieve curve*. Figure 10.2 shows the sieve curves of the grading from Example 10.1 on the total percentage passing and retained basis, respectively. One can see that the TPP and TPR curves are reflections of each other about the 50% horizontal line.

Unfortunately, the curves in Fig. 10.2 do not show clearly enough the all-important fine portion of the grading because the sieve sizes are too close to each other, whereas the coarsest size portion is unnecessarily overextended. One can overcome this problem by using a suitable nonlinear scale, such as $\sqrt{d}$, $\log(nd)$, and so on, that extends the fine portion of the scale and reduces the coarse portion. Such a regularly used scale is the logarithmic scale, more specifically the $\log(10d)$ scale, for the axis of particle sizes instead of the linear scale in Fig. 10.2. This case is illustrated in Fig. 10.3, where the same two sieve curves of the Example 10.1 are presented in this semilog system.

Theoretically, the curves in Figs. 10.2 and 10.3 are the graphical forms of the function $y = f(d)$. This formula provides the total percentage passing or retained quantities as a function of the particle size d. Despite the relative simplicity of the grading curves obtained by the sieve test, the numerical forms of the corresponding formulas representing these curves, even approximate ones, would by uselessly complicated. Besides, such formulas would provide little additional information. Therefore, such formulas are not used in concrete technology for empirical sieve curves. On the other hand, more or less simple formulas can be developed for theoretical sieve curves that fulfill certain mathematical conditions and/or certain conditions concerning the grading. A simple example is the so-called Fuller parabola. This and several other such formulas, will be discussed in Chapter 12.

The percentage passing-retained (PPR) data are shown graphically in the PPR (or Δy) versus $\log(10d)$

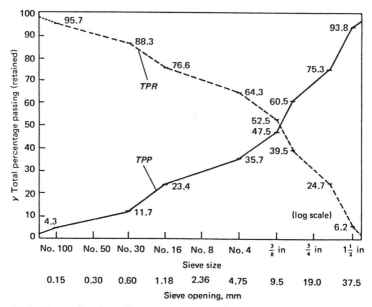

Figure 10.3: Example of grading curves on total percentage passing (TPP) and total percentage retained (TPR) basis, respectively, in the semilogarithmic system of coordinates. (See Example 10.1.)

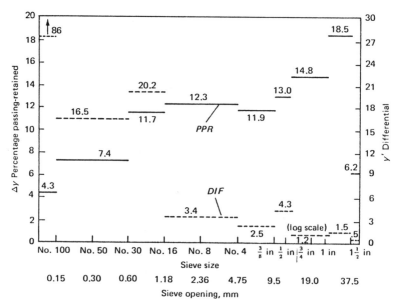

Figure 10.4: Example of grading curves on percentage passing-retained (PPR) basis, and as a differential curve (DIF), respectively, in the semilogarithmic system of coordinates. (See Example 10.1.)

system in Fig. 10.4. The ordinates of this curve can be defines as

$$\Delta y_{i+1,i} = y_{i+1} - y_i \qquad (10.8)$$

where $\Delta y_{i+1,i}$ = ordinate of the PPR curve in the $d_{i+1} - d_i$ size fraction

y_{i+1}, y = ordinates of the TPP curve at sizes d_{i+1} and d_i respectively

d_{i+1}, d_i = consecutive sizes in the sieve series used; for TPP, $d_{i+1} > d_i$.

Another curve designated as DIF and derived from Eq.10.8 is also plotted in Fig. 10.4. The ordinates of this curve approximate the ordinates of the differential curve of the TPP and are calculated as

$$y'_{i+1,i} = \frac{y_{i+1} - y_i}{d_{i+1} - d_i} \qquad (10.9)$$

where $y'_{i+1,i}$ is an ordinate of the DIF curve in the $d_{i+1} - d_i$ size fraction.

Equations 10.8 and 10.9 are also valid when the grading is as TPR, but this case $d_{i+1} < d_i$.

It is, of course, an approximation that the points of the curves in Figs. 10.2 and 10.3, and especially in Fig. 10.4, are connected by straight lines. The error resulting from this approximation can be reduced by applying Sheppard's corrections for grouping [Spiegel 1961].

It was pointed out earlier that the particle-size distribution expressed as percentage by the number of particles is not used in concrete technology. Thus, when a test method, for instance an optical method, provides the grading in terms of the number of particles, this is converted into percentages by weight (or by absolute volume) as follows [Herdan 1960]:

$$y' = f'(d) = ad^3 h'(d) \qquad (10.10)$$

where $f'(d)$, $h'(d)$ = differential function of the cumulative particle-size distribution (TPP or TPR) by weight or absolute volume, and that by the number of particles, respectively

d = particle size parameter depending on the units, particle shape, and whether the $f(d)$ distribution is by weight or by absolute volume

10.7 GRADING REPRESENTATION IN TRIANGULAR DIAGRAM

Another method utilizes trilinear, that is, triangular diagrams for the graphical presentation of grading. This method is suitable primarily for the case when the aggregate is considered to be made up of three size fractions. This is equivalent to the case when the grading curve is made up of only two intermediate points between the minimum and maximum particle sizes. This is obviously a simplification at the expense of accuracy. Fortunately, three properly selected size fractions can provide quite reliable information concerning the concrete-making properties of the grading for the majority of practical cases; this is particularly true when the grading criteria are aptly selected, as discussed in Chapter 12. Whenever a higher accuracy is needed in the grading, the triangular method can be generalized for four or five size fractions with the application of computer graphics.

The triangular diagram is usually, but not necessarily, an equilateral triangle. Each of the three sides of the triangle represents a scale for the quantity of one of the size fractions from 0 to 100%, as shown in Fig. 10.5. A point inside the triangle represents a set of percentages for each of

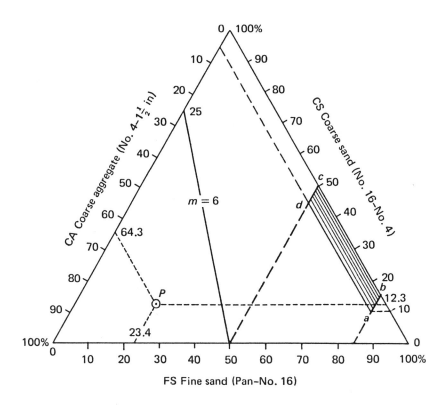

Figure 10.5: Grading representation in triangular diagram. (See Example 10.2.) The assumed values of the fineness moduli are for the three aggregate components as follows: $m_{FS} = 3$, $m_{CS} = 5$, and $m_{CA} = 9$.

the three size fractions, or conversely, each
combination of the three size fractions has a
corresponding point in the triangular diagram. This
point is called the *grading point* and is the
equivalent of the grading curve in the rectangular
system. The position of this point can be defined in
several ways, for instance, by the three distances
between the point and the sides of the triangle
measured on lines parallel to the sides. Only two of
the three distances are independent.

An example for the application of triangular
diagrams has been given in Fig. 2.4, another in Fig.
8.3.Further examples as well as the theory of
triangular diagrams can be found in the literature
[Popovics 1964b]. An additional example below
illustrates the application of a triangular diagram
for grading representation.

Example 10.2 Let the three basic size
fractions of the grading be fine sand with the
size limits of pan and sieve No. 16 (0 and 1.18
mm), coarse sand with No. 16 and No. 4 (1.18 and
4.75 mm), and coarse aggregate with No. 4 and 1
1/2 in (4.75 and 37.5 mm).

Then the grading of Example 10.1 that
consists of 23.4% fine sand, 12.3% coarse sand,
and 64.3% coarse aggregate is represented by the
grading point *P* in Fig. 10.5.

Similarly, all the gradings that contain
fine sand in the quantity 50-85%, coarse sand
in the quantity 10-50%, and, consequently, coarse
aggregate in the quantity 0-5%, are represented
by the area *abcd* in Fig. 10.5. Incidentally,
these are grading limits that are specified for
fine concrete aggregate in ASTM C 33-90. The line
of fineness moduli *m* = 6 in the same figure will
be explained later in Example 11.5.

Triangular diagrams have been used for grading
representation since the early years of concrete
technology [Feret 1892]. This representation is
particularly useful (1) to illustrate the effect of
grading on certain aggregate properties, as has been
shown in Chap. 8 for percent of voids in aggregates;
and (2) to calculate the required blending

proportions for two or more aggregates to comply with quite sophisticated grading specifications, as will be discussed in Chapter 13.

10.8 NUMERICAL CHARACTERIZATION OF GRADING

In the previous paragraphs the grading was represented, or characterized, either graphically, by the grading curve or the grading point, or by the results of the sieve test, that is, by a set of numbers providing the coordinates for the graphical representation. Another kind of grading character- ization is called *numerical characterization*. In this latter case, only one, or perhaps two or three numbers, are used to characterize the concrete- making properties of the grading; therefore, such a numerical method cannot provide information about the details of the particle-size distribution, but rather, characterizes the coarseness or fineness of the grading as a whole.

The numerical characterization of grading has two major advantages: (1) the grading is expressed in the same way as many other concrete properties, such as cement content, water-cement ratio, strength, slump, and so on; therefore, more meaningful com- parisons are feasible between differing gradings on this basis than on the basis of graphical representation; and (2) the numerical grading characterization makes possible the development of formulas for relationships between certain properties of fresh or hardened concrete and the aggregate grading. Therefore, it is a simple but effective tool for the technical or economical optimization of the composition of concrete.

The natural approach for numerical charac- terization seems to be the use of statistical concepts, such as the average particle sizes, and in addition, as a refinement, perhaps a variance of the total grading. After all, grading is the distribution of particle sizes. Therefore, the determinations of several of the most common average particle sizes will be discussed .

Another approach is to establish useful

numerical grading characteristics on an empirical basis. Quite a few empirical numbers have been recommended. Definitions of some of these for one-size grading are presented below. Illustrative numerical values are presented in Table 10.3 for several size fractions with $d = 0.5(D + d_{min})$ substitution. In the formulas, d is the particle size as defined earlier, expressed in millimeters, and d'' is the particle size expressed in inches. Note that, in contrast to bituminous concretes, the voids content of the aggregate (Section 8.4) is not used any more as a grading characteristic in the technology of portland cement concrete.

1. Fineness modulus, m, by Abrams [Abrams 1925]

$$m \approx 3.32 \log d + 3.75 \approx \qquad (10.11)$$

$$\approx 3.32 \log 13.5d \approx \qquad (10.12)$$

$$\approx 3.32 \log d'' + 8.43 \qquad (10.13)$$

2. Specific surface of aggregate s, by Edwards [Edwards 1918]

$$s = \frac{6000}{d} = \qquad (10.14)$$

$$= \frac{236}{d''} \qquad (10.15)$$

The unit of s in these formulas is m^2/m^3 (0.305 ft^2/ft^3).

3. The i index, by Faury [Faury 1958]. These indexes were obtained empirically. No general mathematical form has been provided for the i index in terms of particle size.

4. The A values, by Kluge [Kluge 1949]. These values were also obtained empirically. No formula is available to provide the A values in terms of particle size.

5. Distribution number λ, by Solvey [Solvey 1949]

Table 10.3: Various Numerical Characteristics of Aggregate Fractions[a]

Limits of size fraction		d, mm	d_e, in	s, m²/m³	m	e	λ	ρ	A	i	f_s
Sieve	mm										
3–1½ in	75–37.5	56.25 (100)	2.21 (100)	106.7 (100)	9.56 (100)	.638 (100)	9.33 (100)	2.53 (100)	.020 (100)	0.06 (100)	−2.5 (−100)
1½–¾ in	37.5–19.0	28.25 (50.2)	1.11 (50.2)	212.4 (199)	8.56 (89.5)	1.01 (158)	11.34 (122)	3.57 (141)	.035 (175)	0.12 (200)	−2.0 (−80)
¾–⅜ in	19.0–9.5	14.25 (25.3)	0.561 (25.3)	421.1 (395)	7.58 (79.3)	1.59 (250)	13.95 (150)	5.03 (199)	.055 (275)	0.19 (317)	−1.0 (−40)
⅜ in–No. 4	9.5–4.75	7.12 (12.7)	0.280 (12.7)	842.7 (790)	6.58 (68.2)	2.53 (396)	17.49 (187)	7.12 (281)	.075 (375)	0.27 (450)	1.0 (40)
No. 4–No. 8	4.75–2.36	3.56 (6.32)	0.140 (6.32)	1685 (1580)	5.58 (58.4)	4.02 (631)	22.3 (239)	10.07 (398)	.096 (480)	0.39 (650)	4.0 (160)
No. 8–No. 16	2.36–1.18	1.77 (3.15)	0.0697 (3.15)	3390 (3178)	4.57 (47.8)	6.40 (1003)	29.2 (313)	14.28 (564)	.116 (580)	0.55 (917)	7.0 (280)
No. 16–No. 30	1.18–0.60	0.89 (1.58)	0.0350 (1.58)	6742 (6321)	3.58 (37.4)	10.10 (1584)	39.0 (418)	20.14 (796)	.160 (800)	0.70 (1167)	9.0 (360)
No. 30–No. 50	0.60–0.30	0.45 (0.80)	0.0177 (0.80)	13333 (12500)	2.60 (27.2)	15.94 (2500)	53.5 (573)	28.32 (1119)	.24 (1200)	0.75 (1250)	9.0 (360)
No. 50–No. 100	0.30–0.15	0.225 (0.40)	0.0089 (0.40)	26667 (25000)	1.60 (16.7)	25.30 (3969)	76.8 (819)	40.06 (1583)	.35 (1750)	0.79 (1317)	7.0 (280)
No. 100–pan	0.15–0	0.075 (0.13)	0.0030 (0.13)	?	0	—	?	?	?	1.0 (1667)	2.0 (80)

[a]Values in parenthesis are presented relative to the numerical characteristics of size fractions 3–1½ in (75–37.5 mm).

$$\lambda = \frac{1000}{\log^3 1000d} = \qquad (10.16)$$

$$= \frac{1000}{\log^3 25,400d''} \qquad (10.17)$$

6.   Stiffening coefficient ρ, by Leviant [Leviant 1966]. Although these coefficients were obtained empirically, the following equation provides approximate values of ρ for aggregates of round particle shape:

$$\rho \approx \frac{19}{\sqrt{d}} \approx \qquad (10.18)$$

$$\approx \frac{3.77}{\sqrt{d''}} \qquad (10.19)$$

7. Water requirement values e, by Bolomey [Bolomey 1930]. For aggregates of round particle shape:

$$e = \frac{9}{\left(d_i d_{i+1}\right)^{1/3}} \approx \qquad (10.20)$$

$$\approx \frac{9}{(d^2)^{1/3}} \qquad (10.21)$$

$$= \frac{1.04}{(d''^2)^{1/3}} \qquad (10.22)$$

8. Equivalent mean diameter d_e by Hughes [Hughes 1960]

$$d_e = d'' \qquad (10.23)$$

9. Surface index f_s, by Murdock [Murdock 1960]. These indexes again are empirical. No general mathematical form has been provided for f_s.

Any of the numerical grading characteristics in Table 10.3 can be used to calculate the corresponding numerical characteristic for the complete grading as the weighted average of the numerical characteristics of the size fractions, as follows:

$$h = 0.01 \; \Sigma \; p_i h_i \tag{10.24}$$

where h_i, h = a numerical grading characteristic for

complete grading , respectively

p_i = relative quantity of the ith fraction

as percent of the total aggregate, by absolute volume

Table 10.3 shows that the 10 numerical grading characteristics form three groups: (1) the values of d, d_e and the reciprocal value of s increase approximately proportionally with the particle size increase; (2) the values of e, i, A, λ, ρ, and the reciprocal values of m decrease with the increase of the particle size but, more important, the rate is about half of the rate of change in group (1), m having the lowest rate; and (3) the f_s values increase slightly with the increase of particle size up to size No. 16 (1.18 mm), then decrease, again slightly, assuming negative (?) values for aggregate sizes greater than 3/8 in (9.5 mm).

The primary criterion for the acceptability of any numerical grading characterization is how closely differing gradings of identical numerical characteristic have identical concrete-making properties. Experimental data seem to indicate that the fineness modulus approximates this requirement best, as will be shown in Chapter 12. The specific surface will also be discussed because of its popularity.

10.9 AVERAGE PARTICLE SIZE OF THE COMPLETE GRADING

Another kind of grading characterization can be performed with the average size of the complete particle-size distribution. When the f(d) equation is available, providing the grading of the aggregate, which is the case with certain theoretical gradings, then any of the average particle sizes can be calculated either from the f(d) equation or from the f'(d) equation. Several pertinent formulas are presented below for the case when the grading is expressed on the total percentage passing basis.

1. Arithmetic average

$$\bar{d_a} = 0.01 \int_{d_0}^{D} d\, f'(d)\, dd = \tag{10.25}$$

$$= D - 0.01 \int_{d_0}^{D} f(d)\, dd \tag{10.26}$$

2. Logarithmic average

$$\overline{\log d_l} = 0.01 \int_{d_0}^{D} \log(d)\, f'(d)\, dd = \tag{10.27}$$

$$= \log \frac{D}{d_0} - 0.004343 \int_{d_0}^{D} \frac{f(d)}{d}\, dd \tag{10.28}$$

3.Geometric average

$$\overline{d_g} = \text{antilog } \overline{d_l} = 10^{\overline{d_l}} =$$

$$= \frac{D}{d_0} \exp\left(-0.01 \int_{d_0}^{D} \frac{f(d)}{d} \, dd\right) \quad (10.29)$$

4. Harmonic average

$$\frac{1}{\overline{d_h}} = 0.01 \int_{d_0}^{D} \frac{f'(d)}{d} \, dd = \quad (10.30)$$

$$= \frac{1}{D} + 0.01 \int_{d_0}^{D} \frac{f(d)}{d^2} \, dd \quad (10.31)$$

In these formulas,

$f(d)$ = equation of the grading curve on total percentage passing by weight or absolute volume basis

$f'(d)$ = differential curve of $f(d)$

D = theoretical maximum particle size, that is, $f(D) = 100$

d_{min} = theoretical minimum particle size, that is, $f(d_{min}) = 0$

Note that if, say, arithmetic averages of two different gradings are equal, then any other averages, say, geometric, of the same gradings must be different. Also, for the same distribution:

$$\overline{d_a} \geq \overline{d_g} \geq \overline{d_h} > \overline{d_l}$$

The following example illustrates the use of these formulas.

Example   10.3 Calculate the arithmetic,
logarithmic, geometric, and harmonic average
particle sizes of the grading represented by the
so-called Fuller parabola:

$$f(d) = 100 \sqrt{\frac{d}{D}} \tag{10.32}$$

Solution

$$f'(d) = \frac{50}{\sqrt{dD}}$$

1. Arithmetic average; $d_{min} = 0$
From Eq. (10.25),

$$\overline{d_a} = 0.5 \int_0^D \frac{d}{\sqrt{dD}}\, dd = \frac{D}{3}$$

or from Eq. (10.26),

$$\overline{d_a} = D - \int_0^D \sqrt{\frac{d}{D}}\, dd = \frac{D}{3} \tag{10.33}$$

2. Logarithmic average; $d_{min} = 0.1$ mm.
The logarithm is not defined for $d_{min} = 0$. Thus,
it is assumed here that $d_{min} = 0.1$ mm. This
assumption results in an error in the calculated
value of d_l which, however, is not serious when
Eq. (10.28) is used.
From Eq. (10.28), when the origin of the
semilog system is at $d = 0.1$ mm,

$$\overline{d_l} + 1 = \log(10D) - 0.4343 \int_{0.1}^D \frac{1}{\sqrt{dD}}\, dd =$$

$$= \log (10D) - 0.8686 \left(1 - \frac{1}{\sqrt{10D}}\right) \quad (10.34)$$

3. Geometric average; $d_{min} = 0.1$ mm.
From Eq. (10.29),

$$\overline{d_g} \approx 0.135D \text{ antilog} \frac{0.8686}{\sqrt{10D}} \quad (10.35)$$

4. Harmonic average; $d_{min} = 0.1$ mm.
Here again the assumption of $d_{min} = 0.1$ mm results in an error in the calculation of d_h which, however, is not serious when Eq. (10.30) is used.
From Eq. (10.30),

$$\frac{1}{\overline{d_h}} \approx 0.5 \int_{0.1}^{D} \frac{1}{d\sqrt{dD}} \, dd =$$

$$= \frac{1}{\sqrt{0.1D}} - \frac{1}{D} \quad (10.36)$$

$$\overline{d_h} \approx \frac{D}{\sqrt{10D} - 1} \quad (10.37)$$

When the f(d) or f'(d) equation of the particle-size distribution is not available, which is the usual case, the average particle size still can be determined from the results of the sieve test. Three such methods are presented below.

The first method calculates the average size from the PPR values as follows:

$$\overline{d_c} = 0.01 \sum_{i=1}^{n} \Delta y_i \overline{d_i} \quad (10.38)$$

where d_c = any of the average particle sizes of
the complete grading

Δy_i = relative quantity of the ith size
fraction as percentage passing-
retained by weight or absolute volume

d_i = the pertinent average particle size of
the ith size fraction

n = number of Δy_i size fractions used.

The second method for the determination of any average particle size of the complete grading utilizes the grading curve. This is based on the principle that a definite integral, such as the ones in Eqs. 10.25 - 10.31, can be represented by an area. For instance, Eq. 10.26 is represented in Fig. 10.2 by the area above the TPP grading curve, the boundaries of which are the TPP curve, the y vertical axis, and the 100% horizontal line. If this area is designated by A_{lin}, then the d_a arithmetic average size of the grading in Fig. 10.2 is

$$\overline{d_a} = 0.01 A_{lin} \qquad (10.39)$$

Similarly, the logarithmic average particle size of that grading can be obtained approximately from the A_{log} area above the TPP curve in the semilogarithmic system of Fig. 10.3. Here the boundaries are again the TPP curve, the y vertical axis, and the 100% horizontal line. When the origin of the semilog system of coordinates is at $d = 0.1$ mm, the logarithmic average size is

$$\overline{d_l} \approx 0.01 A_{log} - 1 \qquad (10.40)$$

A graphical method for the determination of the d_g geometric average particle size of a grading is illustrated in Fig. 10.6 based on Eqs. 10.29 and 10.40. The essence of this method is that the vertical line that makes the A_u and A_o areas equal in

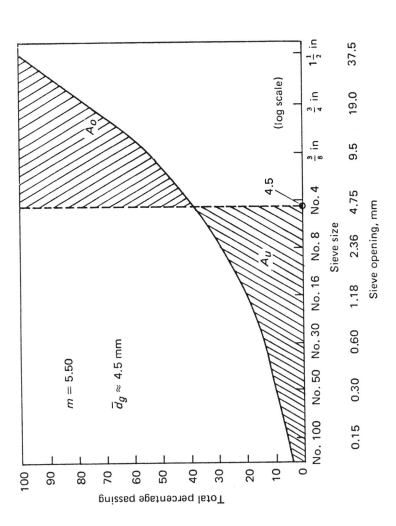

Figure 10.6: Graphical determination of the geometric average particle size $\overline{d}_g$. The vertical line that makes the A_u and A_o areas equal intersects the horizontal axis in this system at $\overline{d}_g$.

the system of Fig. 10.6 will show the approximate value of the d_g on the horizontal d axis.

Equation 10.30 is represented approximately by the area A_h under the DIF curve in the semilogarithmic system in Fig. 10.4. The boundaries of this area are the DIF curve, two vertical lines at the d_{min} and D sizes, respectively, and the horizontal log d axis. Therefore,

$$\frac{1}{d_h} = 0.023A_h'$$

(10.41)

The advantage of using the appropriate area for the determination of any of the average particle sizes is that the area method is applicable regardless of whether the equation of the sieve curve is available.

Methods for the area A determination are discussed in the next chapter.

The third method for the determination of any of the average particle sizes without the knowledge of f(d) or f'(d) is the application of the definition of the particular average for discrete variables for the results of the sieve test or for the coordinates of the sieve curve.

It follows from the definition of the average concept that many different gradings can have the same average particle size. Nevertheless, the average-particle-size concept is significant, since experience has shown that a change in the quantity of a given size fraction in a grading can be counterbalanced within wide limits by appropriate changes in the quantities of other size fractions, so that the concrete-making properties of the aggregate will remain essentially unchanged. This means that the influence of aggregate grading on concrete properties is controlled more by the coarseness of the grading as a whole than by details of the size distribution.

11

Fineness Modulus and Specific Surface

SUMMARY

The oldest and best-supported numerical grading characteristic is the *fineness modulus* for concrete aggregates. The original definition by Abrams, which agrees with the ASTM definition, is as follows: The fineness modulus is obtained by adding the total percentages of an aggregate sample retained on each of a specified Tyler sieve series and dividing the sum by 100. There are several other methods for determination of the fineness modulus, including using the total percentage passing values; from the A_{log} area, which is the area above the TPP, or under the TPR, grading curve in a semilog system; and from the equation of the grading curve; from the fineness moduli and relative quantities of the size fractions. Any of these methods is applicable not only for aggregates but also for mixtures of aggregate and cement.

It is important from a theoretical point of view that the fineness modulus represents an average particle size: It is proportional to the logarithmic average particle size of the grading, and as such, it is a fundamental parameter of the particle-size distribution. The practical basis for the acceptance of the fineness modulus for grading characterization and evaluation is the experimentally justified claim that gradings having the same fineness modulus will require the same quantity of water to produce mix of the same consistency and give concrete of the same strength within certain practical limits. For a given concrete there is an optimum value for the fineness modulus. The pertinent optimum values are given in tables, or may be calculated by formulas. The larger the maximum particle size, the larger the optimum fineness modulus, therefore the larger the optimum logarithmic average particle size, but the smaller becomes the optimum of the relative average particle size. The weakness of the fineness modulus method is that it is somewhat liberal: It may judge a grading as good even when the amount of fine and very fine particles is too much or too little.

The other frequently mentioned numerical grading characteristic is the *specific surface* of the aggregate particles. This can be calculated by methods similar to those applied for the fineness modulus (formulas, area, etc.). The specific surface also has a statistical interpretation: It is proportional to the reciprocal value of the harmonic average of the particle sizes. Experimental results do not support adequately the specific surface method for grading characterization and evaluation of concrete aggregates. However, the weakness of the specific surface method, and that of the fineness modulus method, can be reduced by using both of these numerical characteristics simultaneously for grading evaluation. This is the *D-m-s method*.

11.1   FINENESS MODULUS

Probably the best-known numerical characteristic of grading in concrete technology is the fineness modulus. This will be designated by *m* for aggregates and *M* for aggregate-cement mixtures.

Abrams was the first to develop and recommend the fineness modulus for the grading characterization of concrete aggregate [Abrams 1925]. His definition, which essentially agrees with the definition of ASTM C 125-88, is: The fineness modulus is a factor obtained by adding the total percentages of an aggregate sample retained on each of the following (Tyler) sieve series: No. 100 (0.15 mm), No. 50 (0.30

mm), No. 30 (0.60mm), No. 16 (1.18 mm), No. 8 (2.36 mm), No. 4 (4.75 mm), 3/8 in (9.5 mm),3/4 in (19.0 mm), 1 1/2 in (37.5 mm), and so on, and dividing the sum by 100. The mathematical form of this definition is

$$m = 0.01 \sum_{1}^{n} b_i \qquad (11.1)$$

$$m = n - 0.01 \sum_{1}^{n} y_i \qquad (11.2)$$

where m = fineness modulus

b_i = 100 − y_i = total percentage of the fraction of the aggregate retained on the ith member of the specified Tyler series of sieves

y_i = total percentage of a sample of the aggregate passing through the ith member of the specified Tyler series of sieves above

n = number of nonzero b_i values

When the sieve series used in the sieve analysis is not the one specified above, the missing b_i or y_i values for Eqs. 11.1 and 11.2 can be obtained by graphical interpolation. (See Example 11.1)

Since both of these equations represent numerical integrations, the fineness modulus can also be calculated with a good approximation as follows [Abrams 1918, Hummel 1959]:

$$m \approx \frac{A_{log}}{30.1} \qquad (11.3)$$

where A_{log} is the area above the TPP grading curve in the semilogarithmic system of Fig. 10.3, the boundaries of which are the TPP curve, the y vertical axis at d = 0.1 mm, and the 100% horizontal line.

(The same area is under the grading curve when TPR is used.) The use of Eq. 11.3 is illustrated in Example 11.2.

This area method has certain advantages compared to the summation method of Eqs. 11.1 and 11.2. First, it is visual. In addition, determination of the area does not require the use of any specific series of sieves, or the knowledge of the equation of the sieve curve; therefore it is convenient for computerization when an irregular sieve series was used for the test. Most important, however, a comparison of Eq. 11.3 with Eq. 10.40 and Eqs. 10.27 and 10.28 indicates that the fineness modulus is proportional to the logarithmic average particle size of the grading [Popovics 1952, 1955, 1962]. Therefore:

$$m \approx \frac{\overline{d_l} + 1}{\log 2} \qquad (11.4)$$

$$m = 3.322 \log (10\overline{d_g}) \qquad (11.5)$$

That is [Popovics 1950, 1962a, 1966b], for $d_{min} \leq 0.1$ mm

$$m \approx 0.0332 [100 \log (10D) - 0.4343 \int_{0.1}^{D} \frac{f(d)}{d} dd] \qquad (11.6)$$

$$\approx 0.0332 \int_{0.1}^{D} \log (10d) \, f'(d) \, dd \qquad (11.7)$$

where the symbols are identical with the symbols of Eqs. 10.27 and 10.28.

It follows also that

$$\overline{d_g} = 10^{\overline{d_l}} = \qquad (11.8)$$

$$= 0.1 \times 2^m \qquad\qquad (11.9)$$

and

$$d_l = 0.301m - 1 = \qquad\qquad (11.10)$$

$$= \log \overline{d_g} \qquad\qquad (11.11)$$

Finally, if an aggregate is a blend of several size fractions, the fineness modulus of the aggregate blend can be calculated, in accordance with Eq. 10.24, as follows:

$$m = 0.01 \sum_{1}^{n} p_i m_i \qquad\qquad (11.12)$$

where m = fineness modulus of the aggregate blend
  m_i = fineness modulus of the ith size fraction
  p_i = relative quantity of the ith fraction as
      percent of the total aggregate, by
      absolute volume.
  n = number of size fractions in the aggregate
      blend

Values of m_i for several size fractions are presented in Table 10.3.

Three supplementary comments can be mentioned here:

1. The volume diameter of a particle is not uniquely defined by the limiting sieve sizes; it also depends on the shape of the particles [Powers 1968].Therefore, a given fineness modulus does not correspond to the same average size unless all aggregates compared are composed of particles having the same shape. A crushed material, for instance, will have a larger average volume diameter than an aggregate composed of round particles having the same fineness modulus.

2. As discussed in Chapter 7, gradings of unsieved crushed stones, sands and gravels, soils, dusts and powders, often follow the logarithmic-normal distribution [Loveland 1927, Drinker 1925, Hatch 1957, Kolmogorov 1941, Jaky 1933, Jaray 1955]. Therefore, for such materials the application of the logarithmic average, that is, the fineness modulus concept, is quite fitting.

3. Any of the methods represented by Eqs. 11.1 - 11.12 is applicable not only for aggregates but also for mixtures of aggregate and cement, in which case the fineness modulus M refers to the grading of the "total solid volume" of concrete.

4. When the relative quantities of the components, or fractions, in the blend are expressed by percentages by weight p_{wi}, they should be converted to p_i percentages by absolute volume for proper use of Eq. 11.12. In case of two components, the following formula applies:

$$m = \frac{p_w G_2}{p_w G_2 + (100-p_w) G_1} m_1 + \frac{(100-p_w) G_1}{p_w G_2 + (100-p_w) G_1} m_2 \qquad (11.12a)$$

where G_1 and G_2 = specific gravity of Component 1 and Component 2, respectively

p_w = blending proportion of Component 1 to the total blend, in percent by weight

m_1 and m_2 = fineness modulus of Component 1 and Component 2, respectively

Example 11.1 Determine the fineness modulus of the grading in Example 10.1 by using Eq. 11.1.

The required values of b_i are obtained from Figure 10.3 by graphical interpolation, and presented in Table 11.1. Therefore,

$$m = \frac{576.1}{100} \approx 5.76$$

Example 11.2 Calculate the fineness modulus of the grading in Example 11.1 by using Eq. 11.3.

From Fig. 10.3 again, the area A_{log} and then

Table 11.1

Sieve size	b_i, TPR
No. 100	95.7
No. 50	91.5
No. 30	88.3
No. 16	76.6
No. 8	70.0
No. 4	64.3
$\frac{3}{8}$ in	52.5
$\frac{3}{4}$ in	31.0
$1\frac{1}{2}$ in	6.2
3 in	0
	$\Sigma\, b_i = 576.1$

Table 11.2: Calculation of the Area $A_{\log}$ and the Fineness Modulus from a Grading Curve

Size fraction $d_i - d_{i+1}$, mm	$\log (d_{i+1}/d_i)$	$\dfrac{0.5(y_{i+1} + y_i)}{TPR}$	$\Delta A_{\log}{}^a$
0.1–0.15	0.1761	97.85	17.23
0.15–0.60	0.6021	92.0	55.39
0.60–1.18	0.2937	82.45	24.22
1.18–4.75	0.6048	70.45	42.61
4.75–9.5	0.3010	58.4	17.58
9.5–12.5	0.1192	46.0	5.48
12.5–25.0	0.3010	32.1	9.66
25.0–37.5	0.1761	15.45	2.72
37.5–50.0	0.1299	3.10	0.39
			$A_{\log} = 175.28$

Fineness modulus: $m \approx \dfrac{175.28}{30.1} = 5.82$

[a] $\Delta A_{\log} = 0.5(y_{i+1} + y_i) \log (d_{i+1}/d_i)$, although the partial areas $\Delta A_{\log}$ can also be calculated from the following formula: $\Delta A_{\log} = 0.5(y_{i+1} - y_i) \log (100 d_{i+1}\, d_i)$.

the value of m can be calculated (Table 11.2).

Note that the fineness modulus calculated from A_{log} is necessarily somewhat greater than the value calculated by Eq. 11.1.

Example 11.3 Calculate the fineness modulus of the grading characterized by the Fuller parabola by using Eq. 10.34 in Example 10.3:

$$m \approx 3.32(\overline{d_l} + 1) \approx$$

$$\approx 3.32 \left[\log (10D) - 0.8686 \left(1 - \frac{1}{\sqrt{10D}} \right) \right] \quad (11.13)$$

Example 11.4 By using Eq. 11.6, calculate the fineness moduli of the four gradings that are characterized as (a) one-size, (b) linear, (c) logarithmic and (d) parabolic particle-size distributions.

The pertinent grading equations as well as the calculated moduli are shown in Table 11.3.

Example 11.5 1. A coarse aggregate of fineness modulus $m_{CA} = 9$ is blended with a coarse sand of fineness modulus $m_{CS} = 5$ in 1:3 portion by absolute volume.

Calulate the fineness modulus of the aggregate blend.

Since $p_{CA} = 25\%$ and $p_{CS} = 75\%$ by absolute volume, it follows from Eq. 11.12 that

$$m = 0.25 \times 9 + 0.75 \times 5 = 6.0$$

2. As will be shown later, the modulus $m = 6.0$ can be produced with the 25 and 75% blending proportions. However, when the number of components is greater than two, infinitely many blending proportions form a straight line in the triangular

Table 11.3: Fineness Moduli and Specific Surfaces of Aggregate Fractions for Various Distributions Within the Particle-Size Limits from d_{min} to D^a

Type of particle-size distribution within d_{min} and D^b	Equation of grading $f(d)$ TPP	Fineness modulus m	Specific surface for spherical shape s, m^2/m^{3d}
One size^c	—	$3.32 \log(10d_l) + 0.43$	$\dfrac{6{,}000}{d_l}$
Linear	$100\dfrac{d - d_{min}}{D - d_{min}}$	$3.32\left[\log(10D) + \dfrac{d_{min}}{D - d_{min}}\log\left(\dfrac{D}{d_{min}}\right) - 0.43\right]$	$13{,}820\dfrac{\log(D/d_{min})}{D - d_{min}}$
Logarithmic	$100\dfrac{\log(d/d_{min})}{\log(D/d_{min})}$	$1.66 \log(100Dd_{min})$	$2{,}600\dfrac{1/d_{min} - 1/D}{\log(D/d_{min})}$
Parabolic	$100\dfrac{\sqrt{d} - \sqrt{d_{min}}}{\sqrt{D} - \sqrt{d_{min}}}$	$3.32\left[\log(10D) + \dfrac{\sqrt{d_{min}}}{\sqrt{D} - \sqrt{d_{min}}}\log\left(\dfrac{D}{d_{min}}\right) - 0.87\right]$	$\dfrac{6{,}000}{\sqrt{Dd_{min}}}$

^{a}From [Palotas 1933].
$^b d_{min} = 0.1$ mm should be used in these equations whenever $d_{min} < 0.1$ min.
$^c D = d_{min} = d_1$.
$^d 1\ m^2/m^3 = 0.305\ ft^2/ft^3$.

diagram. An example for this is given in Fig. 10.5, where the three components have fineness moduli of 9 (coarse aggregate), 5 (coarse sand), and 3 (fine sand), respectively; the full range of the blending proportions for those components that provide a fineness modulus of 6 are represented by the line m = 6. Note that the 25% CA and 75% CS proportions above are also shown on the coarse aggregate axis.

11.2 EXPERIMENTAL JUSTIFICATION OF THE FINENESS MODULUS

The theoretical basis of the fineness modulus is that as an average-particle-size concept, it is a fundamental parameter of the particle-size distribution. It is not clear presently why the logarithmic (geometric) average; perhaps because it considers the effect of the fine particles with just the right proportion.

The practical basis for the application of the fineness modulus for grading charcterization and evaluation of concrete aggregate is the claim that gradings having the same fineness modulus "will require the same quantity of water to produce mix of the same plasticity and give concrete of the same strength" as long as the maximum particle size is kept constant and the amount of fine particles is enough, but not too much, to assure adequate workability and denseness of the concrete [Abrams 1925]. This claim is well supported by experimental results not only by Abrams but others as well [Palotas 1933, Hummel 1930, Bolomey 1930]. For instance, Fig. 11.1, after Palotas [Palotas 1933], shows 13 different gradings of identical fineness modulus m = 5.8. Measured values of unit weight, flow, and compressive strength of non-air-entrained concretes made with these gradings and identical cement and water contents are shown in Table 11.4. It can be seen that all the gradings provided practically the same unit weight, flow, and compressive strength, with exception of the obviously unacceptable No.9, which contains only about 3% of particles passing through the 0.1-in (2.5-mm) sieve.

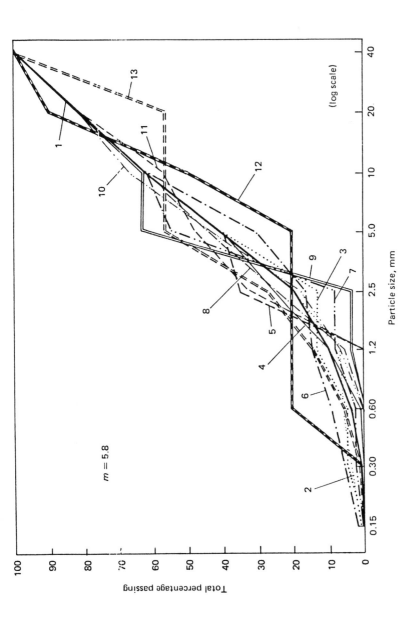

Figure 11.1: Various gradings with identical fineness modulus and maximum particle size [Palotas 1933]. (1 in = 25.4 mm.)

Table 11.4: Unit Weight, Flow, and Compressive Strength of Concretes Made with Various Gradings of Identical Fineness Modulus[a,b]

No. of Grading	Unit weight of concrete, lb/ft³ (kg/dm³)	Flow, in (cm)	Compressive strength, psi (kg/cm²)
1	150.0 (2.40)	18.9 (48)	5190 (365)
2	150.0 (2.40)	18.1 (46)	5162 (363)
3	150.0 (2.40)	18.1 (46)	5005 (352)
4	148.8 (2.38)	18.9 (48)	5261 (370)
5	146.9 (2.35)	20.5 (52)	4906 (345)
6	150.6 (2.41)	17.7 (45)	5034 (354)
7	149.4 (2.39)	19.3 (49)	4991 (351)
8	148.8 (2.38)	20.5 (52)	5162 (363)
9	146.2 (2.34)	18.1 (46)	4337 (305)
10	148.1 (2.37)	19.7 (50)	4977 (350)
11	147.5 (2.36)	20.5 (52)	4892 (344)
12	150.6 (2.41)	19.7 (50)	4849 (341)
13	149.6 (2.39)	18.5 (47)	4735 (333)

[a]From [Palotas 1933].
[b]Fineness modulus: $m = 5.8$. Cement content: 507 lb/yd³ (300 kg/m³) of concrete. W/C = 0.545 by weight.

This and other pertinent measurements provide strong experimental support for the applicability of the fineness modulus, not for details of the particle-size distribution, but for the evaluation of the effects of grading as a whole on certain important properties of concrete. Such test results also show that almost any sand of unorthodox grading can produce a workable and strong concrete provided that it is blended with a suitably graded coarse aggregate in the proper proportion [Powers 1968, Gaynor 1963b, Popovics 1976d].

11.3 OPTIMUM FINENESS MODULI

Another important aspect is that for a given maximum particle size and particle shape of the aggregate, as well as cement content, consistency, and purpose of the concrete, there is an optimum value for the fineness modulus of the aggregate. This is usually the maximum permissible value. That is, when the fineness modulus is less than this optimum value, the needed amount of mixing water is higher; when the fineness modulus is greater than the optimum value, a harsh mixture results, which is again detrimental to the workability of the fresh concrete.

Abrams was the first to publish optimum values of fineness modulus primarily for optimizing concrete strength [Abrams 1925]. Subsequently, several other researchers dealt with the same problem [Swayze 1947, Kellermann 1940, Kennedy 1940], obtaining essentially the same results as Abrams.

Several researchers also tested the optimum values of the so-called combined fineness modulus, which is the modulus of the total solid volume [Swayze 1947, Palotas 1934, Walker 1947]. This modulus represents the logarithmic average particle size of the combination of cement and aggregate. The various results clearly illustrate the fact that "the combined optimum fineness modulus for total solid volumes is essentially constant for a given set of aggregates regardless of the cement content of the mixes involved." Incidentally, this statement supports the experience that the concrete consistency remains nearly constant when the cement content is changed within practical limits [Lyse 1932].

In Table 11.5, after Walker and Bartel [Walker 1947], experimentally obtained values of the combined optimum fineness modulus M_o are shown for combinations of cement, sand, and gravel. From these values Fig. 11.2 was prepared, which illustrates two facts:

1. Values of the combined optimum fineness modulus can be obtained with good approximation from the following formula [Popovics 1961, 1962a]:

$$M_o = 2.45 \log D + 1.05 = \qquad (11.14)$$

$$= 2.45 \log (2.68D) \qquad (11.15)$$

where M_o = combined optimum fineness modulus for combinations of cement, sand and gravel
D = maximum particle size, mm

The limits of validity of Eqs. 11.14 and 11.15 are as follows:

Type of mineral aggregate: sand and gravel of round particle shape
Maximum particle size: 0.1 - 6 in (2.4 - 150 mm)
Cement-aggregate ratio: 1:4 - 1:10 by weight
Consistency:plastic,that is, about 4 in (10 cm) slump
Concrete property to be optimized: strength

Palotas also derived a formula for M_o from his own experimental data [Palotas 1934]. His formula is very much the same as Eq. 11.14.

2. Since fineness modulus is proportional to the logarithmic average particle size of the grading, this can also be expressed using the average-particle-size concept by substituting Eq. 11.4 into Eq. 11.14:

$$\overline{d}_{lo} = 0.737 \log D - 0.68 \qquad (11.16)$$

Table 11.5: Combined Optimum Fineness Moduli and Average Particle Sizes of Total Solid Volume for Different Maximum Sizes

Size of aggregate	Maximum size D, mm	M_O, optimum fineness modulus[a]	$\bar{d}_{go}$, mm[b]	d_{ro}, %
0–No. 28	0.6	1.00	0.200	33.3
0–No. 14	1.2	1.54	0.290	24.2
0–No. 8	2.4	2.08	0.424	17.7
0–No. 4	4.8	2.68	0.642	13.3
0–No. 3	6.3	3.03	0.820	13.0
0–$\frac{3}{8}$ in	9.5	3.42	1.07	11.3
0–$\frac{1}{2}$ in	12.7	3.78	1.37	10.8
0–$\frac{3}{4}$ in	19	4.17	1.80	9.5
0–1 in	25.4	4.54	2.33	9.2
0–1$\frac{1}{2}$ in	38	4.93	3.04	8.0
0–2.1 in	53	5.29	3.92	7.4
0–3 in	76	5.70	5.21	6.85
0–4$\frac{1}{2}$ in	115	6.07	6.73	5.85
0–6 in	152	6.45	8.75	5.75

[a] From [Walker 1947].
[b] $\bar{d}_{go} = 2^{M_O}/10 =$ combined optimum value of the geometric average particle size of the total solid volume.

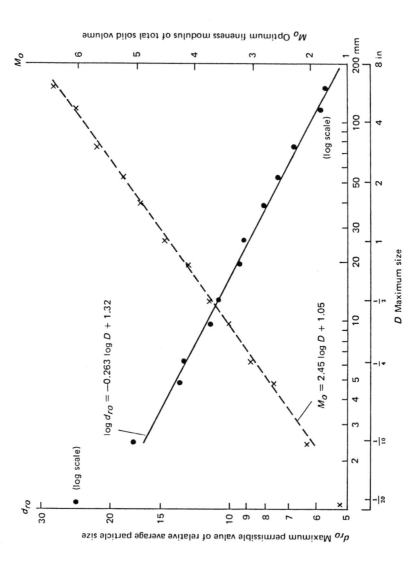

Figure 11.2: Maximum permissible values of the relative average particle size and that of the fineness modulus as a function of the maximum particle size for total solid volume [Popovics 1962].

or from Eq. 10.29:

$$\log \overline{d_{ro}} = -0.263 \log D + 1.32 \qquad (11.17)$$

or, with rounded-off coefficients,

$$\overline{d_{go}} = 0.2D^{0.75} \qquad (11.18)$$

and

$$\overline{d_{ro}} = \frac{20}{D^{0.25}} \qquad (11.19)$$

where d_{lo} = maximum permissible value of the logarithmic average particle size of the grading of the cement-aggregate mixture

   D = maximum particle size, mm

   d_{go} = maximum permissible value of the geometric average particle size of the grading of the cement-aggregate mixture, mm

   d_{ro} = $100d_{go}/D$ = relative size of the d_{go} particle size as percentage of D.

The conditions of validity of Eqs. 11.16 and 11.17 are the same as those for the underlying values of M_o in Table 11.5.

The d_{go} and d_{ro} values calculated from the experimentally obtained M_o values are also presented in Table 11.5. The values in this table, or the corresponding Eqs.11.4 - 11.19 represent the necessary condition for workability of concrete as far as the grading of the total solid volume is concerned. It can be seen that the larger the maximum particle size, the larger the permissible value of modulus M_o and the average particle size d_{go}. However, the permissible maximum values of d_{ro} decrease with increasing D.

Example 11.6 It can be seen from Table 11.5 that if the maximum particle size is $D = 3/4$ in (19 mm), the combined fineness modulus should not exceed

4.17. that is, the geometric average particle size of the dry mixture d_{go} should not exceed the value of 0.071 in (1.80 mm), which is 9.5% of D, regardless of the cement content or the aggregate-cement ratio of the mixture; otherwise the adequate workability of the fresh concrete cannot be assured.

As mentioned earlier, values have been developed on an experimental basis for the optimum, that is, maximum permissible values of fineness moduli for aggregates only, without cement. A group of such m_o values is shown in Table 11.6, derived from Abrams's data by Walker and Bartel. The conditions of validity are also presented. Similar optimum fineness moduli recommended by Palotas [Palotas 1933] on the basis of his experiments are very much the same as the values in Table 11.6.

Equations can be developed for the optimum fineness moduli for mineral aggregates alone. The general equation is [Popovics 1962b]

$$m_o = \frac{G_a(M_o - m_c)}{nG_c} + M_o \qquad (11.20)$$

where m_o, M_o = optimum fineness moduli for the aggregate and for the total solid volume, respectively

$\quad\quad m_c$ = fineness modulus of the cement

$\quad\quad G_a, G_c$ = specific gravity of the aggregate and the cement, respectively

$\quad\quad n$ = aggregate-cement ratio by weight

With the usual values of $G_a = 2.65$, $G_c = 3.15$, and $m_c = -2.5$, the combination of Eqs. 11.14 and 11.20 provides the following practical form:

$$m_o = \left(1 + \frac{0.85}{n}\right)(2.45 \log D + 1.05) + \frac{2.1}{n} \qquad (11.21)$$

The limits of validity for Eq. 11.21 are the same as those for Eq. 11.14 or for the values in Table

Table 11.6: m_o **Maximum Permissible (Optimum) Values of Fineness Modulus for Sand and Gravel[a,b]**

Maximum Size of Aggregate		Pounds of Cement per Cubic Yard (kg/m³) of Concrete							
No.	mm	280 (170)	375 (225)	470 (280)	565 (335)	660 (390)	750 (445)	850 (500)	950 (560)
No. 30	0.60	1.4	1.5	1.6	1.7	1.8	1.9	1.9	2.0
No. 16	1.18	1.9	2.0	2.2	2.3	2.4	2.5	2.6	2.7
No. 8	2.36	2.5	2.6	2.8	2.9	3.0	3.2	3.3	3.4
No. 4	4.75	3.1	3.3	3.4	3.6	3.8	3.9	4.1	4.2
⅜ in.	9.5	3.9	4.1	4.2	4.4	4.6	4.7	4.9	5.0
½ in.	12.5	4.1	4.4	4.6	4.7	4.9	5.0	5.2	5.3
¾ in.	19.0	4.6	4.8	5.0	5.2	5.4	5.5	5.7	5.8
1 in.	25.0	4.9	5.2	5.4	5.5	5.7	5.8	6.0	6.1
1½ in.	37.5	5.4	5.6	5.8	6.0	6.1	6.3	6.5	6.6
2 in.	50.0	5.7	5.9	6.1	6.3	6.5	6.6	6.8	7.0
3 in.	75.0	6.2	6.4	6.6	6.8	7.0	7.1	7.3	7.4

[a] Note: The m_o values shown in the table are valid for aggregates when the fine aggregate is a natural sand and the coarse aggregate is a rounded gravel having voids of about 35% in the dry-rodded condition. 0.1 should be subtracted from the tabulated m_o values for each increase of 5 in the percentage of dry-rodded voids in the coarse aggregate. In other words: for gravel consisting of flat particles, subtract 0.25 from the m_o values in the table; for crushed coarse aggregate, subtract about 0.25 from the m_o values when the particle shape is regular, and up to 0.40, when the particles are flat or elongated; for crushed fine aggregate, subtract about 0.25 from the m_o values. The m_o values above are recommended for non-air-entrained concretes of 1- to 2-in (25- to 50-mm) slump. For air-entrained concretes add 0.1 to the m_o values in the table by reducing the sand content. Also, for concretes of plastic consistency, subtract about 0.25 from the listed m_o values; for no-slump concretes about 0.25 may be m m_o values; for no-slump concretes about 0.25 may be added to the m_o values above.

[b] From [Walker 1947].

11.6. Experimentally obtained m_O values support Eq. 11.21, as can be seen from Fig. 11.3.

Palotas also published an equation on empirical basis for m_o, the optimum fineness moduli of aggregates. This is the following [Palotas 1934]:

$$m_o = 2.66 \log D + 0.0028c + 0.84 \qquad (11.22)$$

where D = maximum particle size, mm
$\quad c$ = cement content of the compacted concrete
$\quad\quad$ kg/m^3

When D is given in inches and c in lb/yd^3,

$$m_{o''} = 1.1552 \ln D'' + 0.001709c + 4.5769 \qquad (11.22a)$$

where ln is natural log. The corrections for Eqs. 11.22 and 11.22a are the same as for values listed in Table 11.6.

Although the forms of Eqs. 11.21 and 11.22 are different, the comparable values calculated from these two equations are close to each other.

11.4 SPECIFIC SURFACE

The other frequently mentioned numerical grading characteristic is the specific surface of the particles. According to the proponents of this concept, the specific surface should control the concrete-making properties of the aggregate because it determines the amount of cement paste needed to cover the aggregate surface. Consequently, the numerical value of the specific surface of the aggregate should be related to the quantities of the two most important concrete components needed, namely cement and water, and be independent of the details of the grading. As will be shown, experimental results have not supported this premise for concrete aggregates in general, perhaps because it does not take the voids content in the aggregate· into consideration. Nevertheless, the specific surface

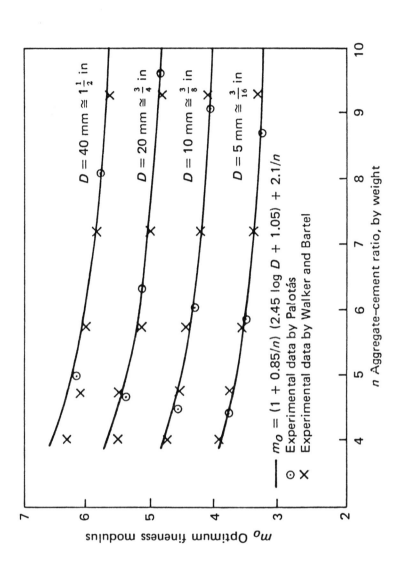

Figure 11.3: Calculated values and experimental values of the optimum fineness modulus for sand and gravel combinations as a function of the maximum particle size and mix proportion [Popovics 1962b, 1966b].

concept has been applied quite successfully in a few other cases, such as the characterization of the fineness, that is, grading of cements, for bleeding of mortars and concretes, and even for the grading of aggregates for bituminous mixes.

The specific surface is defined as the total external surface area of particles having a total quantity of unity. The unit quantity can be a unit of weight or, preferably, a unit of absolute (solid) volume, providing two types of dimensions for the specific surface: (1) ft^2/lb, cm^2/g, m^2/kg, and so on; or (2) ft^2/ft^3, cm^2/cm^3, m^2/m^3, and so on. The conversion from one type of dimensions to the other takes place with the help of the bulk specific gravity of the material.

Of course, the actual size of specific surface is dependent not only on the grading but also on the particle shape. For the purpose of grading characterization it has been customary to assume a spherical shape for the aggregate particles, in which case it is easy to see that the specific surface of a one-size aggregate is

$$s_w = \frac{6}{G_b d}$$

(11.23)

or

$$s = \frac{6000}{d}$$

(11.24)

where s_w, s = specific surface, m^2/kg, and dm^2/m^3, respectively

G_b = bulk specific gravity of the aggregate

d = particle size, mm

$1\ m^2/m^3 = 0.305\ ft^2/ft^3$

Values of s calculated by Eq. 11.24 for several particle sizes have been presented in Table 10.3. It can be seen that halving the particle size results in doubling the specific surface.

Edwards [Edwards 1918] and Young [Young 1919] were the first to summarize the concept of specific surface for concrete technology. Many years after, a general analytical expression was developed for the calculation of the specific surface for a given $f(d)$ particle-size distribution [Popovics 1952, 1955, 1962a, 1962b] as follows:

$$s = b \int_{d_{min}}^{D} \frac{f'(d)}{d} \, dd \qquad (11.25)$$

$$s = b \left(\frac{100}{D} + \int_{d_{min}}^{D} \frac{f(d)}{d^2} \, dd \right) \qquad (11.26)$$

where s = specific surface
      b = factor that depends on the applied units, particle shape, and so on

The other symbols are the same as the symbols of Eqs. 10.30 and 10.31.

If it is assumed that the particles are spherical and smooth, and that the particle size is expressed in millimeters, $f(d)$ in percentage, and the specific surface in square meters per cubic meter (0.305 ft^2/ft^3), then $b = 60$.

A comparison of Eqs. 11.25 and 11.26 to Eqs. 10.30 and 10.31 shows that the specific surface is proportional to the reciprocal value of the harmonic average particle size:

$$s = \frac{b}{d_h} \qquad (11.27)$$

Therefore, it follows from Eq. 10.41 that the specific surface can be determined with good approximation by the area A'_h under the DIF curve in the semilogarithmic system in Fig. 10.4 as follows [Popovics 1952, 1966]:

$$s = 2.30 b A'_h \qquad (11.28)$$

The unit m^2 /m^3 = 1/m shows also that the specific surface is a reciprocal particle-size concept.

Of course, the specific surface of a complete grading can also be calculated from the quantities p_i and specific surfaces s_i of aggregate fractions:

$$s = 0.01 \ \Sigma p_i s_i \qquad (11.29)$$

The assumption of spherical particle shape is an acceptable simplification when the specific surface is used for the numerical characterization of the grading; therefore, this assumption will be used in this chapter unless it is indicated otherwise It is possible, of course. to assume other regular particle shape(s), such as ellipsoid, cube, tetrahedron, and so on, but this would affect (increase) only the value of the factor b in the previous formulas and would still be an approximation. When the actual value of the specific surface is needed, it can be determined, for instance, by suitable permeability measurements [Loudon 1953, Shacklock 1957].

It should be pointed out that the specific surface can be considered as a variance concept. This is so because it is the second moment of the cumulative function h(d) of the particle-size distribution where the distribution is taken by the number of particles. Also, the difference between the geometric average and the harmonic average can be used for the estimation of the standard deviation [Herdan 1960]. A practical way to show numerically how the particle sizes are distributed around the geometric average size d_g is to calculate a *distribution factor u*, as follows:

$$u = s/s_o \qquad (11.30)$$

where s = the calculated specific surface of the aggregate of d_g geometric average particle size

s_o = the calculated specific surface of the grading consisting of indentical particle sizes of d_g; that is $s_o = 6/d_g$

The minimum value of u is 1 which indicates no variation in aggregate sizes, that is a one-size grading. The greater the value of u, the greater the variance of the particle sizes. The u value appears to be a more precise and more general indicator of the "uniformity" of a grading than the coefficient of uniformity used in soil mechanics because the latter takes into consideration only two particle sizes rather than the complete particle size distribution.

Example 11.7 Calculate the specific surface by using Eqs. 11.23 and 11.24, respectively, for one-size aggregates, assuming $G_b = 2.65$.

==

d, mm	0.01	0.1	1.0	10	100
s, m²/m³	600,000	60,000	6000	600	60
s_w m²/kg	226.415	22.642	2.264	0.226	0.0226

Note the drastic increase in the specific surface as the particle size decreases.

Example 11.8 Calculate the specific surface of the grading between the range of 0.1 and D mm characterized by the Fuller parabola by using Eq. 11.27.
By utilizing Eq. 10.36 of Example 10.3:

$$s = \frac{b}{d_h} = \frac{6000(\sqrt{10D} - 1)}{D} \qquad (11.31)$$

Example 11.9 By using Eq. 11.25 and $b = 60$, calculate the specific surface of the four gradings characterized as one-size, linear, logarithmic, and

parabolic, respectively, particle-size distributions.

The pertinent grading equations as well as the
calculated values of the specific surfaces are
shown in Table 11.3.

Example 11.10 Calculate the specific surface of
the grading in Example 11.1 by using Eq. 11.28.

From Fig. 10.4, the area A_h can be determined in
very much the same way as the area A_{log} was calculated in Example 11.2 and Table 11.2. It is assumed
that the minimum particle size is 0.1 mm, therefore

$$A'_h = 35.02$$

thus

$$s = 2.30 \times 60 \times 35.02 = 4833 \text{ m}^2/\text{m}^3$$

11.5 CRITIQUE OF THE SPECIFIC SURFACE AND OTHER NUMERICAL CHARACTERISTICS

The single but decisive objection to the surface area
for grading characterization of concrete aggregate is
that experimental results do not adequately support
its applicability for the grading evaluation of
concrete aggregates. More specifically, it is a
common view that the surface area method
overestimates the effect of fine particles, and
underestimates the effect of large particles with
respect to most concrete-making properties. For
instance, the surface area method cannot explain the
experimental fact that a change in the cement content
of a mixture scarcely influences the consistency
despite the significant change in the size of the
surface area of total solid volume, or that a change
in the maximum particle size from, say, 3/4 in (19 mm)
to 1 1/2 in (38 mm) produces a considerable reduction
in the water requirement of the aggregate although it
hardly affects the value of the specific surface.
Although Newman and Teychenne [Newman 1954] and also
Singh [Singh 1957], concluded from their test results
that aggregates having the same specific surface

produce practically the same consistency and strength if the other variables are kept unchanged, their experiments should not be generalized at all. They used only a single maximum particle size (D = 19 mm = 3/4 in), and a single minimum particle size (0.15 mm) in their experiments. This means that practically no particle passing sieve No. 100 was present in any of their tested gradings. The presenc of these particles would have increased decisively to the specific surface, although not necessarily the concrete making properties. In addition, characteristically enough, an analysis reveals that their test results show better correlations with the fineness modulus of the gradings used than with the specific surface favored by these authors.

It should not be surprising that not all the different numerical-grading characteristics are equally good for the indication of the concrete-making properties of a grading. If, for instance, the numerical characteristics classified as the second group in connection with Table 10.3 are well supported by experiments, then the other numerical characteristics in the different groups cannot be supported very well. But even the best numerical-grading characteristic has definite limitations. For instance, previously discussed experiments demonstrated that different gradings having the same fineness modulus require the same quantity of water to produce mixtures of the same plasticity and give concrete of the same strength regardless of the differences in the details of the particle-size distribution so long as the grading contains enough, but not too much, fine particles. This means that the fineness modulus method provides only a *necessary*, but not a *sufficient* condition for the evaluation of aggregate grading. In other words, if the grading is good, the value of the fineness modulus is always near the optimum. If, however, the fineness modulus of a grading equals the optimum value, this fact does not indicate sufficiently that the grading is excellent or even fair for concrete because a lack or excess of fine particles still can impair the workability.

This experimental fact shows that the fineness

modulus does not reflect adequately the amount of fine particles for concrete-making purposes, which, in turn, explains why concretes made with various gradings of identical fineness modulus may have significantly different water permeability or bleeding under otherwise identical conditions. This is definitely a weakness of the fineness modulus. This weakness has not been reduced significantly by moving the origin of the semilogarithmic system of coordinates for the grading curve form d = 0.1 mm to 0.075 [Hudson 1969], or 0.01 [Palotas 1956] or 0.001 mm either.

11.6 ATTEMPTS TO IMPROVE NUMERICAL GRADING CHARACTERIZATION

Whereas the fineness modulus underestimates the effects of fine particles, the specific surface method overestimates them with respect to many concrete properties. The other numerical-grading characteristics discussed earlier seem to be even less satisfactory.

Therefore, it is safe to say that there is no single number that can characterize perfectly both the fine portion and the coarse portion of an aggregate for concrete technology, not even when the maximum particle size is kept unchanged.

A possible approach to overcome this difficulty is to specify, say, the amount of fine particles in the grading in addition to the value of the numerical characteristic. Another possibility is to apply more than one numerical characteristic simultaneously for grading evaluation. Several such methods have been recommended [Fulton 1956, Zietsman 1957, Hughes 1960, Lecompte 1969]. All these, however, were preceded by a method called the D-m-s method [Popovics 1952, 1966b]. This recommends that for the grading evaluation of a concrete aggregate its maximum particle size D, fineness modulus m, and its specific surface s be used simultaneously. The statistical interpretation of the D-m-s method is that D characterizes the range of the particle-size distribution, m is an average concept, and s is a

variance concept. The equivalent concrete technological explanation is that the values of D and m
control the coarse part of the aggregate, and the
value of s controls the fine part with respect to
concrete-making properties. Thus it is expected that
the $D-m-s$ method provides not only a necessary, but
also a sufficient criterion for the acceptability of
grading of concrete aggregates by using only the
results of a sieve test.

The $D-m-s$ method includes two statements:

1. Different particle-size distributions of indentical maximum size and fineness modulus can be
distinguished by means of specific surface with
respect to concrete-making properties.

2. Concrete with the same properties in both the
plastic and hardened state will be produced by
aggregate gradings that have the same numerical
values of maximum size, fineness modulus, and
specific surface, regardless of the details of the
grading and if other proportions of the mixture are
the same.

Statement 1 is illustrated for compressive strength
in Fig. 11.4. It is reasonable to say that in the
region of specific surface where the strength-versus-
specific surface curves are flat, that is, within
approximately 500 and 1300 in^2/lb in this example,
the percent of voids in the aggregate is not
excessive. This will be discussed more in Chap. 12.

Statement 2, that is, the validity of the $D-m-s$
method, is supported better by experimental data than
the other methods, especially when the gradings are
extreme, such as one-gap or multiple-gap gradings.
For instance, Fig 11.5 shows three highly different
particle-size distributions having identical $D-m-s$
values. The calculation of the specific surfaces was
based on the assumptions that

1. The shapes of the particles are spherical.
2. The specific gravity of aggregate is 2.65.
3. The specific surface of the particles passing

through the 0.1 mm sieve is 40 m^2/kg, regardless of
the actual grading of these very fine particles. (If
the amount of these very fine particles cannot be

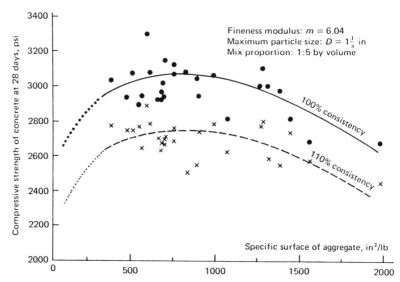

Figure 11.4: Compressive strength of concrete as a function of the specific surface of the aggregate used. The concretes were made with aggregates of differing gradings having identical fineness modulus. The aggregate type and the aggregate-cement ratio were the same for all mixtures. The same water quantity was used in concretes of "100% consistency," and 10% more for those of "110% consistency." Data were taken from Abrams [1925].

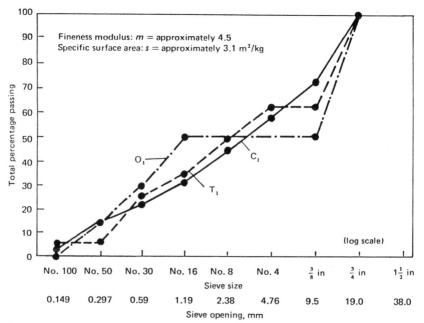

Figure 11.5: Gradings having the same D, m, and s values [Popovics 1961, 1964].

estimated adequately, then either the specific surface of 30 m^2/kg can be used for the particles passing the 0.15 mm (No. 100) sieve, or 20 m^2/kg for the particles passing the 0.2 mm sieve.)

With these gradings several groups of concrete were made and the slump, flow, bleeding, segregation, unit weight, and compressive strength of these concretes were determined. The results show that the properties of concretes made with the C_1 and T_1, gradings were practically identical in both the plastic and hardened states, the O_1 grading resulting in slightly drier consistencies and lower strength. C, T, and O gradings with 1-in maximum particle size showed similar trends [Popovics 1961, 1964c] . The internal structures of concretes made with these gradings are shown in Fig. 13.4, and the appearances of these graded aggregates in Fig. 13.5.

12

Grading Evaluation and Specification

SUMMARY

Grading evaluation is the comparison of one or more grading characteristics either to a specification, to an optimum value, or to another grading. Most frequently *TPP* or *PPR* grading curves, or the fineness modulus, are used. *Grading specification* means the establishment of condition(s) with which one or more grading characteristics should comply to achieve specificed concrete properties, and to assure satisfactory uniformity in the gradings of subsequent batches. Regardless what method is used for this, the maximum particle size should always be specified.

The simplest, and crudest method is to specify the *proportion of fine aggregate* in the total aggregate. Additional conditions, such as the specification of the quantities of more size fractions (*PPR*) or more points of the *grading curve* (*TPP* or *TPR*), tighten the grading specification. The grading curve itself can be specified either in a graphical or in a mathematical form. The usual mathematical form of such "theoretical" gradings is power function. Among the power functions, the Fuller parabola and the Bolomey grading are the best known. The curves and equations can refer either to aggregate or to the mixture of cement and aggregate.

Grading specification by *limit curves* is very popular because of its simplicity. This method requests that the actual grading curve should fall within the specified limit curves in its total length. In the triangular diagram the equivalency to a pair of limit curves is *a grading area.*

Gradings can also be specified by special conditions. Three examples for this are the *percentage grading,* the *particle interference* method, and the *maximum density* principle. Mathematical analysis shows that the gradings provided by these three special conditions are Fuller gradings.

The grading specification by *fineness modulus* requires that the modulus of the aggregate, or that of the cement-aggregate mixture, be close enough to a specified value. It is also possible to specify additional condition(s) that supplement the specified fineness modulus. Such an additional condition can be the amount of fine particles or the specific surface of the aggregate (*D-m-s* method).

The most general of these methods for grading specification are those that utilize the fineness modulus of the combined grading, although these methods require a little experience and common sense. The limit curve method is very popular because of its simplicity despite the fact that it is not applicable for gap gradings. Grading specification by special conditions, or on the basis of percentage passing-retained are not recommended for practical purposes.

12.1 GRADING EVALUATION

Grading evaluation is the comparison of one or more grading characteristics of an aggregate either to a specification, to a theoretical (optimum) value(s), or to the corresponding set of characteristics of another grading. The usual purpose is to see without trial mixes (1) if the grading meets the specification (this is discussed in the following paragraph); (2) how far the grading is from the pertinent optimum; or (3) how far the concrete-making properties of two or more different gradings are from each other.

A perfect method of grading evaluation should have the following properties (in the subjective order of significance):

1. It should judge as equal in value all the

different gradings, including both continuous and gap gradings, that have identical concrete-making properties, and gradings having differing concrete-making properties as different in value.

2. It should be able to indicate if the grading is optimum, and if not, how far it is from optimum.

3. It should be reliable; that is, it should not judge a poor grading as a good one or a good grading as a poor one.

4. It should preferably be numerical.

5. It should preferably be simple and short enough for practical applications.

6. It should preferably have a theoretical basis.

None of the presently available methods of grading evaluation (*TPP* or *PPR* grading curves, fineness modulus, etc.) has all these properties mainly because of the lack of perfect grading characteristic(s). Nevertheless, most methods are applicable for certain practical purposes with restricted limits of validity. Perhaps the method based on the simultaneous application of the fineness modulus and the specific surface of the aggregate (the *D-m-s* method) best meets this set of requirements.

This may be a good place to face the old question of whether gap gradings are better or worse than continuous gradings. The extensive literature discussing this question [Li 1970] contains contradictory opinions. It is clear, however, from the majority of the experimental evidence that gap gradings can produce good concretes if the gap(s) is (are) selected properly [Li 1973], although the majority of investigators did not show significant strength increase attributable to gap gradings [Walz 1974, Albrecht 1965, Popovics 1966b]. Thus, the selection between the gap-grading type and the continuous type is made in most cases on an economical basis; namely, the grading type is chosen that is less expensive under the prevailing local circumstances.

12.2 GRADING SPECIFICATION IN GENERAL

Grading specification means the more or less detailed description of condition(s) with which one or more grading characteristics (grading curve, fineness modulus, quantity of a size fraction etc.) should comply to assure satisfactory performance of and uniformity in the grading of subsequent batches. The basic premise is that gradings meeting a given specification have practically identical good concrete-making properties. Grading specifications for concrete aggregates should assure an adequate concrete workability with minimum water content. Voids content is usually not specified for aggregates for hydraulic concretes; however, it is regularly specified that gradings for asphalt concrete should produce pratically minimum voids content.

The possibilities for grading specification range from simple methods of limited accuracy to sophisticated methods of high reliability. The expenses of complying with a grading specification increase rapidly as the specified details of the grading increase, while the return in benefits increases only at a diminishing rate after a certain point. The intelligent selection of a method of grading specification is usually based on a number of factors, such as the applied form of grading charac-terization, the required quality and uniformity of the concrete, the degree of grading uniformity that can be obtained economically under the prevailing conditions, whether the grading is continuous or gap grading, the purpose of the concrete structure, how well trained the supervising and quality-control personnel are to reinforce the grading specification, what the pertinent polices of the contracting agencies are, and so on.

The purpose of a grading specification is to assure, or approximate, a grading that can provide uniformly the optimum of a given concrete property under a set of technical and economical circumstances, including specified, reasonable amounts of cement and water.

The various optimum gradings for structural concrete have the common feature that they provide a dense mortar in which coarse aggregate particles are embedded in the maximum quantity that still can produce a workable and dense concrete under the prevailing circumstances. This optimum can be determined from the behavior of the fresh concrete, from the strength and entrapped-air content of the hardened concrete, and from the visual inspection of the internal structure of the compacted concrete. This latter method is discussed in detail in Chapter 13.

Two comments seem appropriate here to resolve certain misconceptions about "optimum" gradings:

1. It is unrealistic to expect that there is any single grading that can optimize all, or most, concrete properties simultaneously. Although it is a common feature of all optimum gradings that they provide good workability under the given circumstances, there are numerous more or less different "optimum gradings." For one thing, gradings that are optimum for one concrete property, say, the strength of concrete, may not, and usually do not, provide the optimum for another concrete property, such as impermeability. Second, the optimum of a grading for, say, concrete strength is also influenced by the method of consolidation, the type of aggregate, and the amount of cement to be used. Third, differing particle-size distributions, under otherwise identical conditions, can have practically identical concrete-making properties, that is, can produce the same optimum of the concrete property in question.

2. There is no single grading for fine aggregate that can provide an optimum combined grading, in any sense of the term optimum, with every coarse aggregate. Fine aggregates complying with accepted grading specifications, such as, ASTM C 33-90, can make good concretes with conventional coarse aggregates. However, combined with unusual coarse aggregates, in gap-gradings for instance, the same fine aggregates may be less than optimum. The effects of using a finer sand on the properties of concrete can usually be compensated for by using a smaller proportion of it [Powers 1968, Gaynor 1963b]. In other words, the optimum grading, as well as the optimum quantity of a sand in concrete aggregate, depend on, among other things, the grading and type of coarse aggregate used. The same is true in a reverse sense for the optimum grading of coarse aggregate.

Several methods will be discussed below for grading specification.

12.3 SPECIFICATION OF MAXIMUM PARTICLE SIZE

Regardless what method is used, the maximum particle size of aggregate D should always be specified. The specified value of D depends on the minimum dimension of the concrete structure, the denseness of the reinforcement, the purpose of the concrete structure, the required properties of the concrete, the method of concrete consolidation, and so on.

Except for high-strength concretes, an increase in the maximum particle size in the concrete can provide higher concrete strength while keeping the consistency unchanged. Also, a marked saving in cement can be obtained as the maximum size of the aggregate is increased up to about 3 in (75 mm), because in this way the amount of aggregate in the concrete is also increased [U.S.Bureau of Rec. 1966]. Therefore, in many cases it is desirable to use the largest maximum particle size permitted by the prevailing circumstances. This largest particle size of the aggregate should not be larger, for instance, than one-fifth of the narrowest dimension of the form in which the concrete is to be used, or larger than three-fourths of the minimum clear distance between reinforcing bars [ACI 1971c]. In making concrete test specimens, the maximum size of aggregate preferably should not be larger than one fourth of the smallest dimension of the mold, although one-third is allowed in ASTM C 31-90.

12.4 SPECIFICATION OF THE SAND AND FINE SAND CONTENTS

Probably the simplest method is to specify the proportion of the fine aggregate, that is, sand (particles passing sieve No.4), in the total aggregate. It can be prescribed, for instance, that the fine aggregate and coarse aggregate contents be 40 and 60%, respectively, by absolute volume or by weight. This specification can be equally well written in the form that the fine aggregate-coarse aggregate ratio be 1:1.5 by absolute volume or by weight. For practical purposes, it is better to specify an upper and a lower limit, say, 38 and 42%, for the fine aggregate content. that is, 1.38 and 1.63 for the coarse aggregate-fine aggregate ratio.

The graphical form of this specification is the requirement that the grading curve pass through the 40% TPP point, or pass between the 38 and 42% points, respectively, at the size of sieve No. 4 (4.75 mm) without any additional requirement concerning the details of the grading curve or the grading as a whole. This is obviously a very loose grading specification; yet it has been used with fair results in many practical cases of lesser importance as long as the sources of the fine and coarse aggregates remained unchanged.

The basis of this method of grading specification is that if the fine aggregate content is properly specified, the grading is expected to provide an adequately dense and workable concrete with reasonable amounts of cement and water. In other words, the amounts of fine and coarse aggregates should be well balanced for a good grading, but the nature of this balance is dependent on numerous factors. More specifically, the relative amounts of fine aggregate needed for good grading increases (1) with decrease of the maximum particle size; (2) as particle shape becomes less favorable; (3) with decreasing cement contents; (4) with increase of the average particle size (fineness modulus) of the fine aggregate; (5) with increase in the average particle size (fineness modulus) of the coarse aggregate; (6) when not only good strength but also improved watertightness is required from the concrete; and so on. Because of the large number of these variables, it is not practical to try to give a complete set of recommended values for the amount of fine aggregate in concrete aggregate. As a rough guide, however, the coarse aggregate-fine aggregate ratio for structural concretes is usually between 0.5 and 2 by absolute volume. There is no accepted standard in the United States for the fine aggregate content or the coarse aggregate-fine aggregate ratio, primarily because of the inherent uncertainties of this method of grading specification.

The application of additional conditions, of course, can improve the accuracy of the grading specification. For instance, one can specify not only the amount of sand, that is, particles, passing sieve No. 4 (4.75 mm), but also the amount of fine sand, that is, particles, passing sieve No. 16 (1.18 mm) or some other sieve around this size. According to this method, a concrete aggregate should consist of, say, 25% fine sand, 25% coarse sand, and 50% coarse aggregate of 1 in (25 mm) maximum particle size. This type of specification corresponds to two intermediate points for the grading curve to pass through or to a grading point in the triangular diagram of Fig. 10.5. If the specification provides upper and lower limits for the fine sand, coarse sand, and coarse aggregate, then this can be represented by a pair of grading curves each defined by two points at the size of, say, sieve No. 16 (1.18 mm) and sieve No. 4 (4.75 mm) or by a corresponding area in the triangular diagram as shown in Fig. 10.5. The recommended values for the fine and coarse sand contents are influenced by essentially the same factors discussed above in connection with the fine aggregate content. The reason for selecting both specifying conditions in the fine aggregate region is that the grading of fine aggregate controls the concrete-making properties of the aggregate much more than the grading of the coarse portion.

It is satisfying to note from the technical point of view that despite its simplicity the grading specification by the amounts of fine sand and coarse sand works surprisingly well for the majority of concrete structures as long as the sources of the aggregate components remain unchanged. There is no accepted standard in the United States for the fine sand, coarse sand, and coarse aggregate contents of concrete aggregates.

12.5 SPECIFICATION OF THE GRADING CURVE

Further tightening of the grading specification can be

achieved by prescribing more points between the minimum and maximum particle sizes for the grading curve to pass through. The more such points are prescribed, the more details can be specified for the grading.

The specified grading curve can be given in graphical form (or by a set of coordinates of the grading curve) or by a formula that provides the specified y coordinates of the grading curve as a function of particle size. In either case, the specified curves are *empirical* in nature because they, even the equations, are obtained by curve fitting to experimentally proved, good gradings. Gradings specified by grading curves are sometimes called "theoretical gradings." Europeans like to use theoretical gradings for combination of fine and coarse aggregates. In American practice, it is more common to specify the grading of the fine aggregate and that of the coarse aggregate separately, leaving the grading of the combination to the technologist in the field.

Grading curves can be used either for *aggregates only* or for the *mixture of aggregate and cement*. The practical application of the first type of grading curve is simpler; the second type is important because the grading of the total solid volume controls the consistency of the fresh concrete, as discussed in the previous chapter. If the cement quantity to be used is known, the conversion from one of these sieve curve types to the other is easy by using the reasonable assumption that all the cement particles are smaller than, say, the sieve No. 100 (0.15 mm). For instance, if the sieve curves, or the coordinates of the sieve curves, are given on the total percentage-passing basis, then for particle sizes greater than 0.15 mm:

$$y_c = \frac{100(y+t)}{100+t} \qquad (12.1)$$

$$= \frac{ny+100}{n+1} \qquad (12.2)$$

$$y = y_c(1 + 0.01t) - t = \qquad\qquad (12.3)$$

$$y = y_c\left(1 + \frac{1}{n}\right) - \frac{100}{n} \qquad\qquad (12.4)$$

where y_c, y = ordinates of the sieve curve of the
  cement-aggregate mixture and the aggregate
  without cement, respectively, total percentage
  passing, by absolute volume (or by weight)
n = aggregate-cement ratio, by absolute volume
  (or by weight)
t = $100/n$ = cement quantity, percent of the
  aggregate, by absolute volume (or by weight).

When the cement content is given in the form of
the weight of the cement in one unit volume of
compacted concrete, the corresponding n value can be
calculated with good approximation by the following
formula [Popovics 1974c]:

$$n = G\left(\frac{1425}{c} - 0.5\right) \qquad\qquad (12.5)$$

where G = average bulk specific gravity of the
  aggregate
c = cement content of the compacted concrete,
  lb/yd^3 (1 lb/yd^3 = 0.593 kg/m^3)

More accurate formulas for the c-versus-n
relationships are also available in the literature
[Popovics 1962c, 1968h].
The general use of theoretical grading curves is
limited by the fact that they can provide "optimum"
grading only for a certain range of maximum particle
size, and, the grading curves for aggregates only, for
a limited range of cement content. For instance,
grading curves that have been proved good for
traditional reinforced concrete are not recommended
for concretes of prestressed elements.
The general form of the empirical equation for the

most frequently used theoretical grading curves is

$$y = f(d) = g \left(\frac{d}{D}\right)^i + (100 - g) \left(\frac{d}{D}\right)^h \qquad (12.6)$$

$$= gd_r^i + (100 - g) d_r^h \qquad (12.7)$$

where $y = f(d) = $ ordinate of the cumulative grading
curve at particle size d on total
percentage-passing basis
D = maximum particle size
d_r = d/D = relative particle size
g, i, h = parameters controlling the details and
coarseness of the theoritical grading

Several recommended values for the parameters of Eqs. 12.6 and 12.7 are presented in Table 12.1 along with the limits of optimum.

A modified form of Eqs. 12.6 and 12.7 is recommended by Faury for cement-aggregate mixtures [Faury 1958, Vallette 1963, Duriez 1961]. He recommends one equation for the $d_{min} = 0.0065$ mm through $D/2$ size portion of the grading:

$$y = f(d) = \frac{y_{D/2}}{(D/2)^{1/5} - 0.365} (d^{1/5} - 0.365) \qquad (12.8)$$

$$= -\frac{y_{D/2}}{2.386D^{1/5} - 1} + \frac{y_{D/2}}{0.871 - 0.365/D^{1/5}} \left(\frac{d}{D}\right)^{1/5} \qquad (12.9)$$

and another for the $D/2$ through D portion, as follows:

$$y = f(d) = y_{D/2} + \frac{100 - y_{D/2}}{D^{1/5} - (D/2)^{1/5}} \left[d^{1/5} - \left(\frac{D}{2}\right)^{1/5}\right] \qquad (12.10)$$

Table 12.1: Parameters for the Optimization of Grading Eq. (12.6) as Recommended by Various Authorities

$$y = g\left(\frac{d}{D}\right)^i + (100 - g)\left(\frac{d}{D}\right)^h$$

Authority	Parameters			Limits for optimum		Remark
	g	i	h	n	D	
Fuller and Thompson [1907]	0	—	0.5	6	5–100	For aggregates only
				9	20–40	
EMPA [Ros 1950]	50	1	0.5	4	15–30	For aggregates only
				6	3–8	
Popovics [1962b]	$15\left(1 - \frac{3}{n}\right)$	0	$\frac{0.5}{1 - 1.6n}$	4–10	5–100	For aggregates only. See Fig. 12.1
Bolomey [1947]	8–10	0	0.5	6–8	20–80	For cement–aggregate mixtures
	10–12	0	0.5	6–8	20–80	For cement–crushed aggregate mixtures
Caquot and Faury [1937]	$\frac{100}{2.386\,D^{1/5} - 1}$	0	0.20	?	For 0.0065–$D/2$ fraction	The blending proportion of these two size fractions recommended for cement–aggregate mixtures is determined by $yD/2$ (see text)
	-672.5	0	0.20	?	For $D/2$–D fraction	
Popovics [1962b]	15	0	0.5	4–10	5–100	For cement–aggregate mixtures
Popovics [1962b]	20	0	0.56	4–10	2.5–150	For cement–aggregate mixtures
Popovics	$\frac{100}{n + 1}$	0	0.5	6	5–100	For cement–aggregate mixtures. Aggregate has Fuller grading

Aggregate: sand and gravel unless indicated otherwise. Consistency: 3–5 in slump. n = aggregate–cement ratio, by weight. D = maximum particle size, mm.

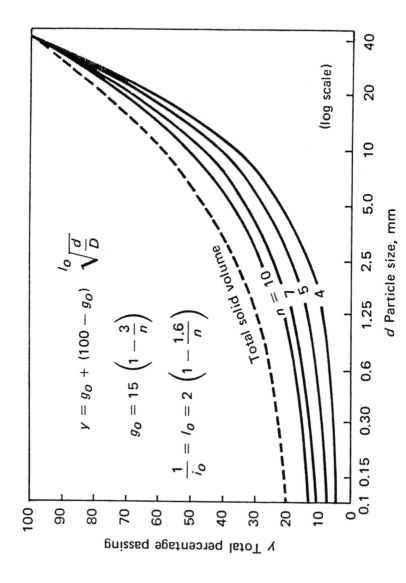

Figure 12.1: "Optimum gradings" of Bolomey type for sand and gravel, and for various aggregate-cement ratios n [Popovics 1962b]. (1 in = 25.4 mm.)

$$= 7.725y_{D/2} - 672.5 + 7.725 \ (100 - y_{D/2}) \ \left(\frac{d}{D}\right)^{1/5} \qquad (12.11)$$

In these equations $y_{D/2}$ is the total percentage passing of the cement-aggregate mixture at the $D/2$ particle size and is as follows:

$$y_{D/2} = A + 17D^{1/5} + C \qquad (12.12)$$

In Eq. 12.12 the parameter A takes the particle shape of the aggregate and the concrete consistency into consideration. For instance, the recommended values for A are 22 for stiff concretes, and 32 for fluid concretes when the aggregate consists of round particles; they are 28 for stiff concretes and 38 for fluid concretes when the aggregate is a crushed material. The parameter C takes the molding resistance of the fresh concrete in the form, that is, the so-called wall effect [Caquot 1937], into consideration and varies usually between 0 and 15. More specifically, greater C values, that is, finer gradings, belong (1) to narrower molds and/or to denser reinforcements; and (2) to milder planned methods of compaction of the concrete. Equations 12.9 and 12.11 show that each Faury grading is a combination of two generalized Bolomey gradings. A simple graphical equivalent of these two gradings is the following: When a Faury grading is plotted in a system of coordinates where the total percentage passing axis y has a linear scale, and the particle-size axis d has a $d^{1/5}$ scale, then the grading curve consists of two connecting straight lines, namely, one from $y = 0$ at the $d = 0.0065$ mm (2.56×10^{-4} in) particle size to $y_{D/2}$ at the $D/2$ particle size and the other from $y_{D/2}$ at the $D/2$ to 100% at D. Examples are presented in Fig. 12.2. This figure also shows good agreement between Faury gradings and Bolomey gradings for larger maximum particle sizes within 0.1 mm $\leq d \leq D$ limits.

Example 12.1 1. When $g = 0$ and $h = 0.5$, then Eq. 12.6 or 12.7 will become the well-known Fuller

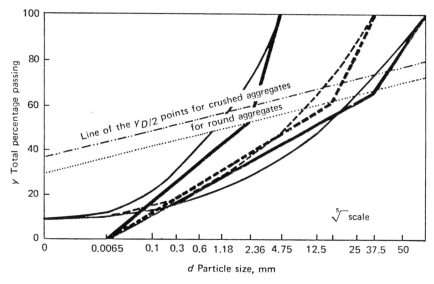

Figure 12.2: Comparison of three Faury gradings (heavy lines) to Bolomey gradings with $g = 9$ (light lines) for cement-aggregate mixtures. Consistency: plastic; wall effect: negligible.

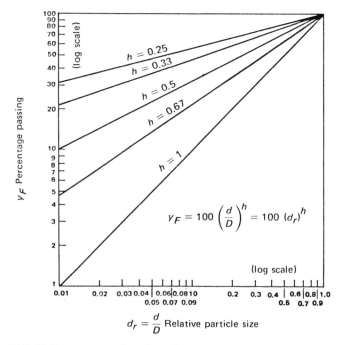

Figure 12.3: Fuller curves of various degrees.

parabola:

$$y = 100\sqrt{\frac{d}{D}} = 100\sqrt{d_r}$$

A grading represented by the equation

$$y = 100\left(\frac{d}{D}\right)^h = 100(d_r)^h \qquad (12.13)$$

where h is a variable, can be called the generalized Fuller grading. The smaller the h parameter is, the more fine particles are in the Fuller grading, as can be seen from Fig. 12.3. 2. A comparison of the m values for this grading as calculated by Eq. 11.13 in Example 11.3 to the optimum m_o values of Table 11.6 shows that the Fuller parabola provides aggregate gradings that are optimum for a wide range of maximum sizes only when the aggregate-cement ratio n is around 6 and for maximum sizes of 3/4 to 1 1/2 in (20 to 40 mm) when $n = 9$, by weight. 3. The grading equation of the total solid volume, when the aggregate of Fuller grading is blended with cement in aggregate-cement ratio n, by weight, is, from Eq. 12.2,

$$y_c = \frac{100}{n + 1} + \left(\frac{100n}{n + 1}\right)\sqrt{\frac{d}{D}} \qquad (12.14)$$

This is a Bolomey-type grading equation ($i = 0$, $h = 0.5$) in Eqs. 12.6 and 12.7 with $g = 100/(n + 1)$. The generalized Bolomey grading can be defined similar to Eq. 12.13.

Another possible form of theoretical grading curves is logarithmic:

$$y = f(d) = \frac{100}{\log^j(kD)}\log^j(kd) \qquad (12.15)$$

where *j* and *k* are parameters controlling the details and coarseness of the theoretical grading. The other symbols are identical with the symbols of Eq. 12.6. Several recommended values for the parameters *j* and *k* are presented in Table 12.2, along with the limits of optimum.

12.6 GRADING SPECIFICATION WITH LIMIT CURVES

The usual objection against the practical application of theoretical grading curves is that it is difficult specify the permissible deviations of the actual grading from theoretical one. This objection can be eliminated by specifying two grading curves, one as an upper limit and the other as a lower limit, with the request that the actual grading curve should fall within the limit curves in its total length. In the triangular diagram, the equivalency to a pair of limit curves is a grading area with the request that the grading point representing the actual grading should not be outside this area.

The limit curves can be given by equations or graphically. For instance, the Fuller parabola can be used as the upper limit and the EMPA curve (Table 12.1) as the lower limit on a total percentage - passing basis for combinations of fine and coarse aggregates, as is done in the Swiss code for reinforced concrete. An example for the graphically given limit curves can be those specified in the German standard DIN 1045 for reinforced concrete, two pairs of which for combinations of fine and coarse aggregates are reproduced in Fig. 12.4. Another example, from the British practice, is presented in Fig. 12.5. The higher the curve number is, the higher the water requirement of the grading, that is, the higher the water content needed for a specific concrete consistency. The lowest curves in these two figures do not provide workable mixtures with low cement contents.

All the gradings whose curves fall within the limits are not equally good, and acceptance of limit curves for grading specification does not imply that

Table 12.2: Parameters for the Optimization of Grading Eq. (12.15) as Recommended by Various Authorities

$$y = \frac{100}{\log^j(kD)} \log^j(kd)$$

Authority	Parameters		Limits for optimum		Remark
	j	k	n	D	
Palotás [1936]	$\dfrac{1 + 0.3m_o}{\log D - 0.3m_o + 1}$	100	4–10	5–80	For aggregates only
Rothfuchs [1964]	2	16	5–8	4–65	For aggregates only
Palotás [1936]	2.4	100	4–10	5–80	For cement-aggregate mixtures
Popovics [1962b]	2.8	600	4–10	2.5–150	For cement-aggregate mixtures

Aggregate: sand and gravel. Consistency: 3–5 in slump. n = aggregate–cement ratio, by weight. m_o = optimum fineness modulus. D = maximum particle size, mm.

460   Concrete Materials

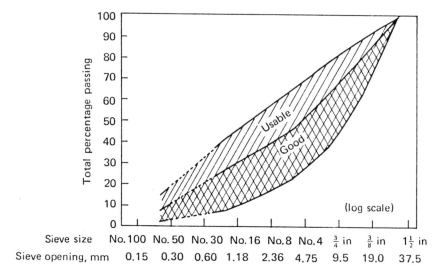

Figure 12.4: Limit curves for continuous gradings of sand and gravel combinations of D = 1.2 in (30 mm) for reinforced concrete (DIN 1045).

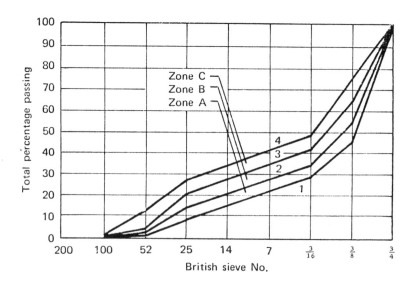

Figure 12.5: British grading curves and zones for aggregates of 3/4 in (19 mm) maximum particle size [Road Research Laboratory 1955].

they are. The fact is, however, that when the limit
curves are set up around an "optimum" grading curve,
and they are not too far apart, the concrete-making
properties of the gradings within the limits do not
differ too much. This is the case, for instance, with
the strength-producing properties of the gradings that
fall in the "good" region in Fig. 12.4. The situation
is different in the "usable" region in the same
figure. This is set up away from the "optimum" curve
in order to permit the use of oversanded gradings when
good gradings are not economically available. In this
region the strength-producing properties of the
grading deteriorate increasingly as the grading curve
moves away from the lower limit. This can be
counterbalanced by increasing the cement and water
contents. Gradings outside the limit curves require,
of course, even higher increases in the water and
cement contents; thus their use is not recommended.

The limit curves given in reinforced concrete
codes, such as those in Fig. 12.4, are suitable
primarily for the production of good strengths with
the usual cement contents and normal-weight
aggregates. The optimum fine aggregate contents of
lightweight aggregates are higher, as a rule, than
those of normal-weight aggregates. When a concrete
property other than strength is important, the
recommended grading can be different. For instance,
gradings for concrete with low permeability should
fall in the good region of Fig. 12.4 but closer to the
upper limit than to the lower limit [Walz 1956]. Even
finer grading is recommended for concretes for
pumping: The line dividing the good and the usable
regions in Fig. 12.4, or a narrow strip around this
line, provides a grading for concretes pumpable under
usual circumstances [Wilson 1974, ACI 1971d].
Prepacked concretes also require special gradings [ACI
1969]. Limit curves that are provided for continuous
gradings are not suitable for the evaluation and
specification of gap gradings, and vice versa.

As with theoretical gradings, limit curves are
used in European practice mostly for the fine
aggregate-coarse aggregate combinations. On the
American continent, ASTM C 33-90 specifies a pair of
limit curves for fine concrete aggregate and 10 pairs
of limit curves for the coarse aggregate depending on

the nominal sizes of the maximum and minimum particle sizes. Figure 12.6 shows examples of this specification. The underlying logic of this method is that since the fine aggregate and the coarse aggregate are usually supplied separately, their gradings can easily be specified separately. The combination of the fine and coarse aggregates should also fulfill the pertinent grading requirements. The fine aggregate limits have also been shown in the triangular system in Fig 10.5. Gradings for sand for masonry mortar are specified in ASTM C 144-89 and are usually finer than the grading of concrete sands.

12.7 GRADING SPECIFICATION ON PERCENTAGE PASSING-RETAINED BASIS

A refinement of the previously discussed method of specifying the amounts of fine sand, coarse sand, and coarse aggregate is when more than three size fractions are required. In practical cases upper and lower limits are specified for the amount of each fraction passing through each given sieve and retained on the next smaller sieve. It is possible to convert from these PPR Limits to upper and lower limits for the grading curve on the total percentage passing (TPP) basis [Dalhouse 1953]. Nevertheless the specification on the passing-retained basis is stricter than the TPP limit curve methods in that the PPR method requires not only that the total percentage passing values fall within the established TPP limits but also that the amount of each size fraction be within the given individual limits. The graphical equivalent of this double requirement is presented at the end of Example 12.2 below.

This type of grading specification is so restrictive that its use is justified only in special cases, such as concrete for pumping [ACI 1985a].

> **Example 12.2** A grading specification is given on the percentage passing-retained basis in the column "specified percentage of the size fraction" of Table 12.3.
> Convert these grading requirements to a pair of limit curves on the total percentage-passing

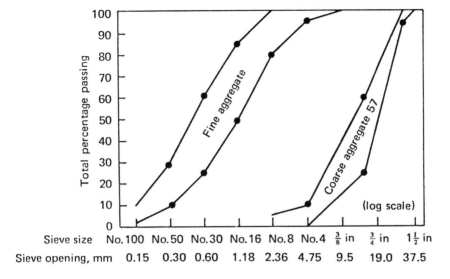

Figure 12.6: ASTM limit curves for the grading of fine concrete aggregate and that of a coarse aggregate of 1-in (25-mm) nominal maximum particle size.

Table 12.3: Conversion of Grading Specification from Percentage Passing-Retained Basis to Total Percentage Passing Basis

Sieve size				Specified percentage of the size fraction	TPP calculation of the points of curves			
Passing		Retained			1 and 2		3 and 4	
		$\frac{3}{4}$ in	19 mm	0		100		
$\frac{3}{4}$ in	19 mm	$\frac{3}{8}$ in	9.5 mm	25–40	75^a	60	75	125
$\frac{3}{8}$ in	9.5 mm	No. 4	4.75 mm	20–30	55^b	30	50	85
No. 4	4.75 mm	No. 8	2.36 mm	12–17	43	13	30	55
No. 8	2.36 mm	No. 16	1.18 mm	6–13	37	0	18	38
No. 16	1.18 mm	No. 30	0.60 mm	5–8	32	–8	12	25
No. 30	0.60 mm	No. 50	0.30 mm	4–7	28	–15	7^c	17
No. 50	0.30 mm	No. 100	0.15 mm	3–5	25	–20	3^d	10
No. 100	0.15 mm	Pan	0	0–5			0^e	5^e

$^a 75 = 100 - 25.$
$^b 55 = 75 - 20.$
$^c 7 = 32 - 25.$
$^d 3 = 28 - 25.$
^{e}Specified value.

basis.

Solution   (Fig. 12.7):

1. Calculate the points of Curve 1 from the lower limits and the points of Curve 2 from the upper limits of the specified percentage of the size fraction as shown in Table 12.3.
2. Calculate the points of Curves 3 and 4, respectively, from these values as shown in Table 12.3.
3. Plot Curves 1, 2, 3, and 4.
4. Establish the upper and lower TPP limit curves by selecting the innermost portions of curves 1, 2, 3, and 4.

Note that only those gradings that fulfill two geometric conditions comply with the given PPR specification: (1) the TPP grading curve should fall within the established limit curves (Curves 3 and 4 in the lower portion) in its total length; and (2) the slope of each segment of this grading curve should not exceed in positive or negative direction the slopes of Curves 1 and 2 in the same zone of particle sizes.

12.8 GRADING SPECIFICATIONS WITH PARTICULAR CONDITIONS

Three cases are discussed here when conditions other than the points of the grading curve are used for grading specification. In each case the grading resulting from the particular condition is compared to the previously discussed generalized Fuller gradings.

12.8.1 Percentage Gradings

The characteristic of "percentage" gradings is that the amount of material Δy_j *retained* on each successively smaller member of a regular sieve series is a fixed percentage p of the quantity retained on the preceding larger sieve Δy_{j+1}. Thus, the following equation holds for every value of j :

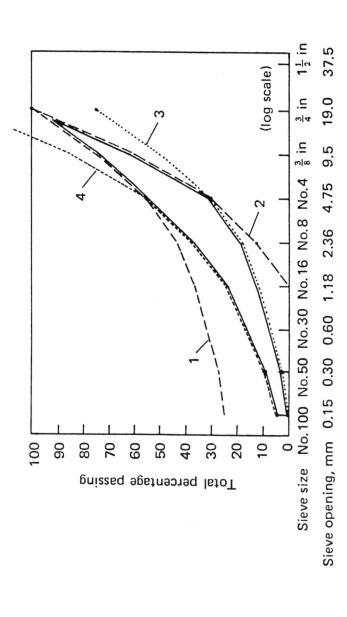

Figure 12.7: A pair of *TPP* limit curves, marked with continuous line, that correspond partially to the grading specification on percentage passing-retained basis presented in Example 12.2.

$$\frac{\Delta y_j}{\Delta y_{j+1}} = \frac{p}{100} = \frac{1}{r}$$ (12.16)

In other words,"70% gradation" is one in which there is 70% as much material of the second largest size as of the largest size, and again 70% as much of the third largest size as of the second, and so on. The r term is often called *size ratio*.

A special form of the percentage grading occurs when each *ordinate* of the sieve curve y_j belonging to each sieve is a fixed percentage of the ordinate belonging to the preceding, larger sieve y_{j+1}. Thus, the following general equation holds for every value of j:

$$\frac{y_j}{y_{j+1}} = \frac{p}{100} = \frac{1}{r}$$ (12.17)

If Eq. 12.16 is fulfilled, then Eq. 12.17 is fulfilled as well. Equation 13.17 is a particular form of Eq. 12.16.

Percentage gradings were used for laboratory purposes, for example, by Kennedy, "in order to avoid questions of particle interference and possible effects of poor gradation" [Kennedy 1940]. They were also used by Glanville et al. [Glanville 1938]. Nevertheless, it was shown by others [Popovics 1962b, Plum 1950] that in the case of a Tyler sieve series, a percentage grading defined by Eq. 12.17 is nothing but a different mode of expression of a generalized Fuller grading. In other words, each percentage grading is a generalized Fuller curve and vice versa: each Fuller curve represents a percentage grading. For a Tyler sieve series with the size ratio of 2, the general equation of a percentage grading is

$$f(d) = 100\left(\frac{d}{D}\right)^{\log r/\log 2}$$ (12.18)

The comparison of Eqs. 12.18 and 12.13 not only

proves the identity of the percentage gradings and Fuller gradings but shows the following equalities as well:

$$h = \frac{\log r}{\log 2} \qquad (12.19)$$

and

$$r = 2^h \qquad (12.20)$$

where $r = 100/p$.

Example   12.3 The customary form of the Fuller gradings is a parabola of second degree ($h = 0.5$). Thus, in this case

$$r = \sqrt{2} = 1.41$$

and

$$p = \frac{100}{1.41} = 70.7\%$$

The practical consequence of the identity of the Fuller gradings and percentage gradings is twofold: (1) The limits of validity presented in the previous paragraph for Fuller gradings are also effective for gradings of the percentage type; and (2) by the consideration of Eqs. 12.19 and 12.20, all methods of calculation that are presented for Fuller gradings can be applied for percentage gradings.

12.8.2 Particle Interference

Weymouth's theory of particle interference [Weymouth 1933] implies that for optimum gradings there always exists an optimum average clear distance between adjacent particles of the same size as they lie in the placed concrete. This optimum distance is such that each particle size has just sufficient space to move into the space between the particles of the next larger size. In other words, to prevent harsh mixtures, the closeness of packing in the concrete should be such that the average clear distance between particles of the same size should not be smaller than the average diameter of the next smaller size group [Powers 1953b]. When the particles are spaced so that the average distance between particles is less than the optimum distance, harshness results because of particle interference.

Based on a proof by Powers, Dunagan has pointed out [Dunagan 1940] that, under some circumstances, the application of Weymouth's principle for continuous gradings results in Fuller gradings. Thus, apart from other inconsistencies of the particle-interference method [Butcher 1956], the limits of validity for Weymouth's theory are, in a sense, the same as those presented in a previous paragraph for Fuller gradings.

12.8.3 Principle of Maximum Density

Furnas has developed mathematical relations for the sizes and amounts of solid particles required to obtain maximum density, that is, minimum percentage of voids, in an aggregate sample. His results indicate that there is a definite amount of the finest size fraction(s) in an aggregate that produces the maximum density. With either an increase or a decrease of this amount, the percent of voids in the aggregate will increase; this could explain, at least partially, the descending branches of the fitted curves in Fig. 11.4. By applying these relations for continuous gradings, Furnas found that "for each class of material there will be a certain constant ratio between the amounts of material of consecutive screen sizes which will give maximum packing for that system. Such a ratio will be designated by the symbol r" [Furnas 1931] .

In this way, the continuous gradings based on the principle of maximum density are reduced under certain conditions to the gradings of the percentage type. Since the percentage gradings and Fuller gradings have been shown to be identical, Furnas's statement not only supports mathematically the popular belief that among the continuous gradings the Fuller gradings can be of maximum density, but his statement also shows that these Furnas gradings should be used for concrete within the limits presented for the Fuller gradings.

Note that the percent of voids in an aggregate sample of generalized Fuller grading and specified maximum particle size is dependent on the particle shape, degree of compaction, and the parameter h of Eq. 12.13. (Fig. 12.8). In other words, the original Fuller grading with $h = 0.5$, or any other constant power, does not provide automatically maximum density, in contradiction to the popular belief. References related to minimum percentage of voids have been mentioned in Chap. 10.

A general remark seems appropriate in closing. The photographs of gap graded as well as continuously graded concretes presented in Chap. 13, or elsewhere in the literature, prove that the internal structure of concrete does not resemble at all the picture advanced by proponents of the maximum denseness principle for grading. This latter is a structure that consists of symmetrically arranged, contacting circles of identical diameter d, representing the coarse aggregate particles, where the remaining holes are filled first with circles of identical diameter $0.15d$ and then with subsequently smaller circles. In actuality, the coarse aggregate particles are distributed in a random manner in the mortar [Wise 1952]. This is true even in the case of concretes where the aggregate is carefully blended to obtain gradings of maximum denseness. Thus, it is little wonder that none of the "ideal gradings" derived mathematically from this unrealistic model of maximum denseness in the aggregates has been proved optimum from the standpoint of concrete technology.

12.9 SPECIFICATIONS WITH FINENESS MODULUS

Specifications with fineness modulus require that the fineness modulus of the fine and coarse aggregate combination (or that of the fine aggregate or the coarse aggregate) be equal, or close, to a prescribed value. When the emphasis is on the strength of concrete, the prescribed values are usually the optimum fineness moduli, such as those presented in Table 11.6 and the related footnotes. It is more convenient for practical purposes to provide upper and

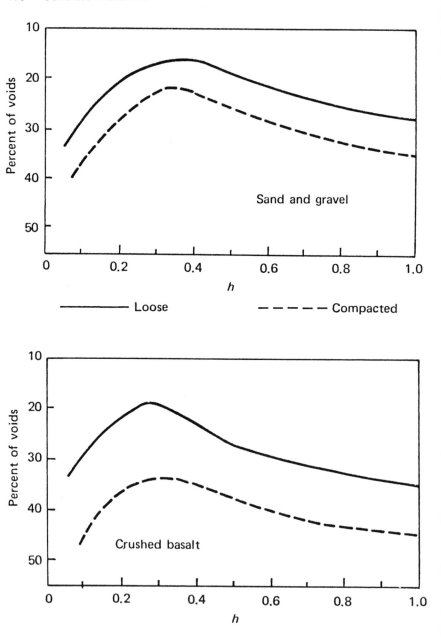

Figure 12.8: Percent of voids in various aggregate samples with $D = 1.2$ in (30 mm) as a function of the exponent h of the generalized Fuller equation $y = 100(d/D)^h$ [Hummel 1959].

lower limits for the acceptable fineness moduli. A typical example is

$$m_a = m_o \left|{\substack{+0.4 \\ -0.6}}\right.$$
(12.21)

where m_a and m_o are the acceptable and the optimum (or specified) fineness moduli, respectively, for fine and coarse aggregate combinations.

It has been pointed out that the fineness modulus is not sensitive enough to the amounts of fine particles; therefore, even gradings with optimum fineness moduli can have too much or too little fine particles for a good concrete. Although such extreme gradings are easily recognizable, and thus can be eliminated, it is also possible to specify additional conditions that supplement the specified fineness modulus for the elimination of useless gradings of acceptable fineness moduli. Such additional conditions can be, for instance, upper and lower limits for the amounts of fine particles or upper and lower limits for the permissible specific surface of the aggregate, as discussed at the end of Chap. 11. It is also evident that such control of the amount of the fine size fraction(s) will keep the percent of voids in the aggregate reasonably low.

In American practice, the fineness modulus is used mostly for the grading evaluation of fine aggregates. ASTM C 33-90, for instance, specifies limit curves for the gradings of fine concrete aggregates (Fig. 12.6) and, in addition, requires that the fineness modulus be not less than 2.3 or more than 3.1. This practice appears paradoxical in view of the relative insensitivity of the fineness modulus to the evaluation of fines. Besides, the simultaneous specification of a sieve curve, or a pair of limit curves, and a fineness modulus is not a particularly useful idea because, as a rule, in such specifications either the curve(s) or the modulus becomes more or less meaningless.

Consequently, this writer believes that there are better ways than common American practice to utilize the fineness modulus for the grading evaluation of concrete aggregates. This belief is supported further in the next paragraph.

12.10 CRITICAL COMPARISON OF VARIOUS METHODS FOR GRADING EVALUATION

A continuous grading that is close enough to the continuous grading curve properly specified for the given case is unequivocally good for the particular purpose. However, there can be many other gradings, among them gap gradings, that are actually just as suitable for making a good concrete as the accepted continuous grading; yet the gap gradings are judged poor or unacceptable by the specified grading curve method since the discrepancies between the actual grading curve and the specified one are usually large. This means that the specified grading curve method provides only the *sufficient* but not the necessary condition for grading evaluation; therefore, it is overly and unnecessarily restrictive.

Essentially the same statement is valid for the methods of grading evaluation based on the principles of percentage passing-retained, of the specified percentage grading, of particle interference, and of the maximum density principle, and in cases where the fine aggregate and coarse aggregate gradings are specified separately. The grading evaluations based on continuous limit curves for fine aggregate-coarse aggregate combinations; on specifying the amounts of fine sand, coarse sand, and coarse aggregate; and on specifying the amounts of fine and coarse aggregates are somewhat less restrictive, although in the case of the latter two methods, this is the consequence of a reduced reliability.

The method of grading evaluation of concrete aggregates by fineness modulus has the opposite weakness; namely, when the grading is good, the value of the fineness modulus is always near the optimum (or specified) value. If, however, the fineness modulus equals the optimum, this does not necessarily mean that the grading is good, because the amount of fine particles, needed for good workability and denseness of the fresh concrete, is not assured by the adequate numerical value of the fineness modulus alone. This means that the fineness modulus method provides only the *necessary* but not the sufficient condition for grading evaluation; therefore it is overly loose. The other numerical grading characteristics, especially the specific surface and the others in the same group (see Table 10.3 and related text) are less suitable for grading evaluation because they have the weaknesses of the fineness modulus without having all of its strengths.

Thus, it may be concluded that the specified grading curve method, as well as the other related methods and to a smaller extent the limit curve method, are too conservative, whereas the fineness modulus method and the other numerical characteristics are too liberal for the grading evaluation of concrete aggregates. It is important to recognize, however, that although one cannot do too much about the restrictiveness of the grading and limit curve methods, it is easy to tighten up the methods based on the fineness modulus by regulating the amounts of fines in the aggregate in one way or another, as discussed at the end of Chapter 11. For this reason and because of the other advantages of the fineness modulus method (it is well supported by experimental data; is numerical: is valid for continuous as well as for gap gradings; is simple; has a statistical interpretation; is good for easily finding the blending proportions for grading corrections; can take into account the effects of maximum particle size, cement content, and particle shape, etc.), this writer prefers the fineness modulus method to any other method for the grading evaluation of fine and coarse aggregate combinations for concrete. The ranking of the methods of grading evaluation in order of personal preference is as follows:

1. Fineness modulus method for the combinations of fine and coarse aggregates, used with common sense
2. Fineness modulus method, as in (1), but supplemented by additional grading condition(s), especially the $D-m-s$ method
3. Limit curve method for combinations of fine and coarse aggregates (primarily for continuous gradings)
4. Specification of the amounts of fine sand, coarse sand, and coarse aggregate (but not for works of primary importance)
5. Specification of the amounts of fine and coarse

aggregates (for less important works and only when the sources of the two aggregate fractions remain unchanged)

6. Specification of the grading of the fine aggregate and that of the coarse aggregate separately

7. Specific surface method

8-9. Specification on the percentage passing-retained basis or by particular conditions, such as maximum density.

13

Internal Structure of Concrete and Its Optimization

SUMMARY

The analysis of the *internal structure* of hardened concrete can reveal if the aggregate grading used is close enough to the optimum. This analysis is based on the measurement of the intercepts of mortar layers among coarse aggregate particles in the finished concrete. The coarser the grading, the closer the coarse aggregate particles are packed. The results indicate that there exists a necessary minimum value for the average mortar layer intercept below which the workability of the concrete is inadequate. This necessary minimum intercept is dependent on several factors, but, significantly, it seems to be independent of the particle shape of the coarse aggregate. For instance, the coarsest permissible, that is, optimum, gradings for any particle shape for traditional concretes are those that provide an average mortar layer intercept of 0.14 in (3.5 mm) in the finished concrete. The primary effect of the particle shape on the grading optimum is that more coarse aggregate particles can be packed without harshness in the concrete when the particle shape is spherical than when it is unfavorable.

The question frequently arises of how to obtain a specified grading by blending aggregates of different gradings. The fulfillment of $n - 1$ grading

475

conditions requires the blending of at least *n* aggregates of different gradings. These conditions can be written in the form of a system of *n* − 1 simultaneous linear equations and inequalities that contain the blending proportions, the grading characteristics of the blending aggregates, and the specified or resulting values of these grading characteristics. This system can be solved either for the blending proportions or for the grading characteristics resulting from the blending. Depending on the number and nature of the grading characteristics as well as on the number of aggregates to be blended, there can be an infinite number of sets of blending proportions, a single set, or no set at all that satisfies all the grading conditions given in the linear system. This numerical method provides exact solutions but becomes cumbersome when the number of grading conditions is greater than three. Fortunately, certain forms of the linear system lend themselves to quick graphical or semigraphical solutions. Their common disadvantage, however, is that they are valid only in special, although important, cases.

For the fulfillment of one grading condition, that is, for blending two aggregates, either the numerical method or the British graphical method is most suitable. The triangular method is convenient for blending three aggregates. For more aggregate fractions, the graphical method by Rothfuchs, the computerized numerical method, or the computerized triangular method is recommended, depending on the type of the grading conditions.

13.1 AGGREGATE GRADING AND THE INTERNAL STRUCTURE OF CONCRETE

The amount of a given fine aggregate and that of a given coarse aggregate are well balanced in a good grading. However, quantitatively this balance is dependent on numerous factors, such as the maximum particle size, particle shape, cement content, method of compaction, and fineness of the sand. In addition to the behavior of the concrete in the fresh as well as in the hardened state, visual inspection of the internal macrostructure of concrete can reveal if the grading used, is close enough to the desired optimum [Popovics 1973b].

An example for this is given in Fig. 13.1, which shows the internal structure of an air-entrained gravel concrete through a cut surface. The fineness modulus of the continuous grading used is 5.7. The maximum particle size is 1 1/2-in (37.5-mm). The cement content of the concrete is 570 lb/yd^3 (340 kg/m^3) of portland cement plus 170 lb/yd^3 (100 kg/m^3) of fly ash. The water-cement ratio is 0.50 by weight; the air content is about 3%. The unit weight of the fresh concrete is 146 lb/ft^3 (2340 kg/m^3), and the slump is about 1 in (25 mm). This concrete has a compressive strength of 5960 psi (41.1 MPa) and a flexural strength of 525 psi (3.6 MPa) at the age of 28 days.

The high strength values indicate that this concrete, and thus the grading, is good. It is a good grading indeed, but not quite optimum. The behavior of this fresh concrete in the laboratory, especially during compaction by rodding performed according to ASTM C 192, indicated that there was a slight excess in the amount of mortar. This observation is supplemented by the fact that the optimum fineness modulus of the aggregate recommended for the maximum strength of the particular mixture is approximately 6.2 (see Table 11.6).

Figure 13.1 shows, therefore, a concrete with a well-graded aggregate: The matrix is a dense mortar in which a number of coarse particles are embedded in a random manner, although perhaps a few more gravel pieces could have been placed in it without overcrowding the internal structure. This evaluation is, of course, highly subjective and only qualitative. To make this approach numerical, a simple analysis of the internal macrostructure of hardened concrete, the intercepts of mortar layers between coarse aggregate particles in a cut, plain concrete surface can be utilized. The coarser the overall grading, the smaller these intercepts become along with the actual thicknesses of the coatings of mortar surrounding the coarse aggregate

Figure 13.1: Concrete made with traditional sand and gravel of a continuous grading. Fineness modulus = 5.7 [Popovics 1973b].

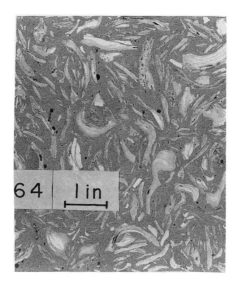

Figure 13.2: Concrete made with 40% beach sand and 60% crushed reef shell by weight. Angularity number = 30. Fineness modulus = 4.95. From Popovics, Reef Shell-Beach Sand Concrete, *Living with Marginal Aggregates,* STP 597, 1976 [Popovics 1976d]. Reprinted by permission of the American Society for Testing and Materials, Copyright.

particles. The mortar intercepts can be measured with the linear traverse method, which is similar to the procedure used in petrography or as described in ASTM C 457 for the air-content determination in the hardened concrete, except that there is no need for a microscope in this case. A random cut is taken through the concrete; a regular grid is placed randomly thereon; and linear intercepts are measured along the lines of the grid with the lengths as intercepts in mortar among coarse aggregate particles. These measured intercepts can be summed and averaged, the result of which is a number called *average mortar intercept.*

The application of the method of mortar intercept to the concrete section in Fig. 13.1 and the sections of other concretes has provided the following conclusions: (1) An average mortar intercept of 0.14 in (3.5 mm) seems to be the necessary minimum in traditional concretes (that is, for concretes with traditional gradings of the fine and coarse aggregates for a smooth, continuous grading with around 1 in (25.4 mm) maximum particle size and with medium cement content) for an adequate workability and maximum strength; and (2) this criterion is valid not only for gravel concrete but also for crushed coarse particles. In other words, this criterion seems to be independent of the particle shape, as illustrated in Fig. 13.2 by another concrete surface containing beach sand as the fine aggregate and crushed reef shell as the coarse aggregate. Thus, one can define the coarsest permissible gradings for any particle shape and for traditional concretes as those that provide a dense mortar and an average mortar layer intercept of 0.14 in (3.5 mm) in the finished concrete. Any amount of traditional mortar less than this minimum in a continuously graded concrete would not provide enough lubrication for the coarse aggregate particles for a workability that is adequate for hand compaction. This does not mean, however, that smaller mortar layer intercepts always result in inadequate workability. For one thing, a reduction in workability caused by a moderate reduction in the mortar quantity can be overcome by intensive mechanical compaction, such as vibration. This is in accordance with the experience that coarser gradings can be used in the concrete when it is compacted by vibration. Second, less mortar can also do the job when its lubricating ability is improved. This can be done to a certain extent either by increasing the cement content or by using a finer sand.

This latter statement is illustrated in Fig. 13.2. This shows a concrete that contains 62%, by absolute volume, of reef shell of 1 1/2-in (37.5-mm) maximum particle size and of angularity number AN = 30, and 32% beach sand practically all the particles of which passed sieve No. 30 (600 μm) and were retained on sieve No. 100 (150 μm) ($m \approx 1.5$). Trial mixes demonstrated that this is the maximum amount of shell that still provides a reasonable workability for this concrete [Popovics 1976d]. The average intercept among the shell particles is 0.120 in (3.04 mm), the low value of which is attributed to the too-fine grading of the applied (beach) sand. Note, however, that when crushed stone (AN = 10) or gravel (AN = 5) (Fig. 13.3) is used to make concrete with a composition similar to the reef shell concrete above, the average mortar intercept values are 0.148 in (3.75 mm) and 0.159 in (4.03 mm), respectively, despite the identical gradings. This means that more coarse aggregate particles can be packed without interference into a concrete when the particle shape is spherical than into an identical concrete when the shape is unfavorable.

The internal structure of a concrete depends not only on the coarseness of the grading and on the particle shape of the aggregate but also on the details of the particle-size distribution employed. This is demonstrated in Fig. 13.4, which is a portion of a larger investigation [Popovics 1964c]. Here three non-air-entrained concretes of differing but identically coarse gradings and of otherwise identical compositions are presented. The cement contents are 520 lb/yd^3 (310 kg/m^3), the water-cement ratios are 0.62 by weight. Concrete No. 10 was made with a continuous grading (C_1), Concrete No. 15 with a one-gap grading (O_1), and Concrete No. 18 with a two-gap grading (T_1). Details of the particle-size distributions are shown in Fig 11.4.

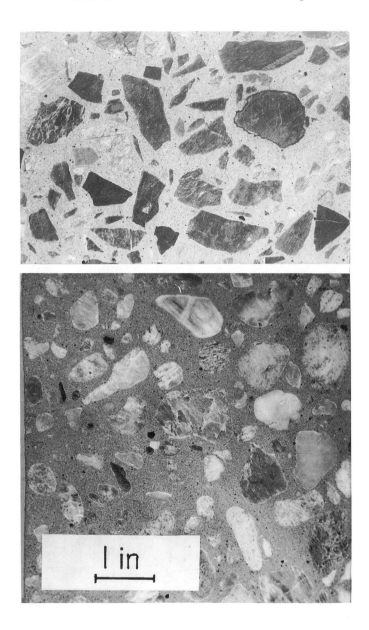

Figure 13.3: Crushed stone and gravel concretes. The gradings and compositions are similar to those of the reef shell concrete in Figure 13.2. Angularity numbers are 10 and 5, respectively [Popovics 1973b].

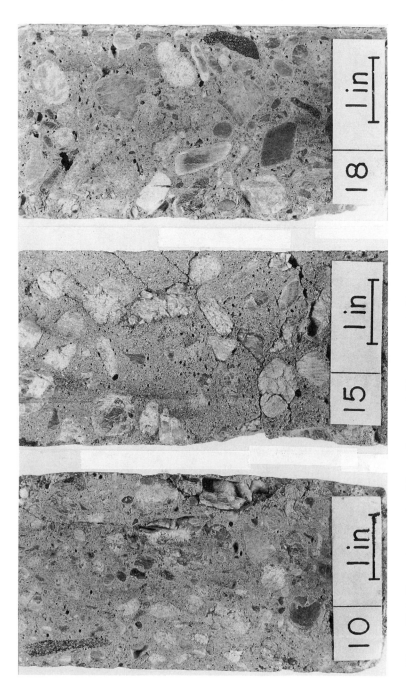

Figure 13.4: Concretes made with continuous (left), one-gap (middle), and two-gap (right) gradings of identical coarseness according to the sieve curves in Figure 11.5 [Popovics 1964c].

These three gradings were set up so that the three important characteristics of coarseness, namely, the maximum particle size, the fineness modulus, and the calculated specific surface of the aggregate, were kept practically constant despite the obvious differences in the particle-size distributions.

The differences in the internal structures in Fig. 13.4 are quite obvious. Especially, the concrete made with the grading of one large gap (O_1) appears different; the grading seems coarser (which is actually not true), and excessive mortar seems to be present (which is true). This latter condition again supports the statement that a thinner average mortar layer is acceptable when a finer sand is used. The appearance of the three graded aggregates is shown in Fig. 13.5.

The linear traverse method can provide the proportion of the cement paste or mortar in the concrete with fair accuracy [Kelly 1958, Axon 1962]. It is also conceivable to obtain the equivalent of an aggregate sieve analysis [Terrier 1967, Stroeven 1973] or the specific surface [DeHoff 1968, Holliday 1966] of the aggregate from linear traverse measurements, similar to the methods used for the determination of the parameters of the air-void system in hardened concrete (ASTM C 457). But the uncertainties are formidable even for spherical particles. If the aggregate is assumed cuboidal, the situation becomes more complex because lines passing through a cube can be longer or shorter than the cube side. Irregular shapes and mixed shapes present even more difficult problems, Thus, the reliability of such grading analysis is highly questionable [Figg 1971].

13.2 NEED FOR BLENDING AGGREGATES

Regardless whether the optimum of a grading is defined by the average mortar intercept or by other methods discussed in Chapter 12, the question frequently arises of how to obtain this optimum grading, or any other specified grading, by blending aggregates of different gradings. In addition, there are other cases where aggregate blending can be

Figure 13.5: Appearance of the continuous (left), one-gap (middle), and two-gap (right) gradings presented in Figure 11.5 [Popovics 1964c].

useful, such as for the production of an aggregate of prescribed specific gravity, for the decrease of the reactivity of an aggregate, for the improvement of the durability, for the increase of the skid resistance, for the reduction of the aggregate price, or for any combination of these requirements. In these cases, either (1) the blending proportions as well as certain properties of the aggregate components are given, and the pertinent properties of the aggregate blend should be determined by calculation; or (2) certain properties of the components are given; the required properties of the blend are specified; and the needed blending proportions should be determined for the fulfillment of the specification.

Here, calculations related to blending proportions will be discussed, first in general terms, then grading problems will be used for the illustration of the specific methods primarily used for the determination of the needed blending proportions.

13.3 GENERAL THEORY OF BLENDING

The common mathematical basis of the various graphical, semigraphical, and numerical methods for finding the needed blending proportions can be stated this way [Popovics 1973c]: In the special case when the specified conditions are given in the form of a system of $p - 1$ linear equations, their fulfillment requires the combination of at least p different aggregates. In the more general case, however, the conditions can be written in the form of a system of $p - 1$ simultaneous linear equations and inequalities, including tolerances, containing the needed blending proportions as unknowns. To these a pth equation should be added to take care of the obvious condition that the sum of the blended aggregates is equal to the total of the combined aggregates. Thus, the general mathematical form of this system is something like this:

$$c_{11}x_1 + c_{12}x_2 + \ldots + c_{1n}x_n = b_1$$

$$c_{21}x_1 + c_{22}x_2 + \ldots + c_{2n}x_n = b_2 \pm \Delta b_2$$

.

.

.

$$c_{i1}x_1 + c_{i2}x_2 + \ldots + c_{in}x_n \geq b_i$$

(13.1)

.

.

.

$$c_{(p-1)1}x_1 + c_{(p-1)2}x_2 + \ldots + c_{(p-1)n}x_n < b_{(p-1)}$$

$$x_1 + x_2 + \ldots + x_n = 1$$

where c_{ij} = the ith characteristic of the jth
        aggregate to be blended
x_j = the sought blending proportion for
        the jth aggregate
b_i = the given i^{th} condition for the
        blended aggregate
p = number of specified conditions + 1
n = number of aggregates to be blended

The value of n can be smaller than, equal to, or larger than p. In the first case, System 13.1 may or may not have a solution(s), but as long as $n \geq p$ and the equations and inequalities are not contradictions, there are infinitely many sets of x_j values, but at least one set, that satisfy this system. Out of these, however, only those sets represent blending proportions where none of the included x_j values is less than zero, since a negative blending proportion is physically meaningless. The lack of such a nonnegative set of solutions indicates that it is impossible to fulfill all the given $p - 1$ grading conditions with the given n aggregates.

In other words, System 13.1 is applicable for every kind of blending problem of aggregates. In most cases it provides the exact solution(s) to the grading problems if there are any. The disadvantage of the numerical method is that it becomes increasingly lengthy and cumbersome when n and/or p are larger than 3, unless a computer is used.

Note that the 13.1 system can be treated as a set of constrains in linear programming, for instance for cost minimization. [Nicholls 1976, Cannon 1971, Popovics 1982c]

It follows from the space-filling role of the aggregate particles that the correct way to express their blending proportions is by absolute volume. In this case, the grading of the combined aggregate should also be expressed in terms of absolute volume. Only when the specific gravities of all the given aggregates to be blended can be considered identical may the proportions be expressed by weight.

The mathematically simplest form of System 13.1 is when $p = n$, and each condition is represented by an equation with a single number on the right-hand side. Here, the number of the aggregates available for blending is one greater than the number of the specified conditions. Mathematically, this problem is represented by a system of n linear equations with n unknowns. Since the determinant of the coefficients cannot be zero in blending problems, there is one and only one set of solutions of this system. If none of the solutions is less than zero in this set, the solutions represent the only combination of the n given aggregates that can fulfill all the $n - 1$ conditions.

When there are more equations than unknowns in the system, it is mathematically impossible, in general, to find an exact set of solutions. In other words, when the number of aggregates to be blended is less than the number of the specified conditions plus one, all the conditions cannot be fulfilled in an exact fashion. It still may be possible, however, to find a set(s) of blending proportions by the least squares method that provides the "best" approximation for the blended aggregate to all the specified conditions [Neumann 1964] .

13.4 IMPROVEMENT OF GRADING

It often occurs in practice that the gradings of
available aggregates do not meet the specifications.
In such cases, two or more aggregates of differing
suitable gradings should be blended in order to
obtain an acceptable grading. The grading
requirements in certain cases, such as pumping or
high impermeability of the concrete, can be quite
complex and strict, in which case four or even more
aggregates may be needed for the production of a
suitable grading.

A set(s) of blending proportions with given
aggregates for an approximation of the required
grading can be obtained from System 13.1, or from
one of its particular numerical or graphical forms,
provided that such a set does exist. Trial-and-error
methods can also be used, although these can be
cumbersome. But regardless of what method is used,
the recommended first step is to plot the grading
curves or grading points of the aggregates to be
blended as well as the specification limits. This
helps one see if a blend can be found at all with
the available aggregates to meet the specification,
and if so, where the critical sizes are.

A typical case for the application of a
particular form of system 13.1 is when the system
consists of n linear equations with $p = n$ unknowns,
and the conditions specify $n - 1$ points within the 0
and 100% points through which the cumulative sieve
curve of the combined grading must pass. A numerical
illustration for three aggregates is given in
example 13.1.

Example 13.1 Produce a grading that passes
through the 20% point at the No. 30 particle
size and the 40% point at the No.4 particle size.
Three mineral aggregates, A, B, C, are given for
blending. Their gradings are shown in Fig. 13.6.
The three blending proportions needed can be
calculated from the following system of linear
equations, where the coefficients are taken from
Fig. 13.6:

$$0.53x_A + 0.10x_B + 0.05x_C = 0.20$$

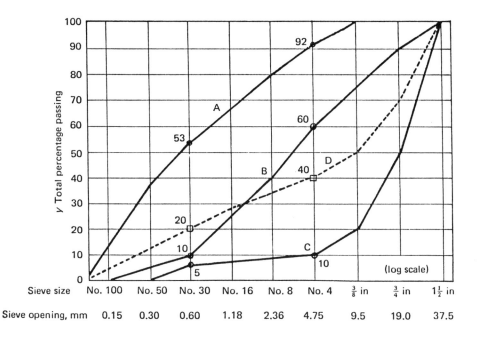

Figure 13.6: Aggregate gradings for Example 13.1 and 13.11. Grading D was obtained by blending 30% aggregate A, 10% aggregate B, and 60% aggregate C.

Table 13.1: Calculation of the Grading of an Aggregate Blend

d_i, mm	Aggregate A y_{Ai}	$0.303y_{Ai}$	Aggregate B y_{Bi}	$0.102y_{Bi}$	Aggregate C y_{Ci}	$0.595y_{Ci}$	Blended aggregate y_{Di}
0.1	3	0.91	0	0	0	0	0.91
0.15	13	3.94	0	0	0	0	3.94
0.30	37	11.21	5	0.51	0	0	11.72
0.60	53	16.06	10	1.02	5	2.97	20.05
1.18	67	20.30	25	2.55	8	4.76	27.61
2.36	80	24.24	40	4.08	9	5.36	33.68
4.75	92	27.88	60	6.12	10	5.95	39.95
9.5	100	30.30	75	7.65	20	11.90	49.85
19.0	100	30.30	90	9.18	50	29.75	69.23
37.5	100	30.30	100	10.20	100	59.50	100.00

$$0.92x_A + 0.60x_B + 0.10x_C = 0.40$$
$$x_A + x_B + x_C = 1.00 \qquad \text{(13.2)}$$

From this,

$$x_A = 0.303 \qquad x_B = 0.102 \qquad x_C = 0.595$$

that is, approximately 30% should be blended from aggregate A, 10% from aggregate B, and 60% from aggregate C.
An easy but important check of the calculations above is

$$0.303 + 0.102 + 0.595 = 1.000$$

Example 13.2 Calculate several points of the grading curve for the blended aggregate obtained in Example 13.1.
The ordinate y_{Di} of the blended grading curve that belongs to the particle size d_i can be calculated as follows:

$$y_{Di} = 0.303y_{Ai} + 0.102y_{Bi} + 0.595y_{Ci}$$

where y_{Ab} y_{Bi}, and y_{Ci} are the ordinates belonging to the same size d_i for gradings A, B and C, respectively.
The details of the calculation are shown in Table 13.1 on a total percentage passing basis. The calculated points of the blended grading are also presented in Fig. 13.6 as curve D. Note that this curve D does pass through the specified points $y_{0.60} = 20\%$ and $y_{4.75} = 40\%$, as required.

13.5 BLENDING OF TWO AGGREGATES

When only one grading condition should be met, at least two aggregates are to be blended. If, for example. the combined sieve curve must pass through the point y_0, the blending proportions can be calculated directly from the following two formulas provided that y_0 is between y_1 and y_2:

$$x_1 = \frac{y_2 - y_0}{y_2 - y_1} \qquad\qquad (13.3)$$

$$x_2 = \frac{y_1 - y_0}{y_1 - y_2} \qquad\qquad (13.4)$$

with the availability of the check:

$$x_1 + x_2 = 1 \qquad\qquad (13.5)$$

where y_1 and y_2 are the pertinent ordinates of the sieve curves of aggregates 1 and 2, respectively.

Equations 13.3 and 13.4 are obviously the solutions of the corresponding system of two linear equations but otherwise similar to Eq. 13.2. Thus, they take into account only one point through which the combined grading curve is required to pass.

Example 13.3 Produce a grading in which the amount of particles passing through a sieve No. 4 (4.75 mm) is 40%. Aggregates B and C in Fig.13.6 are available for blending.
By taking the appropriate y values from Fig. 13.6 and substituting them into Eqs. 13.3 and 13.4:

$$x_B = \frac{10 - 40}{10 - 60} = 0.60$$

$$x_C = \frac{60 - 40}{60 - 10} = 0.40$$

The graphical interpretation of Eqs. 13.3 and 13.4 reveals that (1) the sieve curve of any combination of two aggregates must fall in its total length within the sieve curves of the two aggregates; and (2) the distance between these two sieve curves at any particle size is divided by the combined sieve curve in the ratio of the x_1 and x_2 blending proportions employed.(Linear interpolation)

This consideration may help one select a suitable value of y_0 in a blending problem. It can also be seen in Fig. 13.6 that there is no combination of aggregates A and B that could fulfill the 40% condition given in Example 13.3.

Note also that there is a range of solutions of the linear system, instead of a unique solution, when (1) the number of aggregates to be blended is greater than the number of grading conditions plus one; and/or (2) some or all of the grading conditions are given in the form of inequalities or in the form of ranges rather than as single numbers.

The range of solutions provides a certain flexibility for the engineer in selecting the most suitable set of blending proportions based on economic or other considerations.

Example 13.4 Calculate again the blending proportions for the case discussed in Example 13.3 but with the condition now that $y_0 = 40 \pm 5$.

The upper and lower limits of the sought x_B blending proportion as well as the corresponding value x_C can be calculated again by substituting successively $y_0 = 45$ and $y_0 = 35$ into Eqs. 13.3 and 13.4. The results are

$$(x_B)_{up} = 0.70 \qquad (x_C)_{up} = 0.30$$

and

$$(x_B)_{lo} = 0.50 \qquad (x_C)_{lo} = 0.50$$

That is, any quantity of aggregate B between 50 and 70% combined with aggregate C in the corresponding quantity of $100(1 - X_B)$ percent yields a combined grading in which the amount of particles passing sieve No. 4 (4.75 mm) is the required 40 ± 5%. From this range, one pair of blending proportions can be selected on the basis of local conditions, such as the price and availability of the two aggregates, and so on. Example 13.3 is a special case of the more general blending problem discussed above.

A system similar to Eq. 13.2, and consequently formulas similar to Eqs. 13.3 and 13.4, can be used for the calculation of blending proportions when the fineness modulus, or the specific surface, is used for the grading characterization rather than sieve curves. This is shown for two aggregates in the following example.

Example 13.5 In what proportions should a fine aggregate with a fineness modulus of $m_f = 2.2$ be blended with a coarse aggregate of $m_c = 8.0$ to obtain a combined grading with a fineness modulus of $m_o = 5.4$?

By formulas similar to Eqs. 13.3 and 13.4:

$$x_f = \frac{8.0 - 5.4}{8.0 - 2.2} = 0.45$$

$$x_c = \frac{2.2 - 5.4}{2.2 - 8.0} = 0.55$$

It should be noted that although the blending proportions x_B and x_C in Example 13.3 fulfill the requirement for the grading curve of the blended aggregate to pass through the point $y_0 = 40\%$, the rest of the grading curve may or may not be within acceptable limits. This can be checked by plotting the complete grading curve of the blended aggregate. When the fineness modulus method is used, as in Example 13.5, theoretically there is no need for such checking, since the fineness modulus cannot consider details of grading. From a practical standpoint, however, the plotting of the blending grading is still desirable, primarily to check the amounts of fine particles.

A special case of the grading imperfection is when the *fine aggregate* is too coarse or too fine. A possible correction is the separation of the fine aggregate into two or more size fractions and recombining them in suitable proportions, as discussed above. Also, when the fine aggregate is too coarse, or has unfavorable particle shape, this can be compensated for to a certain extent by (a) using additional mineral admixture; (b) increased cement content; (c) air entrainment; and (d) a water thickener. When the fine aggregate is too fine, this can be compensated for to a certain extent by (a) reducing its quantity; (b) by a reduction of the cement and/or mineral admixture content.

13.6 GRAPHICAL METHODS FOR BLENDING PROPORTIONS

Certain forms of the linear system presented as System 13.1 may lend themselves conveniently to graphical or semigraphical solutions. Such methods may be advantageous beyond their visuality, because they may be faster than numerical methods, they may provide the totality of the solutions, and they may be applicable for computer graphics. Their common disadvantage is that they are only valid for more orless special cases.

Their limited accuracy or, in certain cases, the approximate nature of graphical methods, is not particularly harmful because fluctuations in the gradings of the aggregates to be blended make a blending proportion precision better than 1% meaningless anyway.

Three graphical methods, including the triangular method, will be discussed below.

13.7 GRAPHICAL METHOD BY ROTHFUCHS

An approximate graphical method for the determination of blending proportions, very much the same as the one described below, was probably first offered by Rothfuchs [Rothfuchs 1962]. There are two restrictions concerning the applicability of the method: (1) It is suitable only for grading problems when all the gradings are characterized by sieve curves; and (2) no more than two of the n given individual sieve curves may overlap significantly at any point, viewed from the horizontal axis (which is a rare situation in the case of three or more aggregates to be blended).

But even with these restrictions, the method cannot provide exact solutions for two reasons: (1) The actual curves of the aggregates to be combined are not used, but straight lines are substituted for their curves; and (2) the larger the overlapping of two such grading straight lines, the poorer the approximation of the obtained solution.

The main advantage of this method is its simplicity and that this simplicity is hardly affected by the number of grading conditions or the number of aggregates to be blended. Thus, it is particularly useful for large p and n values.

The mathematical form of this method is the Gaussian algorithm. The use of the method is illustrated in Example 13.6 for four aggregates.

Example 13.6 Determine the blending proportions for the four aggregates shown by continuous lines in Fig. 13.7, so that the combined grading approximates the specified sieve curve shown by the dotted line.

Solution The procedure for solution is as follows: The sieve curves of the four aggregates to be blended are approximated by fitting straight lines. These are represented by dashed lines in Fig. 13.7. The opposite ends of these straight lines are joined together as shown by the dot-and-dash lines. The blending proportions sought can then be determined as the differences of the ordinates of the successive points, marked by circles, where the joining dot-and-dash lines intersect the dotted line representing the required grading. In our example, as shown on the right-hand side of Fig. 13.7, 4% of aggregate 1, 15% of aggregate 2, 24% of aggregate 3, and 57% of aggregate 4 should be blended for the approximation of the specified grading.

The difference between the specified sieve curve and the actual sieve curve of the determined combination of the four aggregates is also illustrated in Fig. 13.7 by the shaded areas.

If the combined grading is specified by a pair of limit curves rather than a single curve, there are, as a rule, infinitely many solutions to the blending problem. Here the acceptable ranges of the blending proportions can be obtained by applying the method above for the upper limit and then repeating the procedure for the lower limit. On the other hand, if

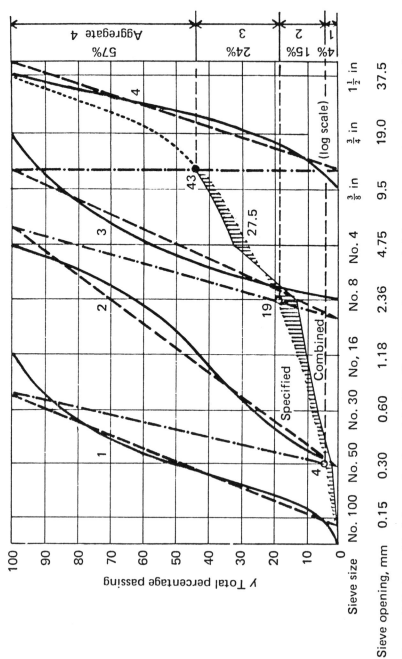

Figure 13.7: A graphical determination of the blending proportions needed for aggregates 1 to 4 to approximate the specified grading.

any of the dot-and-dash lines do not intersect the sieve curve specified for the combined grading, the grading problem has no solution.

A modification of the Rothfuchs method is the following [Lee 1973]: Instead of approximating the grading curves of the aggregates to be blended to straight lines, vertical lines are drawn between the opposite ends of adjacent curves, so that, if the two gradings overlap, the percentage retained on the upper curve equals the percentage passing on the lower curve; and if there is a gap between the two grading curves, the horizontal distances from the opposite ends of the curves to the vertical line are equal. Otherwise, the rest of the graphical procedure remains unchanged and the proportions for the blend can be read off again from the points where these vertical lines cross the specified grading curve.

13.8 BRITISH METHOD

Another simple graphical method is recommended by the British Road Research Laboratory [Road Res. Lab. 1950] for the determination of blending proportions for a combined sieve curve to pass through $p - 1$ specified points. The mathematical form of this method is again the Gaussian elimination algorithm; therefore, no double or triple overlapping of the sieve curves is permitted here, either. In other words, the grading problem in Example 13.6 can be solved with this method, but the one in Example 13.1 cannot. Under these conditions, this method provides the exact solution to the given system of linear equations within the accuracy of a graphical procedure. On the other hand, the method becomes more and more complicated with increasing values of p.

Various applications of the method are demonstrated in Examples 13.7-13.10.

Example 13.7 Determine the blending proportions for aggregates 1 and 2 of Fig. 13.7 by using the British graphical method, so that the amount of particles in the combined grading passing through a No. 50 (300-μm) sieve is 14% (~ 100 X 4/27.5).

Solution The steps of the solution are as
follows (Fig. 13.8): A rectangular diagram is
prepared with percentage scales along three sides
as shown in the figure. The amount of material in
aggregate 1 that passes through a No. 50 (300-μm)
sieve (50%) is marked off along the left-hand
vertical axis. The amount of material in
aggregate 2 that passes through a No. 50 (300-μm)
sieve (0%) is marked off along the right-hand
axis and is joined by a straight line to the
point on the left-hand axis (thick sloping line
called "sieve-size line"). This sieve-size line,
representing the No.50 size, intersects the
horizontal line representing the required
percentage of material passing the No.50 sieve in
the combined grading (14%). This intersection is
marked again by a circle. The percentage of
aggregate 1 sought is indicated on the top scale
by the vertical line (line of combination) drawn
through this point of intersection. In this
example it is approximately 28%.
When aggregates 3 and 4 are to be combined so
that the amount of particles in the combined
grading passing through a 1/2-in (12.5-mm) sieve
is 22% (100 X 15.5/72.5), the same graphical
method provides approximately 20% and 80%
blending proportions for aggregates 3 and 4,
respectively (Fig. 13.9). Note that this
combination does not contain particles passing
through the 3/8-in (9.5-mm) sieve.

This graphical method also has the advantage that
it can provide the complete combined grading with
little additional work. This is shown in Example
13.8.

Example 13.8 Determine several points of the
sieve curve representing the combination of
aggregates 1 and 2 that was obtained in Example
13.7.

Solution The steps of this determination are as
follows (Fig. 13.8): The sieve sizes for the

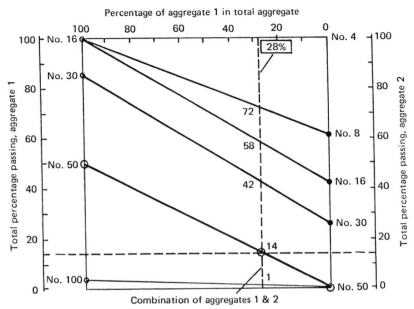

Figure 13.8: An example of the application of the British method for the graphical determination of blending proportions for two aggregates to fulfill one grading condition.

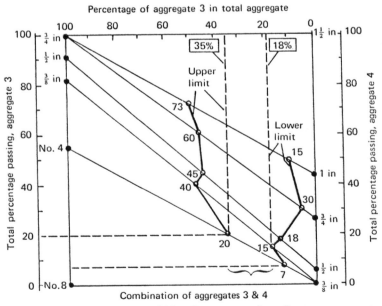

Figure 13.9: Finding the range of blending proportions for two aggregates by the British graphical method when the combined grading is specified by a pair of limit curves.

complete grading of aggregate 1 are marked along
the left-hand axis in the diagram of Fig. 13.8
according to the corresponding percentage passing
values. Aggregate 2 is represented similarly on
the right-hand axis. Each point on the left-hand
axis is joined by a sloping straight line (sieve-
size line) to the point with the same sieve size
on the right-hand axis. Any point of the combined
sieve curve (that is, the percentage of the
combined aggregate passing through any sieve) is
the ordinate of the point of intersection that
the vertical line of combination (as defined in
Example 13.7) makes with the corresponding
sloping sieve-size line. The process is
illustrated in Fig. 13.8 by thin, continuous
lines. It can be seen that, in our example, the
points of the combined sieve curve related to the
No. 100, 50, 30, 16, 8, and 4 sieve sizes are
approximately 1, 14, 42, 58, 72, and 100%,
respectively.

The British graphical method is also applicable
to the case when the aggregates should be blended so
that the sieve curve of the combined grading falls
within a pair of specified limit curves. This is
shown in Example 13.9.

Example 13.9 Determine the blending proportions
for aggregates 3 and 4 of Fig. 13.7 by using the
British method, so that the combined grading
falls within the following limits:

Sieve size	Specified limits, total percentage passing
1 1/2 in (37.5 mm)	100
1 in (25 mm)	50-73
3/4 in (19 mm)	30-60
1/2 in (12.5 mm)	18-45
3/8 in (9.5 mm)	15-40
No. 4 (4.75 mm)	7-20
No. 8 (2.36 mm)	0

Solution The steps of the procedure are as follows (Fig. 13.9): The sieve-size lines are constructed for aggregates 3 and 4 in the same way as discussed in Example 13.8. The specified upper and lower limits are marked off on each sieve-size line as shown in Fig. 13.9. The points representing the lower limit are connected and so are the points representing the upper limit. Any vertical line of combination that can be placed between these limits without intersecting either of the limits represents a blending proportion for aggregates 3 and 4 that produces a combined grading falling between the specified limits. It can be seen in Fig. 13.9 that in this example any proportion for aggregate 3 within 18 and 35% combined with the corresponding amount (within 82 and 65%) of aggregate 4 will yield gradings that comply with the specification.

If the upper limit and lower limit curves intersect each other, the blending problem, as stated, has no solution.

When more than two aggregates are to be combined, the blending proportions can be obtained by repeated application of the British method: Two appropriately selected aggregates should be blended first and then this combination should be treated as one material to be combined with another, and so forth.

An example for four aggregates is presented in Example 13.10.

Example 13.10 Determine again the blending proportions for the case discussed in Example 13.6 but by using the British graphical method.

Solution The procedure is the following: An appropriate combination of aggregates 1 and 2 is found. Then an appropriate combination of aggregates 3 and 4 is found. These steps have been taken in   Example 13.7. Thus, these two combinations should be blended again so that the amount of particles passing throughsieve No. 8 (2.36 mm) in the new combination is 19%. Since it has also been established in Example 13.8 that the amount of particles

passing through a No. 8   sieve in the
combination of aggregates 1 and 2 is 72%,
whereas it is 0% in the combination of
aggregates 3 and 4, the needed blending
proportions for the combination of
combinations can be determined as   shown in
Fig.13.10 by a heavy line. Accordingly,
approximately 26% should be taken from the
combination of aggregates 1 and 2 and 74%
from the combination of aggregates 3 and 4.
Therefore, the needed percentages of the four
aggregates in the total aggregate can be:

Aggregate 1: 100 X 0.26 X 0.28 = 7.3%

Aggregate 2: 100 X 0.26 X 0.72 = 18.7%

Aggregate 3: 100 X 0.74 X 0.20 = 14.8%

Aggregate 4: 100 X 0.74 X 0.80 = 59.2%

Total: 100.0%

There are noticeable differences between these
blending proportions and those determined in Example
13.6, but these new proportions are practically
identical with the exact values that could have been
calculated by the numerical method from the same
grading conditions. As a result of this identity,
the combined grading of Example 13.10 passes through
the No. 50, No. 8, and 1/2-in points of the
specified sieve curve (Fig. 13.7), whereas the
combined sieve curve of Example 13.6 does not.
Nevertheless, one can ascertain from the comparison
of the line of combination in Fig. 13.10 to the
specified as well as the combined sieve curves in
Fig. 13.7 that the goodness of the overall fit
provided by the two graphical methods is more or
less the same.
    Note also that if the specified sieve-size line
does not intersect the horizontal line representing
the required percentage of passing material in the
combined grading, the grading problem, as stated,
has no solution.

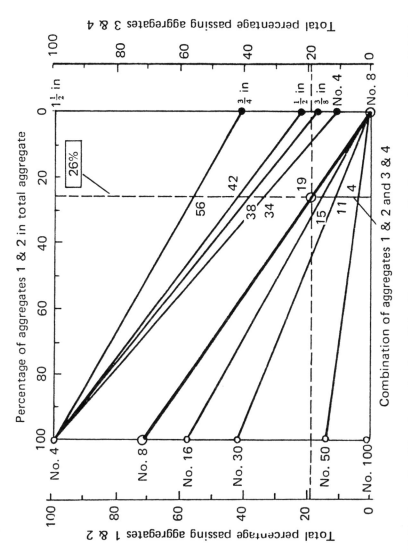

Figure 13.10: Example of the determination of blending proportions for four aggregates by the British graphical method.

13.9 THE TRIANGULAR METHOD

The triangular method is a semigraphical method utilizing triangular charts. It is much more general than the previous two in that it can handle sieve curve, fineness modulus, specific surface, other grading specifications, or combinations of these, both in the form of equalities and inequalities. It is also suitable for blending on the basis of the probability of grading, that is, with a consideration of the expected grading fluctuations in the aggregates to be blended [Sargent 1960]. The triangular method provides the exact solutions within the accuracy limit of graphical methods. Without the application of computer graphics, however, it is restricted essentially to cases where all the gradings included in the blending problem are considered as consisting of not more than three appropriately defined size fractions.

The mathematical basis of the method [Popovics 1964b] is that in the case of transformation of a triangular system of coordinates into another one, the new coordinates x_i, on the one hand, are solutions of a system of three linear equations similar to Eq. 13.2, but, on the other hand, they can also be determined by measuring certain distances p_i in the new system, as shown in Fig. 13.11, and substituting them into the following formulas:

$$x_A = \frac{p_2}{p_1 + p_2}\frac{p_3}{p_3 + p_4}$$

$$x_B = \frac{p_1}{p_1 + p_2}\frac{p_3}{p_3 + p_4}$$

$$x_C = \frac{p_4}{p_3 + p_4}$$

(13.6)

where the check is available that $x_A + x_B + x_C = 1$.

Shaefer was probably the first to recommend these formulas, without an exact mathematical justification, for the determination of blending

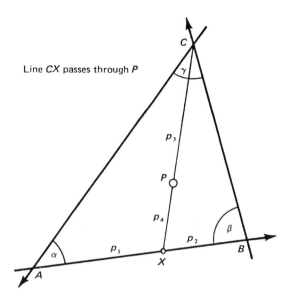

Line CX passes through P

Figure 13.11: Determination of the trilinear coordinates.

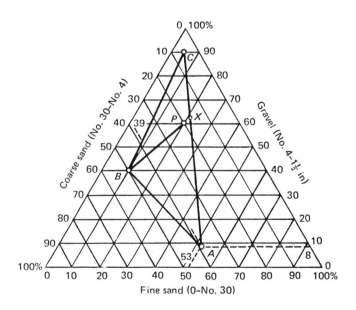

Figure 13.12: The triangular method for the determination of the blending proportions needed for aggregates A, B, and C to produce the grading P.

proportions [Schaefer 1930]. The use of triangular diagrams for the determination of blending proportions is illustrated in the following two examples.

Example 13.11 Determine again the blending proportions for the case discussed in Example 13.1, but by using the triangular diagram method.

Solution The procedure is as follows (Fig. 13.12): All gradings should be considered as consisting of three fractions, the limits of which are determined by the specified points of the combined grading. In our example, the specified points are at sizes No. 30 (600 μm) and No. 4 (4.75 mm). Therefore, the three fractions to deal with are fine sand (pan - No. 30), coarse sand (No. 30 - No. 4), and gravel (No.4 - 1 1/2 in).

An equilateral triangular system of coordinates is prepared with percentage scales along the three sides as shown in Fig. 13.12. Each side is an axis for one of the three aggregate fractions mentioned above. The grading point of aggregate A can be obtained by marking off 53% on the fine sand scale, 39% on the coarse sand scale, and checking if this point cuts out 8% on the gravel scale. This is shown in the figure for point A with dashed lines. The grading points for aggregates B and C, as well as for the specified combined grading (P) are similarly plotted. Points A, B, and C are connected with straight lines to form another triangle. Then another straight line is drawn to pass through the point P and one of the A, B, or C vertices. In our example, point B is selected arbitrarily, and point X is marked off where the line BP intersects the side AC. The distances p are measured from the figure as follows:

$$p_1 = AX = 2.65 \text{ in} \qquad p_3 = BP = 1.50 \text{ in}$$

$$p_2 = CX = 1.35 \text{ in} \qquad p_4 = PX = 0.15 \text{ in}$$

By substituting these values into Eq. 13.6, the
blending proportions sought can be calculated:

$$x_A = \frac{1.35}{4.00}\frac{1.50}{1.65} = 0.307$$

$$x_B = \frac{0.15}{1.65} = 0.091$$

$$x_C = \frac{2.65}{4.00}\frac{1.50}{1.65} = 0.602$$

$$Total = 1.000$$

A comparison of these results to the solutions of
Eq. 13.2 in Example 13.1 shows good agreement.

Note that only such blending problems have
solutions where the point P is within the triangle
ABC.

Example   13.12 In this example, the triangular
method is applied to a more complicated blending
problem.
Assume that materials a, b, and c are given.
Their compositions are shown in Table 13.2.
Determine the blending proportions required to
yield the specified grading shown also in Table
13.2.
If the grading of the available materials are
plotted as points *a, b,* and *c* in Fig. 13.13, then
the points of the shaded area represent the full
range of all possible blending proportions that
will comply with the grading requirements. It can
be seen that in this case there are infinitely
many such mix proportions. The construction of
the limits of the shaded area is as follows: The
positions of the two parallel, sloping dotted
lines are determined by the specified values (5.5
± 0.25) of the fineness modulus of the combined
grading; the two dot-and-dash lines are
determined by the specified limits for the
respective gravel and fine sand contents; the
three dashed lines are determined by the actual

Table 13.2: Grading Data for Materials a, b, and c

Material	G, gravel, % (coarser than No. 4 sieve)	S, sand, % (No. 4– No. 16 sieve)	F, fine sand, % (finer than No. 16 sieve)	Fineness modulus
a	3	27	70	2.3
b	38	49	13	5.05
c	90	7	3	7.05
Specified grading	< 65	No specification	> 10	5.5 ± 0.25

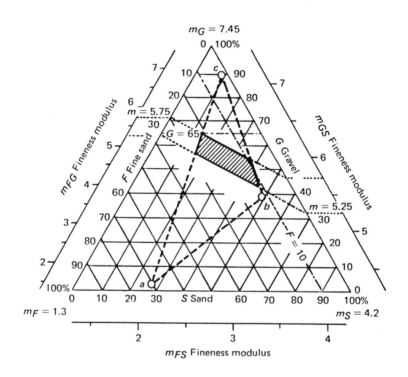

Figure 13.13: Example for proportioning three mineral aggregates by the triangular method to meet simultaneous specifications concerning the fine sand content, gravel content, and the fineness modulus of the combined grading.

gradings of aggregates a, b, and c to be blended. One of the blending proportions can be determined by the semigraphical method from any point of the shaded area and the triangle *abc*. This has been illustrated in the preceding example. By the successive application of the same method to each corner point of the shaded area, the lower and upper limits of the desired blending proportions can be obtained for all three mineral aggregates. Thus, the full range of the blending proportions that will comply with the specified grading requirements is determined. These lower and upper limits are shown in Table 13.3. It can be seen that the amount of material b may vary within 0 and 92%, and the amount of material c may vary within 7.8 and 72.0%. If it has been decided to take, for instance, 50% of material b, then according to Table 13.3, the amount of material c may vary within 33 and 43%, and the amount of material a may vary within 7 and 17% to comply with the specified grading requirements.

In closing, two remarks should be made:

1. The blending proportions determined in Examples 13.11 and 13.12 meet the specifications in terms of the three size fractions defined by the No. 30 (600-μm), No. 4 (4.75-mm), and 1 1/2-in (37.5-mm) sieves; nevertheless, the rest of the grading may or may not be within acceptable limits. Although this is not likely when the three size fractions are chosen intelligently, it can be checked by plotting the complete grading curve of the blended aggregate.
2. A method using a combination of the British method and the triangular diagram for the determination of blending proportions has also been recommended [Sargent 1960]. In this method, the possibilities for blending two aggregates at once to meet certain grading limits are evaluated on the rectangular chart. The ranges for each size are then plotted on a triangular diagram as an area [Lee 1973].

Table 13.3: Limits for the Permissible Blending Proportions of Materials a, b, and c

Material	Blending proportion, %		
b			
Lower	0	4	80
Upper	4	61	92
c			
Lower		$-0.6b + 63$	61
Upper	$-0.43b + 72$	$-0.6b + 73$	$-0.84b + 87$
a			$100 - b$
Lower		$100 - b - c$	
Upper			0

13.10 CRITICAL COMPARISON OF THE METHODS FOR BLENDING PROPORTIONS

The general mathematical form of the blending problem is System 13.1, which contains the blending proportions sought as unknowns. Depending on the number and the nature of the grading conditions, there can be infinitely many sets of blending proportions, a single set, or no set at all that satisfies all the given conditions. The numerical method is applicable for every kind of blending problem of aggregates. In most cases, it provides exact solutions, but it becomes increasingly cumbersome when the number of grading conditions and aggregates to be blended is more than three, unless a computer is used.

The graphical method by Rothfuchs is simple even for large numbers of aggregates to be blended, but it is applicable only in special, although important cases in practice. Also, the solutions it provides are usually approximate in nature.

The restrictions concerning the applicability of the British graphical method are essentially the same as those for the first graphical method. The British method becomes more complicated with increasing numbers of grading conditions and aggregates, but the solutions obtained are exact solutions. It also provides the grading of the combined aggregate with little additional work.

The triangular method is simple and still more general than the other two graphical methods. It also provides exact solutions. It is an ideal method to meet two or more grading conditions, when it is adequate to consider all the gradings included in the blending problem as consisting of not more than three size fractions. In cases more complex than this, however, it may become overly complicated, unless computer graphics are used.

14

Lightweight and Heavyweight Aggregates

SUMMARY

The primary factor controlling the unit weight of a concrete is the average specific gravity of its aggregate. This is the reason that lightweight and heavyweight aggregates are important.

There are two main advantages of *lightweight concretes*: lower unit weight and lower thermal conductivity. Structural lightweight aggregate concretes usually have a compressive strength in excess of 2500 psi (17.25 MPa) and an air-dry unit weight not exceeding 115 lb/ft^3 (1850 kg/m^3) at the age of 28 days. Insulating concretes have much lower unit weights, with strengths between 100 and 1000 psi (0.69 and 6.9 MPa). The only way to reduce the bulk specific gravity of an aggregate is by the inclusion of air into the particles. Sometimes this is done by the action of nature, resulting in natural lightweight aggregates, such as pumice. In other cases the air inclusion is produced by man, resulting in manufactured lightweight aggregates, such as foamed blast-furnace slag, expanded perlite or vermiculite, sintered fly ash, and so on.

Possible problems related to lightweight aggregate concretes are poorer workability and finishability, increased tendency for segregation, uncertainty in the value of the effective water-cement ratio, reduced frost resistance and strength of the concrete. The required quality of various

512

lightweight aggregates is specified in several American standards. Although most of the pertinent test methods are similar to those developed for normalweight aggregates, the results obtained with lightweight aggregates show poorer correlations with the properties of the concrete made with them.

Concretes made with *heavyweight aggregates* are useful primarily for increased attenuation of x-ray and gamma-ray radiations. Increased attenuation of neutron radiation can be achieved by using a concrete with high hydrogen atom content. Heavy aggregates are produced of minerals that have high specific gravity, such as magnetite, ilmenite, limonite, ferrophosphorus, or of steel shot, and so on. Concretes having unit weights over 200 lb/ft^3 (3200 kg/m^3) have been obtained with ore aggregates and over 400 lb/ft^3 (6400 kg/m^3) with steel aggregates. The primary danger in using heavyweight aggregates in concrete is the increased tendency for segregation. Heavyweight concretes may also show poorer workability and finishability as well as reduced frost resistance and strength. Physical requirements and recommended test methods for heavy aggregates do not differ much from those for normal-weight aggregates.

Both lightweight concretes and heavyweight concretes require increased carefulness in proportioning and handling in the fresh state. Otherwise, concrete making with these aggregates is similar to that with normal-weight aggregates.

14.1   INTRODUCTION

It is not unusual anymore that concretes are requested that are lighter or heavier than the traditional, that is, the normal-weight, concretes. The relationship between the composition and the unit weight of an approximately 4-in (10-cm) slump concrete, expressed in traditional units and in metric units, is expressed by the following formulas [Popovics 1964d, 1974c]:

$$U = 0.043c + 10 + G[54 - 0.8(m-6)^2 - 0.015c - 0.6a] \qquad (14.1)$$

or

$$U_m = 1.2c_m + 160 + G[865 - 12(m-6)^2 - 0.5c_m - 10a] \qquad (14.2)$$

where U, U_m = unit weight of the fresh concrete,
  lb/ft^3 and kg/m^3, respectively
  c, c_m = cement content of the compacted
  concrete, lb/yd^3 and kg/m^3, respectively
  G = average bulk specific gravity of the
  combination of fine and coarsaggregates
  m = fineness modulus of the combination of
  fine and coarse aggregates
  a = air content of the fresh concrete, %

These formulas may be reduced to the following simple form without serious loss in accuracy:

$$U = G(45 - 0.6a) + 33 \qquad (14.3)$$

and

$$U_m = G(722 - 9.6a) + 530 \qquad (14.4)$$

These formulas show that the primary factor controlling the unit weight of a compacted concrete is the average specific gravity of its aggregate. This and special properties are the reasons that lightweight and heavyweight aggregates are important for the concrete technologist.

14.2 LIGHTWEIGHT AGGREGATES

There are two main advantages of lightweight concretes as compared to normal-weight concretes, namely, lower unit weight and, not independently, lower thermal conductivity. Whereas in structural concretes mostly the lower unit weight is utilized, the emphasis is on lower thermal conductivity in concretes for insulating purposes (low-density concretes). There are also cases, walls for instance, where both of these properties are important; such concretes may be called moderate-strength concretes. The term *structural lightweight aggregate concrete* can be defined as a concrete having a 28-day compressive strength in excess of 2500 psi (17.25 MPa) and a 28-day air-dry unit weight not exceeding 115 lb/ft^3 (1850 kg/m^3) [ACI 1984]. Insulating concretes have lower unit weights, frequently as low as 50 lb/ft^3 (800 kg/m^3), along with strengths between 100 and 1000 psi (0.69 and 6.9 MPa).

The oldest and probably the most common method to produce a lightweight concrete is to use a lightweight aggregate. A magnificent example for the application of this kind of lightweight concrete is the 143-ft (43-m) diameter dome of the Pantheon in Rome, Italy, built in the second century A.D. A general illustration for the applications of various lightweight aggregates in the three types of lightweight concrete is presented in Fig. 14.1.

The term *lightweight aggregate* means an aggregate of low bulk specific gravity [ACI 1985d]. For structural lightweight concretes this is limited to about one-third to two-thirds that of normal-weight aggregates. The unit weight of the lightweight aggregate is also low, although the relationship between these two weight characteristics is influenced by the grading and the particle shape. The weight characteristic and the strength of the particles are the two most important technical properties of a lightweight aggregate. Smaller particles, as a rule, have higher specific gravities than coarse particles of the same kind of material.

The only practical way to reduce the bulk specific gravity of a stone or stonelike material is by the inclusion of air. Sometimes this is done by the actions of nature, resulting in natural lightweight aggregates. In other cases the air inclusion is produced by man, resulting in manufactured lightweight aggregates [Harrison 1974, Petersen 1974].

The higher the internal porosity of the aggregate, the lighter it is, and as a rule, the lower its heat conductivity, but at the same time

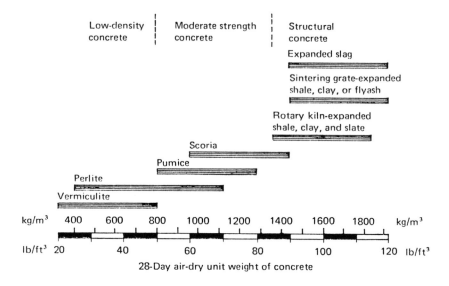

Figure 14.1: Approximate unit weight and uses classification of lightweight aggregate concretes. From ACI Committee 213, *ACI Journal,* American Concrete Institute [ACI 1984].

Table 14.1: Usual Properties of Lightweight Coarse Aggregates

Aggregate	Bulk specific gravity, SSD	Unit weight, lb/ft³ (kg/m³)	Water absorption, % by weight
Pumice	1.25–1.65	30–55 (480–880)	20–30
Foamed blast-furnace slag	1.15–2.20	25–75 (400–1200)	8–15
Expanded perlite	0.90–1.05	~ 10 (~ 160)	10–30
Expanded vermiculite	0.85–1.05	~ 10 (~ 160)	10–30
Expanded clay, shale, and slate	1.1–2.1	35–60 (560–960)	2–15
Sintered fly ash	~ 1.7	37–48 (590–770)	14–24
Saw dust	0.35–0.6	8–20 (128–320)	10–35
Polystyrene foam	0.05	0.6–1.2 (10–20)	~ 50

Note: Fine particles of lightweight aggregates usually have higher bulk specific gravity and unit weight than coarse particles of the same kind of material.

the lower its strength. Therefore, lightweight concretes for insulating purposes have low strengths, whereas a structural lightweight concrete has a higher heat conductivity, which, however, is still considerably lower than that of normal-weight concretes. In general, the higher the strength-unit weight ratio of a lightweight concrete, the better the overall quality of the aggregate used.

Incidentally, a recent, successful application of lightweight aggregates is in paving concretes for the improvement of skid resistance [Gallaway 1970]. Here again, the porous internal structure of the aggregate particles is utilized.

14.3 TYPES OF LIGHTWEIGHT AGGREGATES

14.3.1 Pumice

One of the commonly used natural lightweight aggregates is *pumice*, which is a light, glassy stone, more specifically, a spongy lava. The pores are formed by gases trying to escape when the lava reaches the surface of the earth but is still molten. The internal structure shows fairly evenly distributed, small, more or less interconnected cells. Much pumice is strong enough to produce a good lightweight concrete for structural purposes, such as reinforced roof and floor slabs, highway pavements, walls, and panels, primarily in the precast industry. About 2000 years' experience has proved the pumice-type material to be a good concrete aggregate (Table 14.1).

Other natural lightweight aggregate materials are scoria, tuff, and volcanic cinder.

14.3.2 Foamed Blast-Furnace Slag

Essentially, a similar basic mechanism is used as mentioned above in the production of most manufactured lightweight aggregates, for instance, in *foamed blast-furnace slag*. This slag is a by-product of iron manufacturing, in which silica and alumina constituents combine with lime to form a molten slag collected on the top of the iron in the blast furnace. The slag issues from the furnace as a molten stream at 2550-2900°F (1400-1600°C). If this is allowed to cool slowly, it solidifies to a gray, crystalline, stonelike material, known as air-cooled slag, which can be used as aggregate for base courses or for normal-weight concretes provided that they meet certain quality requirements such as described in B.S. 1047. Cooling the molten slag with a large excess of water produces a material called *granulated slag,* which is a lighter, more friable material. Chilling the slag with a controlled amount of water, applied in such a way as to trap the steam in the mass, gives a porous product of pumicelike character, called foamed slag or expanded slag. This product is then crushed and screened to sizes suitable as lightweight aggregate. The foaming process can change advantageously the mineralogical composition of the slag, removing the unsoundness in almost all slags as well as rendering them stable by the elimination of the β- to γ-dicalcium silicate inversion.

Foamed slag was probably produced first in Germany in 1911 . The lighter types can be used as aggregates in concrete blocks and in insulating concretes, roof screeds, and so on, and the heavier ones for reinforced concrete panels and slabs, in-place wall concreting, other reinforced concrete, and so on. Both extensive laboratory investigation and considerable practical experience have given foamed slag a reputation as a good concrete aggregate provided that it meets certain quality requirements such as specified in B.S. 877 [Short 1963] .

14.3.3 Expanded Perlite

Another manufactured lightweight aggregate is *expanded perlite*. This is one of the lightest inorganic aggregates. Perlite, the basic material, belongs to the riolite group of effusive rocks. It is a dense, glassy rock with a high (2-6%) water content and an internal structure of onionlike concentric rings.

The fundamental mechanism of solid rock expansion is that the rock reaches fusion to such a limited extent that the pores in the rock become plugged by melted material, but at the same time the material remains viscous enough to keep the developed streams and/or gases inside under pressure. This then expands the particle, developing a porous internal structure that is retained on cooling. If the rock reaches the fusion point at a temperature as low as 1300-1470°F (700-800°C), then by heating it further, the whole rock particle becomes plastic; therefore, the expansion takes place more or less equally in all three dimensions. This is what happens in the case of perlite. For best results, the heating should take place rapidly but so that the intensive steam formation coincides with the fusion of the perlite. In this way, the minimum amount of steam can escape from the particle; thus the maximum amount of perlite expansion develops. For instance, an American perlite expanded its volume 15-fold when heated to 1650° F (900° C) in 18 sec, whereas the volume increase was only 4 fold when the heating time was 5 min. Other perlites show similar trends [Ujhelyi 1963].

The extent of perlite expansion is also influenced by the particle size of the perlite before heating. The optimum of these particle sizes depends on the properties of the raw material, but about 0.02 in (0.5 mm) has been a reasonable particle size for several perlites. The production of perlite aggregate started around 1940 in the United States. It is used in large quantities (several million cubic yards per year) both in the United States and in Western, as well as Eastern, Europe, with portland cement, plaster of Paris, magnesium oxychloride cement, water glass, and so on, primarily for insulation purposes and fire protection of steel.

14.3.4 Vermiculite

The expansion of *vermiculite* is similar to that of perlite except that vermiculite expands mostly in one direction as a result of its laminar structure.

The technical properties of such aggregates are similar to those of expanded perlite. Expanded, or exfoliated, vermiculite is used mostly for insulating concretes.

14.3.5 Clays, Shales, and Slates

When certain *clays, shales,* and *slates* are heated to a semiplastic stage, they expand to as much as severalfold their original volume, as a result of the formation of gas within the mass of the material at the fusion temperature, and develop a cellular internal structure. For the production of such aggregate, the raw material should contain mineral constituent(s) that will produce gas at reasonable temperature. If such constituents are not naturally present in the raw material, they may be incorporated during the manufacture. Sometimes the raw material is reduced to the desired sizes before heating, but crushing after expansion may also be applied. The first method produces particles with a smooth, semiimpervious outer shell, or "coating." Coated particles particles have nearly spherical shapes, with lower water absorptions than comparable uncoated particles [Ledbetter 1973].

In the United States, expanded shale was used as aggregate for concrete ships as early as World War I. Since then this type of lightweight aggregate has been produced all over the world under various brand names, including Haydite, Aglite, Rocklite, Leca, and Keramzit.

Heavier expanded clay, shale, and slate aggregates can be used for structural lightweight concretes, walls, even for prestressed concrete elements; the lighter types for blocks, and so on. Compressive strengths as high as 8365 psi (57.7 MPa) were reported at the age of 28 days with a concrete made with an expanded clay aggregate [Shideler 1957] and about 7000 psi (48.3 MPa) with other concretes made with expanded shale or clay aggregates [Reichard 1964].

14.3.6 Fly Ash

Fly ash, the residue obtained from the combustion of powdered coal, when sintered, can also provide good lightweight aggregates. Fly ash serves as an economical raw material because it is a by-product, and because it does not have to be pulverized, it usually contains carbon in sufficient quantity (3-10%) to reduce fuel cost. It is obtained without mining, often from plants near populated areas which have the greatest potential demands for lightweight aggregates [Capp 1970].

The type of fly ash used in manufacturing lightweight aggregates is generally that produced from bituminous coal. The usual method is first to prepare unfired pellets or extruded granules of the fly ash with the addition of water. This is followed by firing on a traveling grate or on a sintering strand to about 2300° F (1260°C), a temperature that softens and agglomerates the pellets into larger particles. The internal structure of a sintered fly ash aggregate particle is multicellular, where the voids in the particle are caused by evaporation of the pellet water and elimination of carbon during the sintering process. Fly ash particles are properly classified as "sintered" products, in contradistinction to the "expanded" materials such as those produced by heating perlite or clay [Minnick 1970].

Fly ash aggregate may be irregular, spherical, or cylindrical in shape, may vary considerably in size, and is normally brown or black in color. Material coming off the end of the grate, if in clinker form, is crushed to the desired sizes and screened.

Sintered fly ash aggregates have been produced since the early 1950s. Concrete strengths in excess of 6000 psi (41.5 MPa) have been produced with the heavier types [Pfeiffer 1971], thus such aggregates are acceptable for use in structural concretes [Kunze 1974a]. Various other types are also used for insulation concretes and for concrete blocks because the thermal conductivity of fly ash lightweight aggregate concrete is quite favorable.

14.3.7 Organic Materials

Certain *organic materials* are also suitable for lightweight aggregates. Such natural materials include crop wastes, such as rice husks, but a more important organic by-product for aggregate is *sawdust*.

Although sawdust consists largely of cellulose, it also contains soluble sugars, acids, resins, waxes, and other organic substances in varying degrees. Some of these have inhibiting effects on the setting and hardening of portland cement. The use of cement other than portland, such as magnesium oxychloride cement, can be useful; nevertheless, many patents have been registered on methods of pretreating sawdust in order to avoid these troubles. Such treatments may involve partial oxidation of the wood, prevention of solvent action, for instance by waterproofing the sawdust, neutralization with lime, and so on [Short 1963]. As a rule, softwoods yield sawdust more suitable for mixtures with portland cement than hardwoods.

Sawdust has been used as an aggregate for more than 50 years for floor finishes, wall and roof units, tiles, and so on. Sometimes, in building blocks for instance, it is blended with sand to reduce the drying shrinkage.

Lightweight aggregates (Styropor, etc.) have been produced from *synthetic organic materials*, such as foamed polystyrene resins. These aggregates can be extremely light and are very good insulators, but they are expensive and it is not simple to make concrete with them. Frequently surface treatment of such aggregates is needed to help the uniform distribution of the particles and/or to improve the bond to the aggregate [Manns 1976]. Thus, they are used only in special cases.

14.4 POSSIBLE PROBLEMS RELATED TO LIGHTWEIGHT AGGREGATES

Some of the lightweight aggregates, especially the fine portions of crushed aggregates, have highly

angular, unfavorable particle shape [Popovics 1977].
This has harmful effects on the workability,
finishing, and perhaps on bleeding of the concrete,
effects that are analogous to those observed with
normal-weight aggregates of unfavorable particle
shape. These can be reduced by air entrainment (up
to 10%), increased cement content, use of mineral
admixtures, or partial substitution of fine, light
particles by normal-weight concrete sand or that
recommended for masonry mortar.

The lighter bulk specific gravity of the
aggregate can also cause problems because it can
produce segregation of the coarse particles from the
concrete mass during mixing, shipping, placing, and
compaction. For instance, during the vibration of
the fresh lightweight aggregate concrete, the coarse
particles have a tendency to move upward. The danger
of segregation can be reduced by careful
proportioning and by proper handling of the fresh
concrete.

Bulk specific gravity of the various size
fractions of a lightweight aggregate usually
increases, sometimes considerably, as particle size
decreases. Therefore, it is recommended to express
the grading and to determine the fineness modulus of
such aggregates in percent by absolute volume
instead of the customary percent by weight [ACI
1981c]. The use of the absolute volume basis for the
aggregate-cement ratio is also recommended when the
richness of a lightweight concrete is to be compared
to that of a normal-weight concrete.

The high absorption value and the high rate of
absorption of most lightweight aggregates [Landgren
1964a] (Table 8.1) can also be a problem if not
checked frequently and counterbalanced in the
proportioning. They also produce an uncertainty in
the calculation of that portion of the total water
in the mixture that is applicable to the effective
water-cement ratio.

The high water absorption can be a problem in
connection with the frost resistance of lightweight
aggregate concretes. If, however, the degree of
saturation in the aggregate is kept low, the frost
resistance of such concretes is usually satisfactory

[Buth 1968a].

Large variations in the shape and surface characteristics, as well as in the early absorption, of different lightweight aggregates account for the wide range in the amounts of mixing water needed to produce a concrete of a given consistency [Hummel 1954a].

Although there is no reliable correlation between aggregate strength and concrete strength [ACI 1984], problems related to inadequate strength occur more frequently with lightweight aggregates than with normal-weight aggregates. Lighter aggregates are usually weaker, but this relationship is influenced by several other factors too, as shown in Fig. 14.2.

14.5 REQUIREMENTS AND TEST METHODS FOR LIGHTWEIGHT AGGREGATES

The required quality of various lightweight aggregates is specified in several American standards. According to ASTM C 330-89, the dry loose unit weight of the fine portion of a lightweight aggregate for structural purposes should not be more than 70 lb/ft^3 (1120 kg/m^3), that of the coarse portion not more than 55 lb/ft3 (880 kg/m^3), and that of the combined fine and coarse aggregate not more than 65 lb/ft^3 (1040 kg/m^3). It is also specified that the unit weight of successful shipments should not differ more than 10% from the sample submitted for acceptance. The grading and grading uniformity are also specified, and so are the permissible maximum amounts of deleterious materials, such as organic impurities, staining materials, clay lumps, popout materials, and loss of ignition.

Similar specifications are provided for lightweight aggregates for concrete masonry units in ASTM C 331-89. ASTM C 332-87 distinguishes two groups of lightweight aggregates for insulating concrete. In the first group, the dry loose unit weight of expanded perlite should be between 7.5 and 12 lb/ft^3 (120 and 196 kg/m^3) and that of expanded vermiculite should be between 6 and 10

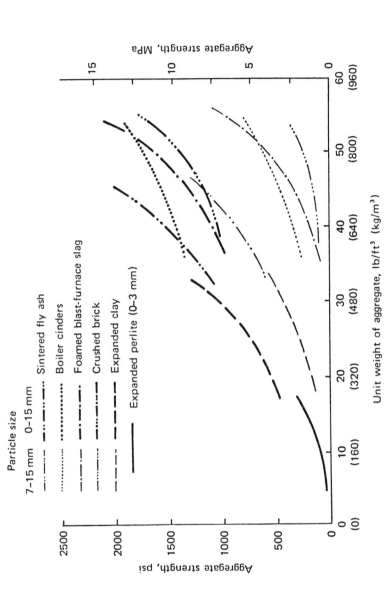

Figure 14.2: Relationship between strength and unit weight of lightweight aggregate particles [Ujhelyi 1963]. As a rule, aggregates with a unit weight up to 15 lb/ft³ (240 kg/m³) are good for insulating concretes; with unit weights between 15 and 40 lb/ft³ (240 to 640 kg/m³) are suitable for walls; and with unit weights between 40 and 65 lb/ft³ (640 to 1,040 kg/m³) for structural lightweight concrete.

lb/ft^3 (96   and   160   kg/m^3).   Unit   weight
specifications for the second group (blast-furnace
slag, pumice, sintered fly ash, etc.), as well as
grading specifications, are similar to those for
lightweight aggregates for structural concretes or
for masonry units. ASTM C 35-89a also provides
quality requirements for lightweight aggregates in
the first group above. B.S.3797, as well as German
DIN 4226, contains general specifications concerning
the quality of lightweight aggregates, including
gradings.

There are no ASTM specifications for the required
values of specific gravity, porosity, or strength of
lightweight aggregates.

The unit weight and the presence of various
deleterious materials in a lightweight aggregate can
be determined by the same or similar ASTM test
methods as specified for normal-weight aggregates
and discussed earlier. The methods for the
determination of water absorption and for specified
gravity used for normal-weight aggregates are not
accurate enough for lightweight aggregates,
primarily because of the peculiar surface characteristics of lightweight aggregate particles. Many
different methods have been recommended in the
literature, but these provide more or less differing
values for the absorption and for the bulk specific
gravity, saturated surface-dry basis [Kunze 1973].
In the United States, the specific gravity factor
method is the most popular, in which quantity
proportional to the bulk specific gravity is
determined in a pycnometer [Landgren 1965]. It
should also be mentioned that it is more meaningful
to express the water absorption of lightweight
aggregates, especially for insulating concretes, in
percent by volume rather than in the customary unit
of percent by weight.

The strength of lightweight aggregate particles
cannot be determined by methods similar to those
that have been discussed in connection with the
strength of normal-weight aggregates. The Los
Angeles abrasion test does not seem suitable for
lightweight aggregates. The two British methods, the
crushing test and the impact test, as well as their

modifications, provide more reliable information concerning particle strength [Hummel 1954b, Nelson 1958], but even these results do not show adequate correlations with the strengths of lightweight concretes made with the tested aggregates.

Because of these uncertainties, the general opinion is that presently one can evaluate the differences between lightweight aggregates only by comparing concretes made with the different aggregates but otherwise identical compositions [ACI 1984].

14.6 HEAVYWEIGHT AGGREGATES

Large-scale production of penetrating radiation and radioactive materials, as a result of the use of nuclear reactors, particle accelerators, industrial radiography, and x-ray and gamma-ray therapy, entails the use of shielding material for the protection of operating personnel against the biological hazards of such radiations. Concrete, both of normal and heavyweight types, is a versatile and popular material for permanent shielding installations because it economically combines radiation absorption properties with good mechanical characteristics.

There are two basic types of radiation to be shielded against: (1) x-ray and gamma radiations; and (2) neutron radiation. An increased attenuation of *x-ray* and *gamma* radiations can be obtained for a given thickness by using heavyweight concrete, that is, a concrete with unit weight considerably higher than the usual 150 lb/ft^3 (2400 kg/m^3) value. Increased attenuation of *neutron* radiation can be achieved for a given thickness by using a concrete with a higher hydrogen atom content or water content. Since the portland cement paste in a mature concrete usually contains less chemically bound water than is necessary for adequate neutron attenuation, the necessary extra fixed water should be supplied by special aggregates.

Many elements, including hydrogen, give off gamma radiation when neutrons are captured. This secondary radiation causes a need for shielding also. Fortunately, concrete of sufficiently high water content is effective against both neutron and gamma radiations [Callan 1962].

The simplest and most economical material to provide adequate shielding against the radiations mentioned above is the traditional, that is, normal-weight concrete in sufficient thickness. In many cases, however, the designer is forced by space restrictions to use reduced thicknesses of shielding concrete. In such cases, concretes made with heavy aggregate(s) and/or aggregate(s) containing fixed water, or some other lightweight material (boron), can be used advantageously [ACI 1972].

Heavy aggregates are produced of materials that have high specific gravity. Of around 60 minerals with specific gravity greater than 3.5, about 10 are available commercially. Of these baryte, magnetite, ilmenite, limonite, and goethite are most suitable [Portland Cem. Ass. 1975a]. Others, such as hematite, taconite, arsenophrite, chromite, psilo-melana, and galena, are of interest but have not been widely used. Ferrophosphorus and ferrosilicon may also be considered. In addition, steel and iron aggregates are available in the forms of shot, punching, scrap, and so on [Davis 1958].

The specific gravity and the amount of fixed water are the two most important technical characteristics of heavy aggregates for shielding concretes. Here again, smaller particles usually have higher specific gravities than coarser part-icles of the same kind of material. Physical pro-perties and radiation absorption factors of the most commonly used heavy aggregates are shown in Table 14.2. Further details concerning the constituents are presented in ASTM C 638-84. Concretes having unit weights over 200 lb/ft^3 (3200 kg/m^3) have been produced with ore aggregates, and over 400 lb/ft^3 (6400 kg/m^3) with steel aggregates.

Table 14.2: Physical Properties and Radiation Absorption Factors of Heavy Aggregates[a]

Heavy aggregate	Source	Primary identification	Specific gravity, saturated surface-dry		Composition, % by weight		Radiation absorption, cm^2/g	
			Coarse pieces	Fine sand	Iron	Fixed water	Fast neutrons	Gamma rays (3 MeV)
Limonite	Michigan	$2FeO_3 \cdot 3H_2O$	3.75	3.80	58	9		
Goethite	Utah	$Fe_2O_3 \cdot H_2O$	3.45	3.70	55	11	0.0372	0.0362
Magnetite	Nevada	Fe_3O_4, etc.	4.62	4.68	64	1	0.0258	0.0359
Magnetite	Montana	Hydrous iron[b]	4.30	4.34	60	2-5		
Barite	Tennessee	92% $BaSO_4$	4.20	4.24	1-10	0		
Barite	Nevada	90% $BaSO_4$	4.28	4.31	1	0	0.0236	0.0363
Ferrophosphorus	Tennessee, Missouri, and Montana	Fe_3P, Fe_2P, FeP	6.30	6.28	70	0	0.0230	0.0359
Steel aggregate	Punchings	Sheared bars	7.78	–	99	0	0.0214	0.0359
Steel shot	Chilled	SAE standard	–	7.50	98	0		

[a]From Davis, *ACI Journal*, American Concrete Institute [Davis 1962].
[b]This ore is primarily magnetite, with some hematite (Fe_2O_3) and limonite.

14.7   POSSIBLE   PROBLEMS   RELATED   TO   HEAVYWEIGHT   AGGREGATES

Concrete making with heavy aggregates is in most cases similar to that with crushed aggregates, and by applying sound principles, concrete strengths over 10,000 psi (69 MPa) can be produced with them [Mather, K. 1965]. Nevertheless, there are several differences in using heavy aggregates as compared to normal-weight aggregates. For instance, when the density of the mortar in a concrete is lighter than the density of the coarse aggregate, which is the case when normal-weight sand is used with heavy coarse aggregate, then the coarse particles have a tendency to segregate from the concrete mass during mixing, shipping, placing, and compaction. The danger of segregation can be reduced by careful proportioning, by using a fine aggregate heavier than the coarse aggregate, by using special admixtures, by proper handling of the fresh concrete [ACI 1985c], or by the application of prepacked concrete. Segregation, both external and internal, is even more harmful in shielding concretes than in strictly structural concretes.

Heavy aggregates are not always available as well-graded materials. This may necessitate the use of gap gradings or other unusual gradings. When the specific gravities of the various aggregate size fractions differ considerably, such as in a blend of normal-weight fine aggregate and heavy coarse aggregate, it is recommended to express the grading and determine the fineness modulus on the basis of percent by absolute volume. The use of the absolute volume basis for the aggregate-cement ratio is also recommended when the richness of a heavyweight concrete is to be compared to that of a normal-weight concrete.

Some of the heavy materials, quite a few limonites for instance, may have low strength. This may cause problems when the heavyweight concrete must serve not only for shielding but also for structural purposes. There are also heavy aggregates that are unsuitable for use in concrete exposed to excessive weathering or abrasive forces.

The particle shape of certain heavy aggregates can be quite unfavorable and impair the workability and finishability of the fresh concrete. Iron and steel punchings, scrap, and so on, are particularly disadvantageous from this point of view. The use of air entrainment, increased cement content, mineral admixtures, and so on, can help the workability. Although these reduce simultaneously the unit weight of the heavy concrete, satisfactory workability is a must because dangerous radiation leaks occur if the shielding concrete is not compacted to a uniformly adequate degree. Note that a, say, 3-in (7.5-cm) slump represents poorer workability in the case of a heavyweight concrete than in the case of a normal-weight concrete. Appropriate allowances should also be made for the fact that mineral ores are less uniform in quality within a deposit than normal-weight aggregates.

It is recommended that iron or steel be rusted before it is used as a coarse aggregate, otherwise the bond between the cement paste and the aggregate surface may be poor. Another factor that may interfere with the bond is magnetic attraction: Fine magnetite particles attached to the steel particles can interfere with the coating of the coarse particles with cement paste. One solution to this problem may be the introduction of a part of the cement and water into the mixer to coat the steel coarse aggregate with cement paste before the magnetite fine aggregate is added to the mixture [Narver 1962].

When a mineral containing boron (colemanite, borocalcite,etc.) is added to the concrete aggregate (for the improvement of neutron shielding), this usually interferes with the setting and hardening of the concrete. The application of a suitable admixture may reduce this problem.

Ferrophosphorous and ferrosilicon materials (heavyweight slags) should be used only after thorough scrutiny. Reactions in concretes containing these aggregates have been known to produce undesirable hydrogen gas in large volume.

14.8 REQUIREMENTS AND TEST METHODS FOR HEAVYWEIGHT AGGREGATES

According to ASTM C 637-84, the physical requirements for heavy aggregates do not differ too much from those for normal-weight aggregates. They should be clean, strong, inert, and free of excessive amounts of deleterious materials. Manufactured aggregates should be reasonably free of dust, oil, and other coatings, except for rust. They should also be reasonably free of flat and needlelike particles.

The grading requirements for conventionally placed heavy concretes are essentially the same as those for normal-weight aggregates. Aggregates for prepacked concrete should not contain fine or medium-size particles [ACI 1969]. The grading, the specific gravity, and the fixed water content of the aggregates should not fluctuate significantly in successive shipments. A Los Angeles abrasion loss not greater than 50% is also recommended.

As far as the composition is concerned, the presence of 4% or more chemically bound, or fixed, water in aggregates for shielding concretes is highly desirable. Otherwise, the composition of heavy aggregates becomes important when low levels of natural gamma radiation activity are required from them. In such cases, of primary concern are the concentrations of potassium, uranium, and thorium [Polivka 1966].

Test methods for heavy aggregates are by and large the same as those specified for normal-weight aggregates. Concrete-making properties of a heavy-weight aggregate with an unknown service record should be thoroughly checked with trial mixes before large-scale application.

15

Handling and Selection of Aggregates

SUMMARY

Every reasonable precaution should be exercised in aggregate handling to see that the particles reach the concrete mixer as uniformly graded and with as uniform moisture content as it is economically feasible to accomplish. Segregation in coarse aggregate is minimized when it is separated into appropriate size fractions to be batched separately. Stockpiling of aggregate should be kept to a minimum. Coarse aggregates in coned or tent-shaped piles have shown particularly excessive segregation. Storage bins should have the smallest practical horizontal cross section and be kept as full as practical. There is a much greater tendency for segregation of an uncrushed gravel than for crushed aggregates. Excessive segregation can be corrected by rehandling and storing the aggregate by one of the recommended methods. Wet sand should be drained until it reaches a practically uniform moisture content prior to its transfer to the batch-plant bins.

The ideal aggregate is seldom available. The problem is, therefore to decide the level of performance that is required in a given situation and select the aggregate with which it is economically attainable. Service record is a most valuable aid in guiding this selection. When a usable service record is not available, the selection of aggregate should be based on the results of appropriate laboratory tests. Not only those aggregate properties that have a direct effect on the properties of concrete should be considered in the evaluation, but also those that have indirect effects.

15.1 AGGREGATE HANDLING

Adequate uniformity of concrete from batch to batch is very important [Kennedy 1975]. For instance, variations in grading of coarse aggregates over the extreme range actually found to exist at operating two-bin concrete plants caused a change of about 3 in (75 mm) in the slump [Hudson 1969]. Therefore, every reasonable precaution should be exercised to see that aggregates reach their ultimate destination in the concrete mixer as uniformly graded and with as uniform moisture content as it is economically feasible to accomplish.

Assuming that the aggregate comes from a well-operated source, the delivery of uniform aggregates to the mixer is a matter of handling in a manner that will retain uniformity [Blanks 1955 Waller 1966]. Handling procedures are discussed in the ACI "Recommended Practice for Measuring, Mixing, Transporting and Placing Concrete" [ACI 1985a]. Figures 15.1 and 15.2 from that report are reproduced here along with the principal recommendations pertinent to aggregates:

1. Segregation in coarse aggregate is minimized when it is separated into appropriate individual size fractions to be batched separately (Fig. 15.2). The amounts of oversize and undersize should be held to a practical minimum and should be uniform.
2. Stockpiling of aggregate should be kept to a minimum because even under good conditions segregation of the coarser particles tends to occur. When stockpiling is necessary, the piles should be built in thin, horizontal or gently sloping layers and not by end-dumping. Trucks,

(a)

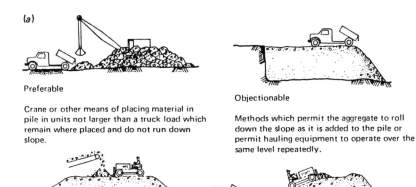

Preferable

Crane or other means of placing material in pile in units not larger than a truck load which remain where placed and do not run down slope.

Objectionable

Methods which permit the aggregate to roll down the slope as it is added to the pile or permit hauling equipment to operate over the same level repeatedly.

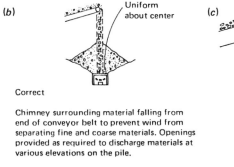

Limited acceptability—Generally objectionable

Pile built radially in horizontal layers by bull-dozer working from materials as dropped from conveyor belt. A rock ladder may be needed in setup.

Bulldozer stacking progressive layers on slope not flatter than 3:1. Unless materials strongly resist breakage, these methods are also objec-tionable.

(b)

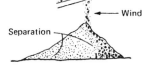

Correct

Chimney surrounding material falling from end of conveyor belt to prevent separating fine and coarse materials. Openings provided as required to discharge materials at various elevations on the pile.

(c)

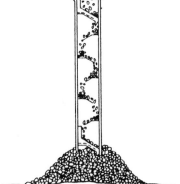

Incorrect

Free fall of material from high end of stacker permitting wind to separate fine from coarse material.

Unfinished or fine aggregate storage (dry materials)

When stockpiling large sized aggregates from elevated conveyors, breakage is minimized by use of a rock ladder.

Finished aggregate storage

Note: If excessive fines can not be avoided in coarse aggregate fractions by stockpiling methods used, finish screening prior to transfer to batch plant bins will be required.

Figure 15.1: Correct and incorrect methods of handling and storing aggregates. Incorrect methods tend to cause segregation and breakage. From ACI Committee 304, American Concrete Institute [ACI 1985a].

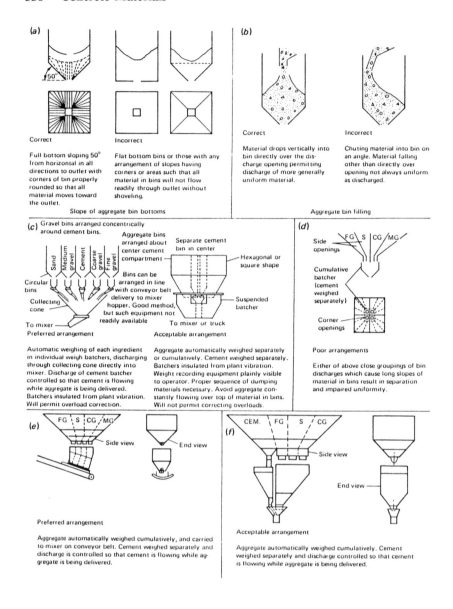

Figure 15.2: Correct and incorrect methods of batching. Incorrect methods tend to impair the uniformity of the aggregate and concrete. From ACI Committee 304, American Concrete Institute [ACI 1985a].

bulldozers, and other vehicles should be kept
off the stockpiles, as they cause breakage and
contamination (Fig.15.1a). Coarse aggregates in
coned or tent-shaped piles have shown particularly excessive segregation.
3. Two sizes of sand cannot be blended satisfactorily by placing them alternately in stockpiles or trucks. It is preferable to handle and
batch them separately.
4. Wind should not be permitted to segregate dry
sand (Fig.15.1b).
5. Storage bins should have the smallest pract-
cal horizontal cross section. The bottoms should
slope at an angle not less than 50° from the
horizontal toward a center outlet (Fig. 15.2a)
Bins should be kept as full as practical to
minimize breakage and changes in grading as
materials are withdrawn. The aggregate should be
filled into the bins vertically over the outlet
(Fig. 15.2b).
6. Wet sand should be drained until it reaches a
practically uniform moisture content prior to
its transfer to the batch-plant bins.
7. During batching operations, aggregates should
be handled in a manner that maintains their
desired grading (Fig. 15.2).

A pertinent investigation also revealed the
following [Miller-Warden Ass. 1967b]:

1. There is a much greater tendency of
segregation for the larger aggregate particles
of an uncrushed gravel than for crushed
aggregates.
2. An acceptable, and presently the most
economical, method of forming and reclaiming
stockpiles from aggregates delivered in trucks
is to discharge the loads in such a way that
they are tightly joined and to reclaim the
aggregate with a front-end loader. When the
aggregate is not delivered in trucks, the least
expensive acceptable results are obtained by
forming the stockpiles in layers with a
clambucket and reclaiming the aggregate with a

front-end loader. Rescreening of the aggregate
immediately prior to feeding it into the
batching bins provides more uniformly graded
aggregate, but this process is overly expensive.

Excessive segregation can be corrected by rehandling and storing the aggregate by one of the
recommended methods.

15.2 SELECTION OF AGGREGATES

A high-quality aggregate consists of particles that
are strong, durable, clean, favorably graded, and
not flat or elongated; that do not slake when wetted
and dried; whose surface texture is somewhat rough;
and that contain no constituents that interfere with
cement hydration or react with cement hydration
products to produce excessive expansion [ACI 1961].
Some other properties, such as specified thermal
properties, may also be required in particular
cases. The ideal aggregate is seldom available. The
problem is then to decide the level of performance
that is required in a given situation and select the
aggregate with which it is economically attainable.
   To the extent practical, the selection of an
aggregate should be based on knowledge of its
behavior in concrete, including uniformity. Service
record, when available in sufficient detail, is a
most valuable aid in guiding judgment. To be
meaningful, the record should cover structure(s)
with concrete proportions and exposures similar to
those anticipated for the proposed work.
Petrographic and other suitable tests should be used
to determine whether the aggregate in the structure
and that proposed for use are sufficiently similar
to make the service record meaningful for the
aggregate evaluation. A structure completely sound
after 10 years or more representative service can be
assumed to constitute an "endorsement" of all
materials used in it, including the aggregates. It
should not be overlooked, however, that unfavorable
service of old concrete may not always be a
sufficient basis for condemning an aggregate.

Meaningful service records must be dependable and the damaging evidence of a recurring nature before rejection is made. For instance, popouts provide a dependable criterion for identifying an undesirable component in the aggregate. On the other hand, there is the possibility that potentially reactive aggregates may have a satisfactory service record if they have been used only with low-alkali cements. Also, an old concrete without entrained air may not accurately indicate performance of concrete with the benefits of proper air entrainment. Where deterioration is associated with a minor constituent, benefaction may make a previously undesirable aggregate usable.

When an applicable service record is not available, the selection of the aggregate should be based on the results of appropriate laboratory tests. Bear in mind, however, that most laboratory tests have limitations that do not make them absolutely reliable.

The economy of concrete also plays an important role in aggregate selection. Therefore, an aggregate with an unfavorable property, for instance with an unfavorable particle shape, should not necessarily be rejected in favor of a more expensive aggregate with better particle shape if the cost of additional cement required for the first aggregate is less than the extra cost of the second aggregate and the use of additional cement will not be detrimental. It is also economical to require only those aggregate properties that are pertinent to its use in the given project.

The effects of aggregate properties on the properties of concrete are summarized in Table 15.1. Some of these aggregate properties have a direct influence on the properties of concrete, whereas others have indirect effects. For instance, the strength of weak aggregates has a direct effect on concrete strength, whereas the aggregate grading with a given maximum particle size influences the concrete strength indirectly through the workability and water requirement of the fresh concrete. The influence of the maximum particle size on concrete strength is an example of the combination of direct

Table 15.1: Aggregate Properties Affecting the Properties of Concrete[a]

Concrete property	Relevant aggregate property
Durability	
Resistance to freezing and thawing	Soundness
	Porosity
	Pore structure
	Permeability
	Degree of saturation
	Tensile strength
	Texture and structure
	Grading
	Maximum size
	Presence of clay
Resistance to wetting and drying	Pore structure
	Modulus of elasticity
Resistance to heating and cooling	Coefficient of thermal expansion
Abrasion resistance	Hardness
Cement-aggregate reaction	Presence of particular constituents
	Maximum size
Strength	Strengt
	Surface texture
	Cleanness
	Grading
	Particle shape
	Maximum size

(continued)

Table 15.1: (continued)

Shrinkage	Modulus of elasticity Particle shape Grading Cleanness Maximum size Presence of clay
Coefficient of thermal expansion	Coefficient of thermal expansion Modulus of elasticity
Thermal conductivity	Thermal conductivity
Specific heat	Specific heat
Unit weight	Specific gravity Particle shape Grading Maximum size
Modulus of elasticity	Modulus of elasticity Poisson's ratio
Slipperiness	Tendency to polish
Economy	Particle shape Grading Maximum size Amount of processing required Availability

==

[a]From [ACI 1961].

and indirect effects.

The aggregate properties listed in Table 15.1 and methods for their determinations have been discussed earlier.

15.3 FUTURE OF AGGREGATES

In many areas supplies and reserves of naturally occurring aggregates will be depleted. Increased emphasis will therefore be placed on manufactured aggregates, many of which are lightweight. The lightweight aggregate concretes will be improved to provide higher strength-weight ratios and greater durability [ACI ad hoc Com. 1971b].

The most notable advancements, however, will come in the development of supplementary aggregates from waste materials. In addition to fly ash, which has already yielded high-quality sintered lightweight aggregate, mineral wool produced from coal ash with lime; magnesia, and sulfur extracted and limestone injected to control SO_2 emissions, will be usable for concrete with modern refuse incinerators; glassy residues similar to fly ash and slag will be obtained and put to similar uses.

As new aggregates are developed, emphasis will be placed on manufacturing them to have specific properties to meet their intended uses (for instance, to improve skid resistance). Marine deposits will be increasingly used to meet needs for aggregates. Aggregates will be used as a means of increasing the strength and ductility of the concrete.

Bibliography

a'Court, C. L., "Mix Design and Abrasion Resistance of Concrete," *Symposium on Mix Design and Quality Control of Concrete*, Cement and Concrete Association, London (1954)

Abdun-Nur, E. A., *Fly Ash in Concrete*, Highway Research Board Bulletin 284, Washington, D.C. (1961)

Abdun-Nur, E. A., "Adapting Statistical Methods to Concrete Production," *Proceedings of National Conference on Statistical Quality Control Methodology in Highway and Airfeld Construction*, Charlottesville, Va. (1966)

Abdun-Nur, E. A., "Sampling of Aggregates for Concrete," *Progress in Concrete Technology*, V. M. Malhotra, Editor, Energy, Mines and Resources, Ottawa, Canada, MRP/MSL 80-89 TR (1980)

Abrams, D. A., *Design of Concrete Mixtures*, Bulletin 1, Structural Materials Research Laboratory, Chicago (1918)

Abrams, D. A., "Tests of Impure Waters for Mixing Concrete," *Proceedings of the American Concrete Institute*, Vol. 20 (1924)

ACI Committee 621, "Selection and Use of Aggregates for Concrete," *ACI Journal*, Vol. 58, No. 5 (1961)

ACI Committee 201, "Durability of Concrete in Service," *ACI Journal*, Proc. Vol. 59, No. 12 (1962)

ACI Commmittee 304, "Preplaced Aggregate Concrete for Structural and Mass Concrete," *ACI Journal*, Proc. Vol. 66 (1969)

ACI Committee 223, "Expansive Cement Concretes - Present State of Knowledge," *ACI Journal*, Proc. Vol.67 (1970)

ACI Committee 212, "Guide for Use of Admixtures in Concrete," *ACI Journal*, Proc. Vol. 68, No. 9 (1971a)

ACI Ad Hoc Committee on Concrete-Year 2000, "Concrete—Year 2000," *ACl Journal*, Proc. Vol. 68, No. 8 (1971b)

ACI Committee 301 Specification for Structural Concrete for Buildings (1971c).

ACI Committee, Placing Concrete by Pumping Methods, *ACI Journal*, Proc. Vol. 68, No. 5 (1971)

ACI, *Concrete for Nuclear Reactors*, Vols. 1 and 2, *ACI Publication SP-34*, American Concrete Institute (1972)

ACI, *Expansive Cements*, *ACI Publication SP-38*, American Concrete Institute, Detroit (1973)

ACI, *Superplasticizers in Concrete*, *ACI Publication SP-62*, American Concrete Institute, Detroit (1979)

ACI Committee 517, "Accelerated Curing of Concrete Atmospheric Pressure - State of the Art," *ACI Journal*, Proc. Vol. 77, No. 6 (1980)

ACI, *Development in the Use of Superplasticizers*, *ACI Publication SP-68*, American Concrete Institute (1981a)

ACI Committee 212, "Guide for Use of Admixtures in Concrete," *Concrete International: Design & Construction*, Vol. 3, No. 5 (1981b)

ACI Committee 211, "Recommended Practice for Selecting Proportions for Structural Lightweight Concrete," *ACI 211.2* (1981c)

ACI Committee 223, "Report No. 223.1R, Standard Practice for the Use of Srinkage-Compensating Concrete," *Concrete International: Design & Construction*, Vol. 5, No. 1 (1983)

ACI Committee 213, "Guide for Structural Lightweight Aggregate Concrete," *ACI Journal*, Proc. Vol. 64, No. 7 (1984)

ACI Committee 304, *Recommended Practice for Measuring, Mixing, Transporting and Placing Concrete*, American Concrete Institute (1985a)

ACI Committee 225, "Guide to the Selection and Use of Hydraulic Cements," *ACI Journal*, Proc. Vol. 82, No. 6 (1985b).

ACI Committee 304, "High Density Concrete: Measuring, Mixing, Transporting, and Placing," American Concrete Institute (1985c)

ACI, *Cement and Concrete Terminology*, *Publication SP-19*, American Concrete Institute (1985d)

ACI Committee 226, "Silica Fume in Concrete," *ACI Materials Journal*, Vol. 84, No. 2 (1987a)

ACI Committee 226, "Ground Granulated Blast-Furnace Slag as a Cementitious Constituent in Concrete," *ACI Materials Journal*, Vol. 84, No. 4 (1987b)

ACI Committee 226. "Use of Fly Ash in Concrete," *ACI Materials Journal*, Vol. 84, No. 5 (1987c)

ACI Committee 212, "Chemical Admixtures for Concrete," *ACI Materials Journal*, Vol. 86, No. 3 (1989)

Adams, L. D., "The Measurement of Very Early Hydration Reactions of Portland Cement Clinker by a Thermoelectric Conduction Calorimeter," *Cement and Concrete Research*, Vol. 6, No. 2 (1976)

Aignesberger, A., Fah, N. L., and Rey, T., "Melamine Resin Admixture Effect on Strength of Mortars," *ACI Journal*, Proc. Vol. 68, No. 8 (1971)

Albrecht, W., and Schaffler, H., "Versuche mit Ausfallkornungen" (Experiments with Gap Gradings), *Deutscher Ausschuss fur Stahlbetonbau*, Heft 168, Berlin (1965)

Alexander, K. M., "Activation of Pozzolanic Materials by Alkali," *Australian Journal of Applied Science*, Vol. 6, No. 2 (1955a)

Alexander, K. M., "Activation of Pozzolans by Treatment with Acid," *Australian Journal of Applied Science*, Vol. 6, No. 3 (1955b)

Alexander, K. M., and Wardlaw, J., "Limitations of the Pozzolana - Lime Mortar Strength Test as Method of Comparing Pozzolanic Activity," *Australian Journal of Applied Science*, Vol. 6, No. 3 (1955c)

Alexander, K. M., Taplin, J. H. and Wardlaw, J., "Correlation of Strength and Hydration with Composition of Portland Cement," *Proceedings of the Fifth International Symposium on the Chemistry of Cement*, Part III., *Properties of Cement Paste and Concrete*, Tokyo (1969)

Alexander, K. M., "The Relationship Between Strength and the Composition and Fineness of Cement," *Cement and Concrete Research*, Vol. 2, No.6 (1972)

Alonso, C., and Andrade, C., "Effect of Nitrite as a Corrosion Inhibitor in Contaminated and Chloride-Free Carbonated Mortars," *ACI Materials Journal*, Vol. 87, No. 2 (1990)

American Coal Ash Association, *Proceeding: Eigth International Ash Utilization Symposium*, Vols. 1 and 2. EPRI CS-5362 (1987)

Anderegg, F. 0., "Grading Aggregates - II - The Application of Mathematical Formulas to Mortars," *Industrial and Engineering Chemistry*, Vol. 23, No. 9 (1931)

Andreasen, A. H. M., and Anderson, J., The Relation of Grading to Interstitial Voids in Loosely Granular Products, (In German) *Kolloid-Zeitung*, Vol. 49 (1929)

Andrews, D. A., "Correspondence to Durability of Concrete," *Concrete*, Vol. 5, No.11, London (1971)

Anon., "Requirements of Mixing Water for Concrete: Effects of Common Impurities," *Indian Concrete Journal*, Vol. 37, No. 3 (1963)

Anon., "Special Cements and Their Uses," *Concrete Construction*, Vol. 18, No. 1 (1973)

Anon., Precautionary Measures Against Deleterious Alkali-Aggregate Reaction in Concrete, *Betontechnische Berichte 1974*, Beton - Verlag Gmbh, Dusseldorf (1974)

Anon., "Richtlinien fur die Prufung der Wirksamkeit von Betonzusatzmitteln" (Guidelines for Testing the Effectiveness of Concrete Admixtures), *Betontechnische Berichte 1975*, Beton-Verlag, Dusseldorf (1976)

Arni, H. T., "Resistance to Weathering," *Significance of Tests and Properties of Concrete and Concrete Making Materials, ASTM SP* 169-A, Philadelphia (1966)

ASTM, *Symposium on Use of Pozzolanic Materials in Mortars and Concretes, ASTM STP* 99, Philadelphia (1950)

ASTM, *Fineness of Cement, ASTM STP* 473, Philadelphia (1969a)

ASTM Committee E-1, *Manual on Test Sieving Methods, ASTM STP* 447, Philadelphia (1969b)

ASTM, "Selected References on Hydraulic Cement," 1974 *Annual Book of ASTM Standards*, Part 13, app. II (1974)

ASTM Committee C-9, *Manual of Concrete Testing*, 1976 *Annual Book of ASTM Standards*, Part 14, Philadelphia (1976)

ASTM, *Manual of Cement Testing, 1988 Annual Book of ASTM Standards*, Vol. 04.01 (1988)

ASTM Committee C-9, *Manual of Aggregate and Concrete Testing, 1989 Annual Book of ASTM Standards*, Part 14, Philadelphia (1989)

Axon, E. O., "A Method of Estimating the Original Mix Composition of Hardened Concrete Using Physical Tests," *Proceedings*, ASTM, Vol. 62 (1962)

Bache, H. H., and Isen, J. C., "Modal Determination of Concrete Resistance to Popout Formation," *ACI Journal*, Proc. Vol 65, no. 6 (1968)

Baker, S. D., and Scholer, C. F., "Effect of Variations in Coarse-Aggregate Gradation on Properties of Portland Cement Concrete," *Highway Research Record Number 441, Grading of Concrete Aggregates*, Highway Research Board, Washington, D.C. (1973)

Bakker, W. T., Workshop Proceedings: *Research and Development Needs for Use of Fly Ash in Cement and Concrete*, Electric Power Research Institute, Palo Alto, Cal. (1982).

Balazs, Gy., and Kilian, J., "Effect of Bentonite on Concretes," *Scientific Publications of the Technical University*, Budapest (1957)

Balazs, Gy., and Boros, M., "The Effect of Calcium Chloride on Cement Mortars Under the Conditions of High Pressure Steam Curing," *Proceedings of the Technical University of Building and Transport Engineering*, Vol. XII, No. 6, Budapest (1967)

Balazs, Gy., The Tensile Strength of Cement Paste and Mortar, (In Hungarian), *Tudomanyos Kozlemenyek*, No. 8, Budapest (1974)

Balazs, G., Buda, G., Borjan, J., et al., Testing the Suitability of Stones for Road BuiEiding, (In Hungarian), *Tudomanyos Kozlemenyek*, no. 20, Budapesti Muszaki Egyetem Epitomernoki Kar Epitoanyagok Tanszek, Kozlekedesi Dokumentacios Vallalat, Budapest (1975)

Balazs, Gy., Borjan, J., Jaime, C. S., Liptay, A., and Zimonyi, Gy., Cracking Sensitivity of Cements, (In Hungarian), *Tudomanyos Kozlemenyek*, No. 24, Budapesti Muszaki Egyetem Epiteszmernoki Kar, Budapest (1979)

Barnes, B. D., Diamond, S., and Dolch, W. L., "Scanning Electron Microscope Characterization of the Surface of ASTM C 109 Standard Ottawa Sand," ASTM *Cement, Concrete and Aggregate*, CCAGDP, Vol. 1, No. 1 (1979)

Barton, W.R., "Cement," *Mineral Facts and Problems*, *Bulletin* 630, U.S.Department of Interior, U.S. Bureau of Mines (1965)

Batchelor, C. S., "Friction Material," *Encyclopedia of Chemical Technology*, Kirk-Othmer (Ed.), Second Edition, Volume 10, Interscience Publishers, New York, London, Sydney (1966)

Bates, P.H. and Klein, A.A., "Properties of the Calcium Silicates and Calcium Aluminate Occurring in Normal Portland Cement," *Technologic Paper* 78, National Bureau of Standards (1917)

Bentur, A., and Ish-Shalom, M., "Properties of Type K Expansive Cement of Pure Components, II. Proposed Mechanism of Ettringite Formation and Expansion in Unrestrained Paste of Pure Expansive Components," *Cement and Concrete Research*, Vol. 4, No. 5 (1974)

Bentur, A., and Ish-Shalom, M., "Properties of Type K Expansive Cement of Pure Components. IV. Hydration of Mixtures of C3S with Pure Expansive Component," *Cement and Concrete Research*, Vol. 5, No. 6 (1975)

Berger, R.L., Frohnsdorff, G.J.C., Harris, P.H. and Johnson,P.D., "Application of X-Ray Diffraction to Routine Mineralogical Analysis of Portland Cement," *Symposium on Structure of Portland Cement Paste and Concrete*, Special Report 90, Highway Research Board, Washington, D.C. (1966)

Berke, N. S., Pfefer, D. W., Weil, T. G., "Protection against chloride-induced corrosion," *Concrete International - Design and Construction*, Vol. 10, No. 12 (1988)

Berman, H.A.,and Newman, E.S., "Calorimetry of Portland Cement - I. Effect of Various Procedures on Determination of Heat of Solution," *Proceedings*, ASTM, Vol. 63 (1963)

Berntsson, L., Chandra, S., Kutti, T., "Principles and Factors Influencing High-Strength Concrete Production," *Concrete International*, Vol. 12, No. 12 (1990)

Berry, E. E. and Malhotra, V. M., *Fly Ash in Concrete*, CANMET SP85-3, Ottawa, Canada (1986)

Bhatty, M. S. Y., "Mechanism of Pozzolanic Reaction and Control of Alkali Aggregate Expansion," *ASTM Cement, Concrete, and Aggregates*, CCAGDP, Vol. 7, No. 2 (1985)

Blaine, R. L., Hunt, C. M., and Tomes, L. A., "Use of Internal-Surface-Area Measurements in Research on Freezing and Thawing of Materials," *Proceedings, Highway Research Board*, Vol. 32 (1953)

Blaine, R.L., Arni, H.T. and Clevenger, R.A., "Water Requirements of Portland Cement," *Interrelations Between Cement and Concrete Properties*, Part I, *Building Science Series* 2, National Bureau of Standards, Washington, D.C. (1965)

Blaine, R.L., Arni, H.T. and Evans, D.N., "Variables Associated with Expansion in the Potential Sulfate Expansion Test," *Interrelations Between Cement and Concrete Properties*, Part 2," *Building Science Series* 5, Section 4, National Bureau of Standards, Washington, D.C. (1966a)

Blaine, R.L. and Arni, H.T., "Heat of Hydration of Portland Cement," *Interrelations Between Cement and Concrete Properties*, Part 2, *Building Science Series* 5, Section 5, National Bureau of Standards, Washington, D.C. (1966b)

Blaine, R.L. and Arni, H.T., "Variables Associocated with Small Autoclave Expansion Values of Portland Cements," *Interrelations Between Cement & Concrete Properties*, Part 2, *Building Science Series* 5, Section 6, National Bureau of Standards, (1966c)

Blaine, R. L., Arni, H. T., and DeFore, M. R., "Compressive Strength of Test Mortars," *Interrelations between Cement and Concrete Properties*, Part 3, *Building Science Series* 8, Section 7, U. S. Department of Commerce, National Bureau of Standards, Washington, D. C. (1968)

Blaine, R.L., Arni, H.T. and Evans, D.N., "Shrinkage of Hardened Portland Cement Pastes," *Interrelations Between Cement and Concrete Properties*, Part 4, *Building Science Series* 15, Section 9, National Bureau of Standards, Washington, D.C. (1969)

Blanks, R.F. and Kennedy, H L., *The Technology of Cement and Concrete*, Vol.1, Wiley, New York - Chapman & Hall, London (1955)

Bloem, D. L., and Gaynor, R. D., "Effects of Aggregate Properties on Strength of Concrete," *ACI Journal*, proc. Vol. 60, No. 10 (1963)

Bloem, D. L.,   "Soundness and Deleterious Substances," *Significance of Tests and Properties of Concrete and Concrete Making Materials, ASTM STP 169-A*, Philadelphia (1966)

Blondiau, L., "Essais de fabrication et de resistance de tuyaux de faible diametre en beton de ciment sursulfate" (Investigation of the Fabrication and Strength of Concrete Pipes of Small Diamater Made with Supersulfated Cement), *Revue des Materiaux de Construction*, No. 528 (1959)

Blondiau, L., "Aptitude du ciment sursulfate a la construction des egouts," (Applicability of Supersulfated Cement to Sewer Construction), *Revue des Materiaux de Construction*, Nos. 535 and 536 (1960)

Bobrov, B. S., "Mutual Influence of 3CaO-SiO2 and 4CaO-Al2O3-Fe2O3 in Portland Cement Hydration," Supplementary Paper, II-2, *The VI. International Congress on the Chemistry of Cement*, Moscow (1974)

Bogue, R. H., "Calculation of Compounds in Portland Cement," *Industrial and Engineering Chemistry*, (Analytical Edition), Vol. 1, October 15 (1929)

Bogue, R. H., and Lerch, W., "The Hydration of Portland Cement Compounds," *Industrial and Engineering Chemistry*, Vol. 26, August (1934)

Bogue, R. H., "Cement," *Encyclopedia Americana*, Vol.6, Groliers, New York (1955a)

Bogue, R. H., *The Chemistry of Portland Cement*, 2nd ed., Reinhold, New York (1955b)

Bolomey, J.,   "Module de finesse d'Abrams et calcul de l'eau de gachage des betons" (Fineness Modulus of Abrams and the Calculation of the Mixing Water of Concrete), *Festschrift 1880-1930 der Eidgenossischen Materialprufungsanstalt* (1930)

Bolomey, J., (The Grading of Aggregate and its Influence on the Characteristics of Concrete), *Revue Mater. Construct. Trav Publ.*, ed. C (1947)

Bombled, I. P. and Klavens, O., "Comportement rheologique des pates, mortiers et betons: mesure, evolution, influence de certains parameters" (Rheological Behavior of Pastes, Mortars and Concretes: Measurement, Evoloution, Influence of Certain Parameters), *Revue de Materiaux de Construction, Ciments et Betons*, No. 617 (1967)

Bonzel, J., and Dahms, J., *Alkalireaktion im Beton* (Alkali-Aggregate Reaction in Concrete), *Betontechnische Berichte 1973*, Beton-Verlag, Dusseldorf (1974a)

Bonzel, J., and Siebel, E., "Fliessbeton und seine Anwendungsmoglichkeiten" (Highly Fluidified Concrete and Its Field of Application), *Betontechnische Berichte 1974*, Beton-Verlag, Dusseldorf (1974b)

Bonzel, J., and Krumm, L., "Betonzusatze" (Admixtures for Concrete), *Zement-Taschenbuch 1976-77*, Bauverlag, Wiesbaden, Berlin (1976a)

Bonzel, J., and Dahms, J., "Zur Prufung des Frostwiderstandes von Betonzuschlag" (Testing the Frost Resistance of Concrete Aggregate), *Beton*, Vol. 26, no. 5 and 6 (1976b)

Bonzel, J., Krell, J., und Siebel, E., "Alkalireaktion im Beton," (Alkali Reaction in Concrete), *Betontechnishe Berichte 1986 - 88*, Beton Verlag, Dusseldorf (1989)

Braniski, A., "Refractory Barium-Aluminous Cement and Concrete," *Chemistry of Cement*, Proceedings of the Fourth International Symposium, *NBS Monograph 43*, Vol. II, Washington, D.C. (1962a)

Braniski, A., "Analogies et differences des ciments au calcium, strontium et baryum" (Similarities and Differences of Calcium, Strontium and Barium Cements), *Revue des Materiaux de Construction, Ciments & Betons*, No. 560 (1962b)

Bredsdorff, P., Idorn, G. M., Kjaer, A., Plum, N. M., and Poulsen, E., "Chemical Reactions Involving Aggregate", *Chemistry of Cement. NBS Monograph*, Washington (1960)

Bretz, J., "Zerstrorungsfreie Prufung von Bauxitbeton- Bauten in Ungarn" (Nondestructive Testing of High- Alumina Cement Concrete Buildings in Hungary), *Zerstorungsfreie Pruf- und Messtechnik fur Beton und Stahlbeton*, Proceedings of the International Conference, Leipzig (1969)

Brink, R. H., "Rapid Freezing and Thawing Test for Aggregate," *Highway Research Board Bulletin*, No. 201 (1958)

Brink, R. H., and Timms, A. G., "Weight, Density, Absorption, and Surface Moisture," *Significance of Tests and Properties of Concrete and Concrete Making Materials*, ASTM STP 169-A, Philadelphia (1966)

Brown, L. S., "Long-Time Study of Cement Performance in Concrete - Chapter 4. Microscopical Study of Clinkers," *ACI Journal*, Proc. Vol. 44 (1948)

Bruere, G. M., "Air Entrainment in Cement and Silica Pastes," *ACI Journal*, Proc. Vol. 51 (1955)

Bruere, G. M., "Mechanism by Which Air Entraining Agents Affect Viscosities and Bleeding Properties of Cement Pastes," *Australian Journal of Applied Science*, Vol. 9, No. 4 (1958)

Bruere, G. M., "The Effect of Type of Surface-Active Agent on Spacing Factors and Surface Areas of Entrained Bubbles in Cement Pastes," *Australian Journal of Applied Science*, Vol. 11, No. 2 (1960)

Bruere, G. M., "The Relative Importance of Various Physical and Chemical Factors on Bubble Characteristics in Cement Pastes," *Australian Journal of Applied Science*, Vol. 12, No. 1 (1961)

Bruere, G. M., "Importance of Mixing Sequence When Using Set-Retarding Agents with Portland Cement," *Nature*, Vol. 199, No. 4888, London (1963)

Bruere, G. M., "Effects of Mixing Sequence on Mortar Consistencies When Using Water-Reducing Agents," *Symposium on Structure of Portland Cement Paste and Concrete, Special Report* 90, Highway Research Board, Washington, D.C. (1966)

Bruere, G. M., "Some Influences on Admixture Requirements in Concrete," *NRMCA Publication* No. 146, National Ready Mixed Concrete Association, Silver Spring, Md. (1974)

Brunauer, S., Emmett, P. H., and Teller, E., "Adsorption of Gases in Multimolecular Layers," *Journal of the American Chemical Society*, Vol. 60 (1938)

Brunauer, S., Copeland, L.E., Kantro, D.L., Weise, C.H. and Schultz, E.G., "Quantitave Determination of the Four Major Phases in Portland Cement by X-Ray Analysis," *Proceedings*, ASTM, Vol. 59 (1959)

Brunauer, S., "The Role of Tobermorite Gel in Conrete," *Structural Concrete*, Vol. 1, no. 7, London (1963)

Brunauer, S., and Copeland, L. E., "The Chemistry of Concrete," *Scientific American*, Vol. 210, No. 4 (1964a)

Brunauer, S., and Kantro, D. L., "The Hydration of Tricalcium Silicate and β-Dicalcium Silicate from 5C to 50C," *The Chemistry of Cements*, ed. H. F. W. Taylor, Vol. 1, Chap. 7, Academic Press, London and New York (1964b)

Brunauer, S., Odler, I., and Yudenfreund, M., "The New Model of Hardened Portland Cement Paste," *Highway Research Record* No.328, Highway Research Board, Washington, D.C. (1970)

Buck, A., D., and Dolch, W. L., "Investigation of a Reaction Involving Nondolomitic Limestone Aggregate in Concrete," *Journal of the ACI*, Proc. Vol. 63 (1966)

Buck, A. D., "Control of Reactive Carbonate Rocks in Concrete," *Technical Report* No. C-75-3, USAE Waterways Experiment Station, Vicksburg, Miss. (1975)

Buck, A., D., "Alkali Reactivity of Strained Quartz as a Constituent of Concrete Aggregate," *Cement, Concrete, and Aggregate*, ASTM, CCAGDP, Vol. 5 (1983)

Budnikov, P. P. and Erschler, E.Ya., "Studies of the Processes of Cement Hardening in the Course of Low-Pressure Steam Curing of Concrete," *Special Report* 90, Highway Research Board,   Washington D.C.(1966)

Budnikov, P. P., and Kravchenko, I. V., "Expansive Cements," *Proceedings of the Firth International Symposium on the Chemistry of Cement*, Part IV, *Admixtures and Special Cements*, Tokyo (1969)

Building Research Establishment, "High Alumina Cement Concrete in Buildings," *Concrete*, BRE Building Research Series, Vol. 1, The Construction Press, Lancaster, London, New York (1978).

Bukovatz, J. E., Crumpton, C. F. and Worley, H. E., "Kansas Concrete Pavement Performance as Related to D-Cracking," *Transportation Research Record*, No. 525. Transportation Research Board, Washington, D. C. (1974)

Burton, R. A., "Friction and Wear," *Tribology - Friction, Lubrication, and Wear*, A. Z. Szeri, (Ed.), Hemisphere Publishing Corporation, Washington D.C. etc. (1980)

Butcher, B. J., and Hokins, H. J., Particle Interference in Concrete Mixes, ACI Journal, Proc. Vol. 53 (1956)

Buth, E., Ivey, D. L., and Hirsch, T. J., "Correlation of Concrete Properties with Tests for Clay Content of Aggregate," *Highway Research Record* No. 124, Highway Research Board, Washington, D.C. (1966)

Buth, E. and Ledbetter, W. B., "The Importance of Moisture Absorption Characteristics of Lightweight Coarse Aggregate," *Highway Research Record* No. 226, Highway Research Board, Washington, D.C. (1968a)

Buth, E., Ivey, D. L., and Hirsch, T. J., "Dirty Aggregate, What Difference Does It Make?," *Highway Research Record* No. 226, Highway Research Board, Washington, D.C. (1968b)

Buth, E., and Ledbetter, W. B., "Influence of the Degree of Saturation of Coarse Aggregate on the Resistance of Structural Lightweight Concrete to Freezing and Thawing," *Highway Research Record* No. 328, *Concrete Durability, Cement Paste, Aggregates, and Sealing Compounds*, Highway Research Board, Washington, D.C. (1970)

Butt, Yu. M., Kolbasov, V. M., and Timashev, V. V., "High Temperature Curing of Concrete Under Atmospheric Pressure," *Proceedings of the Fifth International Symposium on the Chemistry of Cement, Part III, Properties of Cement and Concrete*, Tokyo (1969)

Cadle, R. D., *Particle Size*, Reinhold, New York (1965)

Cady, P. D., "Mechanisms of Frost Action in Concrete Aggregates," *Journal of Materials*, JMLSA, Vol 4, No. 2 (1969)

Call, Bayard M., "Slump Loss With Type "K" Shrinkage Compensating Cement Concretes and Admixtures," *Concrete International: Design & Construction*, Vol. 1, No. 1 (1979)

Callan, E. J., "Thermal Expansion of Aggregates and Concrete Durability," *ACI Journal*, Proc. Vol. 48 (1952)

Callan, E.J., "Concrete for Radiation Shielding," *ACI Journal*, Proc. Vol. 50 (1953), *ACI Publication Compilation* No.1 Second ed. (1962)

Calleja, J., "New Techniques in the Study of Setting and Hardening of Hydraulic Materials," *ACI Journal*, Proc. Vol. 48 (1952)

Campbell, R. H., "A Program to Test Cements for Variations in Strength Producing Properties," *ACI Journal*, Proc. Vol. 65, No. 4 (1968)

Campbell, R. H., Harding, W., Misenhimer, E., and Nicholson, L. P., "Surface Popouts: How Are They Affected by Job Conditions?," *ACI Journal*, Proc. Vol. 71, No. 6 (1974)

Campen, W. H., "The Development of a Maximum Density Curve and Its Application to the Grading of Aggregates for Bituminous Mixtures," *Proc. AAPT*, Vol. 11 (1940)

Cannon, J. P. and Krishna Murti, G. R., "Concrete Optimized Mix Proportioning (COMP)," *Cement and Concrete Research*, Vol 1, No. 4 (1971)

Capp, J. P. and Spencer, J. D., *Fly Ash Utilization - A Summary of Applications and Technology*, Bureau of Mines (1970)

Caqout, A., and Faury, J., Plasticite de mise en oeuvre du beton en construction de beton arme. Influences des principaux factors en jeu. (Plasticity of Concrete Mixtures in Reinforced Concrete Construction. Effects o the Principal Factors), *Annuales de l'Institut Technique de Batiment et des Travaux Publics* (1937)

Carrasquillo, R. L. and Snow, P. G., "Effect of Fly Ash on Alkali-Aggregate Reaction in Concrete," *ACI Materials Journal*, Vol. 84, No. 4 (1987)

Cebeci, O. Z., Saatci, A. M., "Domestic Sewage as Mixing Water in Concrete," *ACI Materials Journal*, Vol. 86, No. 5 (1989)

Celani, A., Moggi, P. A., and Rio, A., "The Effect of Tricalcium Aluminate on the Hydration of Tricalcium Silicate and Portland Cement," *Proceedings of the Fifth International Symposium on the Chemistry of Cements*, Part II, *Hydration of Cements*, Tokyo (1969)

Cembureau, *Review of Standards for Cement Other Than Portland Cement*, The Cement Statistical and Technical Association, Malmo (1958)

Cembureau, *Review of the Portland Cement Standards of the World*, The Cement Statistical and Technical Association, Malmo (1961)

Cement Admixtures Association and Cement and Concrete Association, "Superplasticizing Admixtures in Concrete," Cement and Concrete Association, London (1976)

Cement and Concrete Association, *Hydraulic Cement Pastes: Their Structure and Properties,* Proceedings of a Conference held at University of Sheffield, 8-9 April 1976, Cement and Concrete Association, London (1976)

Chaiken, B., and Halstead, W. J., "Correlation Between Chemical and Mortar Bar Tests for Potential Alkali Reactivity of Concrete," *Highway Research Board Bulletin* No. 239, Washington, D.C. (1960)

Chalmers, R. A., "Chemical Analysis of Silicates," *The Chemistry of Cements,* ed. H.F.W. Taylor, Vol. II, Academic Press, London and New York (1964)

Chatterji, S., and Majumdar, A. J., "Studies of the Early Stages of Paste Hydration of High Alumina Cements Hydration of Individual Aluminates," *Indian Concrete Journal,* Vol. 40, No. 2 (1966a)

Chatterji, S., and Majumdar, A. J., "Studies of the Early Stages of Paste Hydration of High Alumina Cements - 3: Hydration of Commercial High-Alumina Cements," *Indian Concrete Journal,* Vol. 40, No. 6 (1966b)

Clanton, J. R., and Hennes, R. G., "Stability and Durability of Plant-Mix Macadam," *Proceedings of the 31st Annual Meeting,* Vol. 31, Highway Research Board, Washington, D.C. (1952)

Clifton, J. R. and Mathey, R.G., "Compilation of Data from Laboratory Studies," *Interrelations Between Cement and Concrete Properties,* Part 6, *NBS Building Science Series* 36, Section 14, Washington D.C. (1971)

Clifton, J. R., Brown, P. W. and Frohnsdroff, G., "Reactivity of Fly Ashes with Cement," *Cement Research Progress 1977,* Chapter 15, American Ceramic Society (1977)

Commissie Voor Uitvoering Van Research, "Admixtures for Concrete," Translation No. 131, Cement and Concrete Association, London (1968)

Concrete Society, "Admixtures for Concrete," *Concrete,* Vol. 2, No. 1, London (1968)

Cook, H. K.,   "Thermal Properties," *Siginficance of Tests and Properties of Concrete and Concrete Making Materials*,   *ASTM STP*   169-A, Philadelphia (1966)

Copeland, L. E., Brunauer, S., Kantro, D.L., Schultz, E G. and Weise, C.H., "Quantitative Determination of the Four Major Phases of Portland Cement by Combined X-Ray and Chemical Analysis," *Analytical Chemistry*, Vol.31 (1959)

Copeland, L. E., Kantro, D.L. and Verbeck, G., "Chemistry of Hydration of Portland Cement," *The Chemistry of Cement*, Proceedings of the Fourth International Symposium, *NBS Monograph* 43, Vol.I, Session   IV,   Washington, D.C. (1960)

Copeland, L. E., and Schulz, E. G., "Electron Optical Investigation of the Hydration Products of Calcium Silicates and Portland Cement," *Journal of the PCA Research and Development Laboratories*, Vol. 4, No. 1 (1962)

Copeland, L. E., and Kantro, D. L., "Chemistry of Hydration of Portland Cement at Ordinary Temperature," *The Chemistry of Cements*, ed. H.   W. Taylor, Vol. 1, Chap. 8, Academic Press, London and New York (1964)

Copeland, L. E., Bodor, E., Chang, T. N., and Weise, C. H., "Reactions of Tobermorite Gel with Aluminates, Ferrites, and Sulfates," *Journal of the PCA Research Development Laboratories*, Vol. 9, No. 1 (1967)

Copeland, L. E. and Kantro, D. L., "Hydration of Portland Cement," *Proceedings of the Fifth International Symposium on the Chemistry of Cement*, Part II, *Hydration of Cements*, Tokyo (1969)

Copeland, L. E., and Verbeck, G. J., "Structure and Properties of Hardened Cement Pastes," A Principal Paper, *VIth International Congress on the Chemistry of Cement*, Moscow (1974)

Cordon, W. A.,   *Freezing and Thawing of Concrete - Mechanisms and Control, American Concrete Institute Monograph   3 (1966)*

Cusens, A. R. and Harris, J., "A Rheological Study of Fresh Cement Pastes," *RILEM Seminar on Fresh Concrete*, Paper 2.80, Leeds (1973)

Czernin, W., *Cement Chemistry and Physics for Civil Engineers*, Crosby Lockwood & Son, London (1962)

Dahl, L. A., "Estimation of Phase Composition of Clinker," *Rock Products*, Vols. 41-42 (1938-1939)

Dalhouse, J. B., Plotting Gradation Specifications for Bituminous Concrete, *Public Roads*, vol. 27, no. 7 (1953)

Dalziel, J. A., "The Effects of Different Portland Cements Upon The Pozzolanicity of Pulverized-Fuel Ashes and The Strength of Blended Cement Mortars," *Technical Report* 555, Cement and Concrete Association (1983)

Davis, H. S., " High-Density Concrete for Shielding Atomic Energy Plants" *ACI Journal*, Proc. Vol. 54, May (1958), *ACI Publication Compilation* No.1, Second Ed. (1962)

Davis, R. E., "A Review of Pozzolanic Materials and Their Use in Concrete," *Symposium on Use of Pozzolanic Materials in Mortars and Concrete*, ASTM *Special Technical Publication* 99 (1950)

De Larrard, F., Buil, M., "Granularite et compacite dans les materiaux de genie civil," (Particle size and compactness in civil engineering materials) *Materials and Structures - Research and Testing*, RILEM, Vol. 20, 116, Paris (1987)

DeHoff, R. T., and Rhines, F. N., *Quantitative Microscopy*, McGraw-Hill (1968)

Dent-Glasser, L. S., "X-Ray Diffraction," *The Chemistry of Cements*, ed. H.F.W. Taylor, Vol. II, chap.19, Academic Press, London and New York (1964)

Detweller, R. J., and Mehta, K. P., "Chemical and Physical Effects of Selica Fume on The Mechanical Behavior of Concrete," *ACI Materials Journal*, Vol. 86, No. 6 (1989)

Dewar, J. D., "The Workability and Compressive Strength of Concrete Made with Sea Water," *Technical Report* TRA/374, Cement and Concrete Association, London (1963)

Diamond, S., "A Critical Comparison of Mercury Porosimetry and Capillary Condensation Pore Size Distributions of Portland Cement Pastes," *Cement and Concrete Research*, Vol. 1, No. 5 (1971)

Diamond, S., "Identification of Hydrated Cement Constituents Using a Scanning Electron Microscope - Energy Dispersive X-Ray Spectrometer Combination," *Cement and Concrete Research*, Vol. 2, No. 5 (1972)

Diamond, S., "Interactions Between Cement Minerals and Hydroxycarboxylic-Acid Retarders: I. Apparent Adsorption of Salicylic Acid on Cement Compounds; II. Tricalcium Aluminate - Salicylic Acid Reaction; III. Infrared Spectral Identification of the Aluminosalicylate Complex; IV. Morphology of the Aluminosalicylate Complex," *Journal of the American Ceramic Society*, Vol. 54, No. 6, Vol. 55, No. 4, Vol. 55, No. 8, Vol. 56, No. 6 (1971 - 1973)

Diamond, S., and Thaulow, N., "A Study of Expansion Due to Alkali-Silica Reaction as Conditioned by the Grain Size of the Reactive Aggregate," *Cement and Concrete Research*, Vol. 4, No. 4 (1974)

Diamond, S., "A Review of Alkali-Silica Reaction and Expansion Mechanisms. 1. Alkalies in Cements and in Concrete Pore Solutions," *Cement and Concrete Research*, Vol. 5, No. 4 (1975)

Dikeou, J. T., Steinberg, M., et. al., *Concrete-Polymer Materials*, Third Topical Report, *REC-ERC-71-6* and *BNL 50275 (T-602)* (1971)

Dise, J. R., "Significance of the Test for Normal Consistency of Hydraulic Cement," *Cement, Comparison of Standards and Significance of Particular Tests*, *ASTM STP* 441, Philadelphia (1968)

Dolar-Mantuani, L., "Petrographic Examination of Natural Concrete Aggregates," *Highway Research Record* No. 120, *Aggregate Characteristics and Examination*, Washington, D.C. (1966)

Dolar-Mantuani, L., "Alkali-Silica-Reactive Rocks in the Canadian Shield," *Highway Research Report* No. 268, Highway Research Board, Washington, D.C. (1969)

Dolar-Mantuani, L., "Late-Expansion Alkali-Reactive Carbonate Rocks," *Highway Research Record* Number 353, *Mineral Aggregates*, Highway Research Board, Washington, D.C. (1971)

Dolar-Mantuani, L., *Handbook of Concrete Aggregates*, Noyes Publication, Park Ridge, N.J. (1983)

Dolch, W. L., "Studies of Limestone Aggregates by Fluid-Flow Methods," *Proceedings*, ASTM, Vol. 59 (1959)

Dolch, W. L., "Porosity," *Significance of Tests and Properties of Concrete and Concrete Making Materials*, *ASTM STP* 169-A, Philadelphia (1966)

Dolch, W. L., "The Air Entrainment Test for Cement, C 185," *Cement, Comparison of Standards and Significance of Particular Tests*, ASTM STP 441, Philadelphia (1968)

Double, D. D. and Hellawell, A., "The Solidification of Cement," *Scientific American*, Vol. 237, No.1 (1977)

Drinker, P., "The Size-Frequency and Identification of Certain Phagocytosed Dusts," *Journal of Industrial Hygiene*, Vol. 7 (1925)

Dunagan, W. M., The Application of Some of the Newer Concepts to the Design of Concrete Mixes, *ACI Journal*, Proc. Vol. 36 (1940)

Duncan, M. A. G., Swenson, E. G., Gillott, J. E., and Foran, M. R., "Alkali-Aggregate Reaction in Nova Scotia. I. Summary of a Five-Year Study", *Cement and Concrete Research*, (1973a)

Duncan, M. A. G., Swenson, E. G., Gillott, J. E., and Foran, M. R., "Alkali-Aggregate Reaction in Nova Scotia. II. Field and Petrographic Studies", *Cement and Concrete Research*, (1973b)

Duncan, M. A. G., Swenson, E. G., Gillott, J. E., and Foran, M. R., "Alkali-Aggregate Reaction in Nova Scotia. III. Laboratory Studies of Volume Change", *Cement and Concrete Research*, (1973c)

Dunn, J. R., and Hudec, P. P., "The Influence of Clay on Water and Ice in Rock Pores," *Physical Research Report RR* 65-5, New York State Department of Public Works (1965)

Dunstan, M. R. H., "Development of High Flyash Content Concrete," *Proceedings*, Institution of Civil Engineers, Part 1, Vol. 74, London (1983)

Duriez, M. and Arrambelle, J., *Nouveau traite de materiaux de construction* (New Treatise on the Materials of Construction), Vol.I, Dunod, Paris (1961)

Duriez, M. and Arrambide, J., Nouveau traite de materiaux de construction (New Treatise on the Materials of Construction), Vol. 2, Dunod, Paris (1962)

Dutron, P., Bang, C. and Kral, S., "Present Methods for Mechanical Testing of Cements," A Principal Paper, *VIth International Congress on the Chemistry of Cement*, Moscow (1974)

Dutt, R. N., and Lee, D., "Upgrading Absorptive Aggregates by Chemical Treatments" (Abridgment), *Highway Research Record* No. 353, *Mineral Aggregates*, Highway Research Board, Washington, D.C. (1971)

Edison (The) Portland Cement Co., *The Romance of Cement*, Livermore & Knight Company, Providence, New York, Boston (1926)

Edwards, L. N., "Proportioning the Materials of Mortars and Concretes by Surface Area of Aggregates," *Proc. ASTM*, Vol. 18, Part II (1918)

Eitel, W., *Silicate Science*, Vol.V, *Ceramics and Hydraulic Binders*, Academic Press, New York and London (1966)

El-Rawi, Nagih M.,   "Correlation of Properties of
Iraqi Limestone," *Highway Research Record* No. 447,
*Materials for Pavement Marking and for Joints in
Concrete Structures*, Highway Research Board,
Washington, D.C. (1973)

Epps, J. A., and Polivka, M., "Effect of Aggregate
Type on the Properties of Shrinkage-Compensating
Concrete," *Highway Research Record* No. 328, *Concrete
Durability, Cement Paste, Aggregates, and Sealing
Compounds*, Highway Research Board, Washington,
D.C.(1970)

Ernst, S. S., Jr., "An Experiment in Water Pollution
Control," *NRMCA Publication* 141 (1972)

Evans, D. N., Blaine, R. L., and Worksman, P.,
"Comparison of Chemical Resistance of Supersulfate
and Special Purpose Cements," *Chemistry of Cement*,
Proceedings of the Fourth International Symposium,
NBS Monograph 43, Vol. II, Washington, D.C. (1960)

Faber, J. H., Capp, J.P. and Spencer, J.D., *Fly Ash
Utilization*, Proceedings: Edison Electric Institute -
National Coal Association - Bureau of Mines
Symposium, Pittsburgh, Pennsylvania,   March 14-16,
1967. U.S. Department of the Interior, Bureau of
Mines, *Information Circular* 8348, Washington,
D.C.(1967)

Faber, J. H., Eckard, W. E., and Spencer, J. D.
(eds.), *Ash Utilization*, Proceedings: Third
International Ash Utilization Symposium, U.S.
Department of the Interior, *Information Circular*
8640, Washington, D.C. (1974)

Fagerlund, G.,   "Degree critique de saturation. Un
outil pour I'estimation de la resistance au gel des
materiaux de construction" (Critical Degree of
Saturation. A Tool for Estimating the Frost
Resistance of Building Materials), *Materials and
Structures Research and Testing* (Paris), Vol 4, No.
23 (1971)

Farmer, V. C., "Infra-Red Spectroscopy of Silicates and Related Compounds," *The Chemistry of Cements*, ed. H.F.W. Taylor, Vol.II, Chap. 23, Academic Press, London and New York (1964)

Fattuhi, N. I., "Influence of Air Temperature on the Setting of Concrete Containing Set Retarding Admixtures," *ASTM Cement, Concrete and Aggregates* , CCAGDP, Vol.7, No.1 (1985)

Faury, J., *Le beton* (Concrete), 3rd ed., Dunod, Paris (1958)

Feldman, R. F., and Sereda, P. J., "Characteristics of Sorption and Expansion Isotherms of Reactive Limestone Aggregates," *ACI Journal*, Proc. Vol. 58 (1961)

Feldman, R. F., and Sereda, P. J., "A Model for Hydrated Portland Cement Paste as Deduced from Sorption-Length Change and Mechanical Properties," *Materials and Structures - Research and Testing*, Vol. 1, No. 6 (1968)

Feldman, R. F., "Sorption and Length-Change Scannings Isotherms of Methanol and Water on Hydrated Portland Cement," *Proceedings of the Fifth International Symposium on the Chemistry of Cement*, Part III, *Properties of Cement Paste and Concrete*, Tokyo (1969)

Feldman, R. F., "Assessment of Experimental Evidence for Models of Hydrated Portland Cement," *Highway Research Record* No.370, *Concrete*, Highway Research Board, Washington, D.C.(1971a)

Feldman, R. F., "The Flow of Helium into the Interlayer Spaces of Hydrated Portland Cement Paste," *Cement and Concrete Research*, Vol. 1, No. 3 (1971b)

Feldman, R. F., "Helium Flow and Density Measurement of the Hydrated Tricalcium Silicate-Water System," *Cement and Concrete Research*, Vol. 2, No. 1 (1972)

Feret, R., *Sur la compacite des mortiers hydrauliques* (About the Denseness of Cement Mortars), *Annales des Ponts et Chaussees* (1892)

Fifth International Symposium on the Chemistry of Cement, Proceedings, Part IV, *Admixtures and Special Cements*, Tokyo (1969)

Figg, J. W., and Bowden, S. R., *The Analysis of Concretes*, Building Research Station, Her Majesty's Stationery Office, London (1971)

Figg, J. W., and Lees, T. P., "Field Testing the Chloride Content of Sea-Dredged Aggregates," *Concrete*, Vol. 9, No. 9, London (1975)

Fletcher, K. E., and Roberts, M. H., "Test Methods to Assess thc Performance of Admixtures in Concrete," *Concrete*, Vol. 5, No. 5, London (1971a)

Fletcher, K. E., and Roberts, M. H., "The Performance in Concrete of Admixtures with Accelerating, Retarding or Water-Reducing Properties," *Concrete*, Vol. 5, No. 6, London (1971b)

Fletcher, K. E., and Roberts, M. H., "Effect of a Polyethylene Oxide Admixture on the Strength of Concrete," *Concrete* , Vol. 5, No. 10, London (1971c)

Ford, C. L.,"Determination of the Apparent Density of Hydraulic Cement in Water Using a Vacuum Pycnometer," *ASTM Bulletin* 231, Philadelphia (1958)

Forslind, E., "Fresh Concrete," *Building Materials, Their Elasticity and Inelasticity*, ed. M. Reiner, Chap. VII, Interscience, New York (1954)

Foster, B. E., "Summary," *Symposium on Effect of Water-Reducing Admixtures and Set-Retarding Admixtures on Properties of Concrete*, ASTM STP 266, Philadelphia (1960)

Foster, B. E. and Blaine R.L., "A Comparison of ISO and AS TM Tests for Cement Strength," *Cement, Comparison of Standards and Significance of Particular Tests*, ASTM STP 441, Philadelphia (1968)

Foth, J., "Experimentelle Bestimmung des optimalen Aufbaus von Korngemischen" (Experimental Determination of the Optimum Grading for Concrete Aggregates), *Betonstein-Zeitung*, Vol. 32, No. 9 (1966)

Fowkes, R. S., and Fritz, J. F., "Theoretical and Experimental Studies on the Packing of Solid Particles: A Survey," *Information Circular 8623, U.S. Department of the Interior, Bureau of Mines* (1974)

French, P. J., Montgomery, R. G. J., and Robson, T. D.,"High Strength Concrete Within the Hour," *Concrete* Vol. 5, No. 8, London (1971)

Frohnsdroff, G., *Blended Cements, ASTM STP* 897, Philadelphia (1986)

Fuller, W. B., and Thompson, S. E., The Laws of Proportioning Concrete, *Transactions, ASCE* (1907)

Fulton, F. S., "The Fineness Modulus and the Grading of Aggregates" *Concrete and Constructional Eng.*(1956)

Fulton, F. S., *Concrete Technology*, A South African Handbook, 3rd ed., Portland Cement Institute, Johannesburg (1964)

Furnas, C. C., "Grading Aggregates - I - Mathematical Relationship for Beds of Broken Solids of Max. Density, *Industrial and Engineering Chemistry*, Vol. 23, No. 9 (1931)

Gallaway, B. M., "Taylor-Made Aggregates for Prolonged High Skid Resistance on Modern Highways," *An Inter- american Approach for the Seventies, Materials Technology - I*, Second Interamerican Conference on Materials Technology, Mexico City (1970)

Gard, I. A., "Electron Microscopy and Diffraction," *The Chemistry of Cements*, ed. H. F. W. Taylor, Vol. II, Chap. 21, Academic Press, London and New York (1964)

Garrity, L. V., and Kriege, H. F., "Studies of Accelerated Soundness Tests," *Proceedings, Highway Research Board*, Vol. 15 (1935)

Gaynor, R. D., "Tests of Water-Reducing Retarders," *Publication* No. 108, National Ready Mixed Concrete Association (1962)

Gaynor, R. D., "Effects of Prolonged Mixing on the Properties of Concrete," *Publication* 111, National Ready Mixed Concrete Association   (1963a)

Gaynor, R. D., Effect of Fine Aggregate on Concrete Mixing Water Requirement, Presented at the 47th Annual Convention of the National Sand and Gravel Association, San Francisco (1963b)

Gaynor, R. D., "Exploratory Tests of Concrete Sands," *National Sand and Gravel Association and National Ready Mixed Concrete Association*, Silver Sprinig, Md. (1968a)

Gaynor, R. D., "Laboratory Freezing and Thawing Tests. A Method of Evaluating Aggregates," *Proceedings of the 49th Annual Tennessee Highway Conference, Bulltin* no. 34, Engineering Experiment Station, University of Tennessee, Knoxville (1968b)

Gaynor, R. D., "Shrinkage Compensated Cement and the Ready-Mixed Concrete Producer," Presentation at the Ready Mixed Concrete Operating Problems Session, Bal Harbour, Fla. (1973)

Gaze, M. E. and Smith, M.A., "High Magnesia Cements 1: Curing at 50o as a Measure of Volume Stability, Concrete," *BRE Building Research Series*, Vol.1, The Construction Press, Lancaster, London, New York (1978a)

Gaze, M. E. and Smith, M.A., "High Magnesia Cements 2: The Effect of Hydraulic and Non-Hydraulic Admixtures on Expansion, Concrete," *BRE Building Research Series*, Vol.1, The Construction Press, Lancaster, London, New York (1978b)

Gebauer, J., and Coughlin, R. W., "Preparation, Properties and Corrosion Resistance of Composites of Cement Mortar and Organic Polymers," *Cement and Concrete Research*, Vol. 1, No. 2 (1971)

Gibson, F. W. (Ed.), *Corrosion, Concrete, and Chlorides ACI Publication* SP-102, American Concrete Institute (1987)

Gilliland, J. L., "A Survey of Portland Cement Specifications of the United States of America, in Cement, Comparison of Standards and Significance of Particular Tests," *ASTM STP* 441, Philadelphia (1968)

Gillmore, Q. A.,*Practical Treatise on Limes, Hydraulic Cements and Mortars*, New York (1874)

Gillott, J. E., Duncan, M. A. G., and Swenson, E. G., "Alkali-Aggregate Reaction in Nova Scotia. IV. Character of the Reaction," *Cement and Concrete Research*, Vol. 3, No. 5 (1973)

Gjorv, 0. E., and Shah, S. P., "Testing Methods for Concrete Durability," *Materials and Structures Research and Testing*, Vol. 4, No. 23 (1971)

Glanville, W. H., Collins, A. R., and Matthews, D. D., The Grtading of Aggregates and Workability of Concrete, Road Research Technical Paper 5, Department of Scientific and Industrial Research and Ministry of Transport, London (1938)

Glasser, F. P., and Taylor, H. F. W., "Aqueous and Hydrothermal Chemistry, Weight Change Curves and Density Determinations," *The Chemistry of Cements*, ed. H. F. W., Taylor, Vol. 11, Chap. 25, Academic Press, London and New York (1964)

Gonnerman, H. F., "Study of Cement Composition in Relation to Strength, Length Changes, Resistance to Sulfate Waters and to Freezing and Thawing, of Mortars and Concrete," *Proceedings, ASTM*, Vol.34, Part II (1934)

Gonnerman, H. F., and Lerch, W., "Changes in Characteristics of Portland Cement as Exhibited by Laboratory Tests Over the Period 1904 to 1950," *ASTM Special Publication* No. 127, Philadelphia (1952)

Gonnerman, H. F., Lerch, W. and Whiteside, T.M., "Investigations of the Hydration Expansion Characteristics of Portland Cements," *Research Department Bulletin* 45, Portland Cement Association, Chicago (1953)

Gonsalves, G. F., and Eisenberg, J. F., "Implication of Statistical Quality Control of Portland Cement Concrete," Final Report, *Report* No. 8, Arizona Department of Transportation, Phoenix, Ariz. (1975)

Gouda,V. K., Mikhail,R.Sh. and Shater.,M.A., "Hardened Portland Blast-Furnace Slag Cement Pastes. III. Corrosion of Steel Reinforcement Versus Pore Structure of the Paste Matrix," *Cement and Concrete Research*, Vol. 5, No.2 (1975)

Graf, O.(ed), *Die Prufung nichtmetallischer Baustoffe* (Testing Non-Metallic Building Materials), 2nd ed., *Handbuch der Wekstoffprufung* (Materials Testing Handbook), ed. E. Siebel, Vol. III, Springer Verlag, Berlin (1957)

Graf, O., Albrecht, W. and Schaffler, H., *Die Eigenshaften des Betons* (Properties of Concrete), 2nd ed., Springer Verlag, Berlin - Gottingen - Heidelberg (1960)

Grattan-Bellew, P., E., (Ed.), *Concrete Alkali-Aggregate Reactions*, Noyes Publications, Park Ridge, N.J. (1987)

Grieb, W. E., Werner, G., and Woolf, D. O., "Tests of Retarding Admixtures for Concrete," *Highway Research Board Bulletin* 310, Washington, D.C. (1962)

Griffin, D. F., and Henry, R. L., "Integral Sodium Chloride Effect on Strength, Water Vapor Transmission, and Efflorescence of Concrete," *ACI Journal*, Proc. Vol. 58 (1961)

Griffin, D. F., and Henry, R. L., "Effect of Salt in Concrete on Compressive Strength, Water Vapor Transmission, and Corrosion of Reinforcing Steel," *Proceedings*, ASTM, Vol. 63 (1963)

Grudemo, A., "Electron Microscopy of Portland Cement Pastes," *The Chemistry of Cements*, ed. H. F. W. Taylor, Vol. 1, Chap. 9, Academic Press, London and New York (1964)

Gruenwald, E., "Lean Concrete Mixes - A Study of the Effect of Cement Content, Cement Fineness on Compressive Strength, Durability and Volume Change of Concrete," *Proceedings*, ASTM, Vol. 39 (1939)

Guinier, A. and Regourd, M. "Structure of Portland Cement Minerals," *Proceedings of the Fifth International Symposium on the Chemistry of Cement*, Part I, *Chemistry of Cement Clinker*, Tokyo (1969)

Guomundsson, G., and Asgeirsson, H., "Some Investigations on Alkali Aggregate Reaction," *Cement and Concrete Research*, Vol. 5, No. 3 (1975)

Gutmann, P. W., "Expansive Ccments in the United States — A Three-Year Review," *Civil Engineering*, ASCE (1967)

Gutt, W., "Manufacture of Portland Cement from Phosphatic Raw Materials," *Proceedings of the Fifth International Symposium on the Chemistry of Cement*, Part I, *Chemistry of Cement Clinker*, Tokyo (1969)

Gutt, W. and Smith, M.A., "Studies of Phosphatic Portland Cements, Concrete," *BRE Building Research Series*, Vol.1, The Construction Press, Lancaster, London, New York (1978a)

Gutt, W., Teychenne, D.C. and Harrison, W. H., "The Use of Lighter-Weight Blastfurnace Slag as Dense Coarse Aggregate in Concrete," *Concrete*, BRE Building Research Series, Vol. 1, The Construction Press, Lancaster, London, New York (1978b)

Hadley, D. W., "Alkali Reactivity of Carbonate Rocks - Expansion and Dedolomitizaton," *Proceedings*, Highway Research Board, Vol. 40, Washington D.C.(1961)

Hadley, D. W., "The Nature of the Paste-Aggregate Interface," *Joint Highway Research Project Interim Report* No. 40, Purdue University, West Lafayette, Ind. (1972)

Haegermann, G., Huberti, G. and Moll, H., *Vom Caementum zum Spannbeton* (From Caementum through Prestressed Concrete), Vol.1, Bauverlag, Wiesbaden, Berlin (1964)

Hale, K. F., and Brown, M. H., "Application of High Voltage Electron Microscopy to the Study of Cement," *Structure, Solid Mechanics and Engineering Design*, Proceedings of the Southampton 1969 Civil Engineering Materials Conference, Part 1, Wiley-lnterscience, London, New York (1971)

Halstead, P. E., "Expanding and Stressing Cements," *The Chemistry of Cements*, ed.H. F. W. Taylor, Vol. 11,Chap. 15, Academic Press, London and New York (1964)

Halstead, W. J., and Chaiken, B., "Chemical Analysis and Sources of Air-Entraining Admixtures for Concrete," *Public Roads*, Vol. 27, No. 12 (1954)

Halstead, W. J., and Chaiken, B.,"Water-Reducing Retarders for Concrete - Chemical and Spectral Analyses," *Highway Research Board Bulletin* 310, Washington, D.C. (1962)

Halstead, W. J., "Energy Involved in Construction Materials and Procedures," *National Cooperative Highway Research Program Synthesis of Highway Practice* 85, Transportation Research Board, Washington, D.C. (1981)

Halstead, W. J., "Use of Fly Ash in Concrete," *National Cooperative Highway Research Pogram Synthesis of Highway Practice* 127, Transportation Research Board, Washington, D.C. (1986)

Hansen, H. I., "Aggregate Production" *Handbook of Heaby Construction*, J. A. Havers and F. W. Stubbs, Jr., (Ed.), Second Edition Section 20, McGraw-Hill Book Company, New York, etc. (1971)

Hansen, T. C., "Physical Structure of Hardened Portland Cement Paste - A Classical Approach," *Materials and Structures - Research and Testing*, Vol. 19, No.114 (1986)

Hansen, W. C., and Livovich, A. F.,"Factors Influencing the Physical Properties of Refractory Concrete," *Ceramic Bulletin*, Vol. 34, No. 9 (1955)

Hansen, W. C., "Actions of Calcium Sulfate and Admixtures in Portland Cement Pastes," *Symposium on Effect Or Water-Reducing Admixtures and Set-Retarding Admixtures on Properties of Concrete, ASTM STP 266,* Philadelphia (1960)

Hansen, W. C., "Solid-Liquid Reactions in Portland Cement Pastes," *Materials Research and Standards,* Vol. 2, No. 6, Philadelphia, (1962)

Hansen, W. C., "Anhydrous Minerals and Organic Materials as Sources of Distress in Concrete," *Highway Research Record,* No. 43, *Properties of Concrete,* Highway Research Board, Washington, D.C. (1963a)

Hansen, W. C., "Porosity of Hardened Portland Cement Paste," *ACI Journal,* Proc. Vol. 60 (1963b)

Hansen, W. C., "Chemical Reactions," *Significance of Tests and Properties of Concrete and Concrete Making Materials, ASTM STP 169-A,* Philadelphia (1966)

Hansen, W. C, "Basic Chemistry of Reactions of Aggregates in Portland-Cement Concrete," *Journal of Materials,* Vol. 2, No. 2 (1967)

Hansen, W. C., "Potential Compound Compsition of Portland Cement Clinker," *Journal of Materials,* JMLSA, Vol.3, No.1 (1968)

Hansen, W. C., "Interactions of Organic Compounds in Portland Cement Pastes," *Journal of Materials,* Vol. 5, No. 4 (1970)

Hansen, W. C., Offut, J. S., Roy, D. M., Grutzeck, M. W., Schatzlein, K. J., *Gypsum and Anhydrate in Portland Cement,* Third Edition, United States Gypsum Company (1988)

Harder, P., "Oilwell Cement - Something Special in Reliability," *Minerals Processing,* Vol. 8, No. 3 (1967)

Harman, J. W., Cady, P. D., and Bolling, N. B., "Slow-Cooling Tests for Frost Susceptibility of Pennsylvania Aggregates," *Highway Research Record* No. 328, *Concrete Durability, Cement Paste, Aggregates, and Sealing Compounds*, Highway   Research Board, Washington, D.C. (1970)

Harr, M. E., *Mechanics of Particulate Media*, McGraw-Hill, Inc., New York (1977)

Harrison, W. H., "Synthetic Aggregate Sources and Resources," *Concrete*, Vol. 8, No. 11 (1974)

Hatch, T., "Determination of "Average Particle Size" from the Screen Analysis of Non-Uniform Particulate Substances, *Journal of the Franklin Institute*, No. 215 (1957)

Hattori, K., "Experience With Mighty Super-Plasticizer In Japan," *Superplasticizers in Concrete*, ACI Publication SP-62, Detroit (1979)

Hayden, R., "Was ist Portlandzement?" (What is Portland Cement?) *Zement-Kalk-Gips*, No.2 (1954)

Heck, W. J., "Study of Alkali-Silica Reactivity Tests to Improve Correlation and Predictability for Aggregates," *Cement, Concrete, and Aggregate, ASTM*, CCAGDP, Vol. 5, No.1 (1983)

Helms, S. B., and Bowman, A. L., "Extension of Testing Techniques for Determining Absorption of Fine Lightweight Aggregate," *Proceedings, ASTM*, Vol. 62, Philadelphia (1962)

Helmuth, R. A., *Fly Ash in Cement and Concrete*, Portland Cement Association, Skokie, IL (1987)

Herdan, G., *Small Particle Statistics*, 2nd. rev. ed., Academic Press, New York - Butterworths, London (1960)

Hester, J. A., and Smith, O. F., "Alkali-Aggregate Phase of Chemical Reactivity in Concrete," Part II, *Cement and Concrete, ASTM Special Technical Publication* 205, Philadelphia (1958)

Hickey, M. E., "Compressive Strength Studies of
Concrete Made with Types I, II, and V Cement with
and Without Calcium Chloride," *Concrete Laboratory
Report* C-648, U.S. Department of the Interior,
Bureau of Reclamation, Denver (1953)

Higginson, E. C., and Kretsinger, D. G., "Prediction
of Concrete Durability from Thermal Tests of
Aggregate," *Proceedings, ASTM*, Vol. 53 (1953)

Higginson, E. C., "Mineral Admixtures," *Significance
of Tests and Properties of Concrete and Concrete
Making Materials, ASTM STP* 169-A, Philadelphia (1966)

Higginson, E. C., "The Effect of Cement Fineness on
Concrete," *Fineness of Cement, ASTM STP* 473,
Philadelphia (1970)

Highway Research Board, *Calcium Chloride in Concrete,
Highway Research Board Bibliography* 13, Washington,
D.C. (1952)

Highway Research Board, "Alkali-Aggregate Reaction,"
Highway Research Board Bulletin 171, Washington, D.C.
(1958a)

Highway Research Board, *The Alkali-Aggregate
Reaction in Concrete, Research Report* 18-C, Highway
Research Board, Washington, D.C. (1958b)

Highway Research Board, "Pavement Slipperiness
Factors and their Measurement," *Highway Research
Board, Bulletin* 186, Washington, D.C. (1958c)

Highway Research Boardb, *Chemical Reactions of
Aggregates in Concrete: Identification of
Deleteriously Reactive Aggregates and Recommended
Practices for Their Use in Concrete, Special Report*
31, Washington, D.C. (1958d)

Highway Research Board, "Report on Cooperative
Freezing and Thawing Tests of Concrete," *SP J7*,
Washington, D.C. (1959a)

Highway Research Board, "Skid Prevention Research,"
Highway Research Board Bulletin 219, Washington, D.
C. (1959b)

Highway Research Board, *Properties of Concrete, Highway Research Record No. 43*, Washington, D.C. (1963)

Highway Research Board, *Symposium on Alkali-Carbonate Rock Reactions, Highway Research Record* Number 45, Washington, D.C. (1964)

Highway Research Board, *Admixtures for Highway Concrete*, An Annnotated Bibliography, Washington, D.C. (1965)

Highway Research Board,   "Aggregates and Skid Resistance of Bituminous Pavements," *Highway Research Record* No. 341, Washington, D.C. (1970)

Highway Research Board, *Admixtures in Concrete, Highway Research Board Special Report* 119, Washington, D.C. (1971)

Highway Research Board, *Guide to Compounds of Interest in Cement and Concrete Research, Highway Research Board Special Report* 127, Washington, D.C. (1972a)

Highway Research Board, *New Tire Studs, Alternate Traction Aids, and Wear-Resistant Pavement, Highway Research Record* No. 418, Washington, D.C. (1972b)

Highway Research Board, *Skid Resistance, National Cooperative Highway Research Program Synthesis of Highway Practice* 14, Washington D.C. (1972c)

Hiltrop, C. L., and Lemish, J.,   "Relationship of Pore-Size Distribution and Other Rock Properties to Serviceability of Some Carbonate Aggregates," *Highway Research Board Bulletin* No. 239 (1960)

Hime, W. G. and Labonde, E. G.,   "Particle Size Distribution of Portland Cement from Wagner Turbidimeter Data," *Journal of the PCA Research and Development Laboratories*, Vol. 7, No. 2 (1965)

Hobbs, D. W., and Gutteridge, W. A., Particle Size and its Influence Upon the Expansion Caused by the Alkali-Silica Reaction," *Magazine of Concrete Research*, Vol. 31, No. 109 (1979)

Hobbs, D. W.,   "Influence of pulverized-fuel ash and granulated blastfurnace slag upon expansion caused by the alkali-silica reaction," *Magazine of Concrete Research*, Vol. 34, No. 119 (1982)

Hoff, G. C., "Use of Self-Stressing Expansive Cements in Large Sections of Grout, Mortar, and Concrete," Report 1, *Technical Report* C-74-6, USAE Waterways Experiment Station, Vicksburg, Missi. (1974)

Hoff, G. C., Houston, B. J., and Sayles, F. H., "Use of Regulated-Set Cement in Cold Weather Environments," *Miscellaneous Paper* C-75-5, USAE Waterways Experiment Station, Vicksburg, Missi. (1975a)

Hoff, G. C., "A Concept for Rapid Repair of Bomb-Damaged Runways Using Regulated-Set Cement," *Technical Report* C-75-2, USAE Waterways Experiment Station, Vicksburg, Missi. (1975b)

Holliday, L., "Geometrical considerations and Phase Relationships," *Composite Materials* (Ed. L. Holliday). Elsevier Publishing Co., Amsterdam, London,New York (1966)

Hope, B. B., and Manning, D. G., "Creep of Concrete Influenced by Accelerators," *ACI Journal*, Proc. Vol. 68, No. 5 (1971)

Hosek, J.,   "Properties of Cement Mortars Modified by Polymer Emulsion," *ACI Journal*, Proc. Vol. 63, No. 12 (1966)

Hraste, M., and Bezjak, A., "A New Approach to the Study of the Influence of Cement Fineness on the Strength of Cement Mortars," *Cement and Concrete Research*, Vol. 4, No.6 (1974)

Hudson, S. B., and Waller, H. F., *Evaluation of Construction Control Procedures, National Cooporative Highway Research Program Report* 69, Highway Research Board, Washington, D.C. (1969)

Hughes, B. P., "Rational Concrete Mix Design," *Proc. Institution of Civil Engineers* , Vol. 17, London (1960)

Hughes, B. P., and Bahramian, B., "A Laboratory Test for Determining the Angularity of Aggregate," *Magazine of Concrete Research*, Vol. 18, No. 56, London (1966)

Hughes, B. P., and Ash, J. E., "The Effect of Mix Proportions and Aggregate Dust upon the Compressive Strength of Concrete," *Magazine of Concrete Research*, Vol. 20, No.63, London (1968)

Hughes, B. P., and Famili, H., "Saturated Air Techniques for Determining the Absorption of Aggregates," *Technical Paper PCS* 61, Concrete Society, London (1970)

Hummel, A., Evaluation of Sieve Analysis and the Fineness Modulus of Abrams (In German), *Zement.* Vol. 19, No. 15 (1930)

Hummel, A. and Wesche, K., "Schuttbeton aus verschiedenen Zuschlagstoffen (Lightweight Concrete with Different Aggregates), *Deutscher Ausschuss fur Stahlbeton*, Heft 114, Wilhelm Ernst, Berlin (1954a)

Hummel, A., "Die Ermittlung der Kornfestigkeit von Ziegelsplitt und anderen Leichtbeton-Zuschlagstoffe" (Determination of the Particle Strength of Crushed Brick and Other Lightweight Aggregates), *Deutscher Ausschuss fur Stahlbeton*, Heft 114, Wilhelm Ernst, Berlin (1954b)

Hummel, A., and Wesche, K., "Von der Erhartung verschiedener Bindemittel bei niederen Warmegraden" (Hardening of Various Cements at Low Temperatures), *Zement-Kalk-Gips*, Vol. 8, No. 9 (1955)

Hummel, A., *Das Beton-ABC* (The Alphabet of Concrete), 12th ed., Wilhelm Ernst, Berlin (1959)

Hveem, F. N., "Sand-Equivalent Test for Control of Materials During Construction," *Proc. Highway Research Board*, Vol. 32 (1953)

Idorn, G. M., *Durability of Concrete Structures in Denmark*, Danish Technical Press, Copenhagen (1967)

Idorn, G. M., "Hydration of Portland Cement Paste at High Temperature under Atmospheric Pressure," *Proceedings of the Fifth International Symposium on the Chemistry of Cement*, Part III, *Properties of Cement Paste and Concrete*, Tokyo (1969)

Ish-Shalom, M., and Bentur, A., "Properties of Type K Expansive Cement of Pure Components. III. Hydration of Pure Expansive Component Under Varying Restraining Conditions," *Cement and Concrete Research*, Vol. 5, No. 2 (1975)

Ivey, D. L., and Hirsch, T. J., "Effects of Chemical Admixtures in Concrete and Mortar," *Research Report* 70-3, Texas Transportation Institute, Texas A & M University (1967)

Iyer, L. S., et al., "Durability Tests on Some Aggregates for Concrete," *Journal of the Construction Division*, ASCE, voL 101, no. C03, Proc. Paper 11574 (1975)

Jackson, F. H. and Tyler, I. L., "Long-Time Study of Cement Performance in Concrete - Chapter 7," *ACI Journal*, Proc. Vol. 47 (1951)

Jackson, F. H., and Timms, A. C., "Evaluation of Air-Entraining Admixtures for Concrete," *Public Roads*, Vol. 27, No. 12 (1954)

Jackson, F. H., "Long-Time Study of Cement Performance in Concrete. Chapter 9—Correlation of the Results of Laboratory Tests witn Field Performance Under Natural Freezing and Thawing," *ACI Journal*, Proc. Vol. 52, No. 2 (1955)

Jackson, F. H., "Long-Time Study of Cement Performace in Concrete - Chapter 11, Condition of Three Test Pavements after 15 Years," *ACI Journal*, Proc. Vol. 54 (1958)

Jaeger, J. C., and Cook, N. C. W., "Friction in Granular Materials," *Structure, Solid Mechanics and Engineering Design*, Proceedings of the Southampton 1969 Civil Engineering Materials Conference, Part 1, Wiley-lnterscience, London-New York (1971)

Jaky, J., Principles of Soil Mechanics and Their
Technical Application, *Vizugyi Kozlemenyek*, Budapest,
No. 1 - 2 (1933)

Jambor, J., "Relation Between Phase Composition,
Overall Porosity and Strength of Hardened
Lime-Pozzolana Pastes," *Magazine of Concrete
Research*, Vol. 15, No. 45, London (1963)

Jaray, "Relationship between the Total Surface Area
and Liquid Limit of Soils," *Gedenkbuch fur Prof. Dr.
J. Jaky*, Akademiai Kiado, Budapest (1955)

Jeffery, J. W., "The Crystal Structures of the
Unhydrous Compounds," *The Chemistry of Cements*, ed.
H.F.W. Taylor, Vol. I, Chap. 4, Academic Press,
London and New York (1964)

Johnston, C. D., "Admixture-Cement Incompatibility:
A Case History," *Concrete International: Design &
Construction*, Vol. 9, No. 4 (1987)

Joisel, A., "Composition des betons hydrauliques,"
(Composition of Concretes) *Annales de l'Institut
Technique du Batiment et des Travaux Publics*, Paris
(1952)

Joisel, A., "Activite chimique des adjuvants:
accelerateurs et retardeurs" (Chemical Activities of
Admixtures: Accelerators and Retarders), RILEM-ABEM
*International Symposium on Admixtures for Mortar and
Concrete*, Rapport II/3, Brussels (1967)

Joisel, A., "Testing Methods of Admixtures,"
Materials and Structures - Testing and Research,
Vol. 1, No. 2, Paris (1968)

Jones, T. R., Jr., and Stephenson, H.K.,
"Proportioning, Control, and Field Practice for
Lightweight Concrete," *ACI Journal,* Proc. Vol. 54,
No.6 (1957)

Jons, E. S. and Osbaeck, B,. "The Effect of Cement
Composition on Strength Described by a Strength-
Porosity Model," *Cement and Concrete Research*, Vol.
12, No. 2 (1982)

Kalousek, G. L., "High-Temperature Steam Curing of
Concrete at High Pressure," *Proceedings of the Fifth
International Symposium on the Chemistry of Cement,*
Part III, *Properties of Cement Paste and Concrete,*
Tokyo (1969)

Kalousek, G. L., "Hydration Processes at the Early
Stages of Cement Hardening," A Principal Paper, *VIth
International Congress on the Chemistry of Cement,*
Moscow (1974)

Kantro, D. L., Brunauer, S., and Weise, C. H.,
"Development of Surface in the Hydration of Calcium
Silicates," *Solid Surfaces and the Gas-Solid
Interface, Advances in Chemistry, Series* 33, American
Chemical Society (1961)

Kantro, D. L., Brunauer, S., and Weise, C. H.,
"Development of Surface in the Hydration of Calcium
Silicates. II. Extension of Investigation to Earlier
and Later Stages of , Hydration," *Journal of Physical
Chemistry,* Vol. 66, No. 10 (1962)

Kantro, D. L. and Brunauer, S., "Surface Area and
Particle Size Distribution Determinations," *The
Chemistry of Cements,* ed. H.F.W. Taylor, Vol. II,
Chap. 24, Academic Press, London and New York (1964)

Kaplan, M. F., "The Effects of the Properties of
Coarse Aggregates on the Workability of Concrete,"
Magazine of Concrete Research, Vol. 10, No. 29,
London (1958)

Kaplan, M. F., "Ultrasonic Pulse Velocity, Dynamic
Modulus of Elasticity, Poisson's Ratio and the
Strength of Conaete Made with Thirteen Different
Coarse Aggregates," *RILEM Bulletin,* No. 1, Paris
(1959)

Kellermann, W. F., "Designing Concrete Mixtures for
Pavements", *Proceedings ASTM,* Vol. 40 (1940)

Kelly. J. W., Polivka, M., and Best, C. H., "A
Physical Method for Determining the compostion of
Hardened Concrete," *ASTM Special Technical
Publication 205,* Philadelphia (1958)

Kennedy, H. L., "Revised Application of Fineness Modulus in Concrete Proportioning," *ACI Journal*, Proc. Vol. 36 (1940)

Kennedy, T. W., Hudson, W. R. and McCullogh, B. F., "State-of-the-Art in Variability of Material Properties for Airport Pavement Systems," *Contract Report* S-75-6, USAE Waterways Experiment Station, Vicksburg (1975)

Kerton, C. P. and Murray, R.J., "Portland Cement Production," *Structure and Performance of Cements*, P. Barnes, Editor, Chapter 5, Applied Science Publishers (1987)

Kester, B. E., "Significance of Tests for Fineness," *Proceedings, ASTM*, Vol.63 (1963)

Kholin, I. I., and Royak, S. M., "Blast-Furnace Cement in the USSR, *Chemisry of Cement*, Proceedings of the Fourth International Symposium, *NBS Monograph* 43, Vol. II, Washington,D.C. (1960)

King, G. J. W., and Dickin, E. A., "The Evaluation of Grain Shapes in Silica Sands from a Simple Flow Test," *Materials and Structures-Research and Testing*, Vol. 5, No. 26, (1972)

King, J. C., and Bush, G. W., "Symposium on Grouting: Grouting of Granular Materials," *Journal of the Soil Mechanics and Foundation Division, ASCE*, Vol. 87, No. SM2 (1962)

Klein, A., "An Improved Hydrometer Method for Use in Fineness Determinations," *Symposium on the New Methods for the Particle Size Determiminations in the Subsieve Range*, Philadelphia (1941)

Klein, A., and Troxell, G. E., "Studies of Calcium Sulfoaluminate Admixtures for Expansive Cements," *Proceedings, ASTM*, Vol. 58, Philadelphia (1958)

Klein, A., Karby, T., and Polivka, M., "Properties of an Expansive Cement for Chemical Prestressing," *ACI Journal*, Proc.Vol. 58 (1961)

Klein, A., and Bertero, V., "Effects of Curing
Temperature and Creep Characteristics of Expansive
Concretes," *Proceedings, ASTM*, Vol. 63, Philadelphia
(1963)

Klein, A., and Bresler, B., "A Review of Research on
Expansive Cement Concretes, Part 1: Introduction,"
The Structure of Concrete, Proceedings of an
International Conference, London (1968)

Kleinlogel, A. and Hajnal-Konyi, K., "Beziehung
zwischen Abbindezeit und Warmeentwicklung des
Zements" (Relationship Between the Time of Setting
and Heat Development of Cement), *Zement*, Vol. 22, No.
2 (1933)

Kleinlogel, A., *Einflusse auf Beton*, (Influences on
Concrete), 4th ed., Wilhelm Ernst, Berlin (1941)

Klieger, P., "Long Term Study of Cement Performance
in Concrete - Chapter 10, Progress Report on Strength
and Elastic Properties of Concrete," *ACI Journal*,
Proc. Vol. 54 (1957)

Klieger, P., "Effect of Mixing and Curing Temperature
on Concrete Strength," *ACI Journal*, Proc. Vol. 54
(1958)

Klieger, P. and Isberner, A.W., "Portland
Blast-Furnace Slag Cements," *Journal of the PCA
Research & Development Laboratories*, Vol.9 (1967)

Klieger, P., "Shrinkage Compensated Concrete: A
Review of Current Technology," *NRMCA Publication* 134,
National Ready Mixed Concrete Association, Silver
Spring, Md. (1971)

Klieger, P., "Air-Entraining Admixtures,"
*Significance of Tests and Properties of Concrete and
Concrete Making Materials*, ASTM STP 169B, Chapter 45,
Philadelphia (1978)

Klieger, P., "Something for Nothing - Almost,"
Concrete International: Design and Construction, Vol.
2, No. 1 (1980)

Kluge, F., "Vorausbestimmung der Wassermenge bei Betonmischungen fur bestimmte Betonguten und Frischbetonkonsistenzen" (Prediction of the Mixing Water for Concrete for Specified Strength and Consistency), *Der Bauingenieur*, Vol. 24, No. 6 (1949)

Kohler, G., and Nage, J., "Comparative Investigations of lnternational Test Methods for Small-Sized Coarse Aggregates," *Highway Research Record* No. 412, Highway Research Board, Washington, D.C. (1972)

Kokubu, M., "Fly Ash and Fly Ash Cement," *Proceedings of the Fifth International Symposium on the Chemistry of Cement*, Part IV, *Admixtures and Special Cements*, Tokyo (1969)

Kokubu, M. and Yamada, J., "Fly Ash Cements," A Principal Paper, *VIth International Congress on the Chemistry of Cement*, Moscow (1974)

Kolmogorov, A. N., Logarithmic-Normal Distribution Law of the Dimensions of Particles at Crushing, *Dokladi Akademii Nauk SSSR*, Vol. 31 (1941)

Kondo, R., and Ueda, S., "Kinetics and Mechanisms of the Hydration of Cements," *Proceedings of the Fifth International Symposium on the Chemistry of Cement*, Part II, *Hydration of Cements*, Tokyo (1969)

Kondo, R., Daimon, M. and Okabayashi, S., "Effects of Grain Size Distribution in Alite Cement on Pore Size Distribution and Strength of Hardened Mortar," *Mechanical Behavior of Materials*, Proceedings of the International Conference on Mechanical Behavior of Materials, Vol.IV, Society of Materials Science, Japan (1972)

Kondo, R., and Daimon, M., "Phase Composition of Hardened Cement Paste," A Principal Paper, *VIth International Congress on the Chemistry of Cement*, Moscow (1974)

Kosina, J., "Einfluss der Hydrofobizitat der Zement auf die betontechnologischen Eigenschaften" (The Effect of the Hydrophobic Properties of Cements on Their Concrete Making Properties), *Wissenschaftliche Zeitschrift*, Heft 4, Hochschule fur Bauwesen, Leipzig (1959)

Kramer,W., "Blast-Furnace Slags and Slag Cements," *Chemistry of Cement*, Proceedings of the Fourth International Symposium, *NBS Monograph* 43, Vol. II, Washington, D.C. (1960)

Krause, K., *Uber Den Einfluss der Belastungsgeschwindigkeit auf den Elastizitatsmodul des Mortels und Betons sowie des Zementsteins und Zuschlags* (Effect of the Rate of Loading on the Modulus of Elasticity of Mortar and Concrete as Well as on That of the Cement Stone and Aggregate), Westdeutscher Verlag, Koln (1975)

Kravchenko, I. V., "Rapid-Hardening and High Strength Portland Cements," A Principal Paper, *VIth International Congress on the Chemistry of Cement*, Moscow (1974)

Kreijger, P. C., "Improvement of Concretes and Mortars by Adding Resins (Topic Ia)," *Materials and Structures-Testing and Research*, Vol. 1, No. 3, Paris (1968)

Krenkel, P. A., "Principles of Sedimentation and Coagulation as Applied to the Clarification of Sand and Gravel Process Water," *NSGA Circular* 118, Silver Spring, Md. (1972)

Krokosky, E. M., "Strength vs. Structure. A Study of Hydraulic Cements," *Materials and Structures - Research and Testing* Vol. 3, No. 17 (1970)

Kroone, B., and Crook, D. N., "Studies of Pore Size Distribution in Mortars," *Magazine of Concrete Research*, Vol. 13, No. 39, London (1961)

Kroone, B., "A Method of Detecting and Determining Lignin Compounds in Mortars and Concretes," *Magazine of Concrete Research* , Vol. 23, No. 75-76, London (1971)

Kucynski, W., "Loi d'absorption d'eau pour les granulats du beton en fonction de deux parametres des caracteristiques granulometriques" (Law of Water Absorption of Concrete Aggregates in Terms of Two Parameters of the Grain Size Characteristics), *Materials and Structures - Research and Testing*, Vol. 7, No.38 (1974)

Kuhl, H., *Zement-Chemie* (Chemistry of Cements), Vol.II, 3rd ed., VEB Verlag Technik, Berlin   (1961a)

Kuhl, H., *Zement-Chemie*, (Chemistry of Cement), Vol.III, VEB Verlag Technik, Berlin (1961b)

Kunze, W. and Wesche, K., "Vegleich verschiedener Verfahren zur Bestimmung der Kornrohdichte von Leichtzuschlagen" (Comparison of Various Methods to Determine the Bulk Density of Lightweight Aggregate Particles), *Betonwerk + Fertigteil Technik*, No. 6 (1973)

Kunze, W., "Gesinterte Flugashepellets als Zuschlag fur Konstruktionsleichtbeton" (Sintered Fly Ash Pellets as Aggregates for Structural Lightweight Concrete), *Betonwerk + Fertigteil-Technik* (1974a)

Kunze, W., "Untersuchungen an Leichtzuschlag aus gesintertem Waschbergmaterial," (Experiments on Lightweight Aggregate Made of Sintered Byproduct of Coal-Wash), *Beton*, Vol. 24, No. 6 (1974b)

L'Hermite, R., "Volume Changes of Concrete," *Chemistry of Cement*, Proceedings of the Fourth International Symposium, Vol.II, *NBS Monograph* 43, Washington, D.C. (1960)

Labahn, O. and Kaminsky, W.A., *Cement Engineer's Handbook*, 3rd English Edition, Bauverlag GmbH, Wiesbaden-Berlin (1971)

Lachaud, R., "Donnees de la diffusion centrale des rayons X sur la texture d'alites hydratees a court terme et a deux temperatures" (Data on the Textures of Alites Hydrated for a Short Period and at Two Temperatures Provided by Small-Angle Scattering of X-Rays), *Materials and Structures - Research and Testing*, Vol. 5, No. 26 (1972)

Lafuma, H., "La recherch fondamentale dans le domain de la chimie des ciments en France" (Fundamental Research on the Chemistry of Cement in France), *Materiaux de Construction, Ciments & Betons*, No. 603 (1965)

Landgren, R., "Determining the Water Absorption of Coarse Lightweight Aggregates for Concrete," *Proceedings, ASTM*, Vol.64 (1964a)

Landgren, R., "Water-Vapor Adsorption-Desorption Characteristics of Selected Lightweight Concrete Aggregates," *Proceedings, ASTM*, Vol. 64 (1964b)

Landgren, R., Hanson, J. A., and Pfeifer, D. W., "An Improved Procedure for Proportioning Mixes of Structural Lightweight Concrete," *Journal of the PCA Research and Development Laboratories*, Vol.7, no. 2 (1965)

Lange, H., and Modry, S., "Beziehungen zwischen den Eigenschaften von porosen Kalkstein Zuschlagstoffen und der Forst Widerstandsfahigkeit des Betons" (Relationship Between the Properties of a Porous Limestone Aggregate and the Frost Resistance of Concrete), *Wissenschaftliche Zeitschrift*, Vol. 3, Hochschule fur Bauwesen Leipzig (1969)

Larson, T. D., Mangusi, 1. L., and Radomski, R. R., "Preliminary Study of the Effects of Water-Reducing Retarders on the Strength, Air Void Characteristics, and Durability of Concrete, *ACI Journal*, Proc. Vol. 60 (1963)

Larson, T. D., "Air Entrainment and Durability Aspects of Fly-Ash Concrete," *Proceedings, ASTM*, Vol. 64, Philadelphia (1964a)

Larson, T. D., Cady, P., Franzen, M., and Reed, J., "A Critical Review of Literature Treating Methods of Identifying Aggregates Subject to Destructive Volume Change When Frozen in Concrete and a Proposed Program of Research," Highway Research Board, *Special Report* 80, Washington, D.C. (1964b)

Larson, T. D., Boettcher, A., Cady, P., Franzen, M., and Reed, J., "Identification of Conrete Aggregates Exhibiting Frost Susceptibility," *National Cooperative Highway Research Program Report* 15, Highway Research Board, Washington, D.C. (1965)

Larson, T. D., and Cady, P., "Identification of Frost-Susceptible Particles un Concrete Aggregate," *National Cooperative Highway Research Program Report* 66, Highway Research Board, Washington, D.C. (1969)

Larson, L. J., Mathiowetz, R. P., and Smith, J. H., "Modification of the Standard Los Angeles Abrasion Test," *Highway Research Record*, No. 353, *Mineral Aggregates*, *Highway Research Board*, Washington, D.C. (1971)

Lauwereins, M. A., "Water Pollution - Chicago Style," *Modern Concrete* (1971)

Lazar, J., "Die Gesetzmassigkeiten der Korngrossenverteilung maschinell zerkleinerten Materialhaufen" (Law Concerning the Particle Size Distribution of Granular Materials Crushed by Machine), *Acta Technica* Vol. XVII, no. 3-4, Budapest (1957)

Lea, F. M. and Desch, C. H., *The Chemistry of Cement and Concrete*, 2nd ed., Edward Arnold, London (1956)

LeChatelier, H., "Recherches experimentales sur la constitution des mortiers hydrauliques" (Experimental Researches on the Constitution of Hydraulic Mortars), *Annales des Mines*, ser. 8, Vol. 11, (1887) (English translation by J. L. Mack, McGraw ,Publishing Co., New York (1905)

Lecompte, P., "The Fineness Modulus and Its Dispersion Index" *ACI Journal*, Proc. Vol. 66 (1969)

Ledbetter, W. B., "Synthetic Aggregates from Clay and Shale: Recommended Criteria for Evaluation," *Highway Research Record* No. 430, Highway Research Board, Washington, D.C. (1973)

Lee, D., "Review of Aggregate Blending Techniques," *Highway Research Record Number 441, Grading of Concrete Aggregates*, Highway Research Board (1973)

Lees, G.,   "The Measurement of Particle Elongation
and Flakiness: A Critical Discussion of British
Standard and Other Methods," *Magazine of Concrete
Research*, Vol. 16, No. 49, London (1964)

Legg, F. E., Jr., "Aggregates," *Concrete
Construction Handbook*, 2nd ed., ed. J. J. Waddell,
chap. 2, McGraw-Hill, New York (1974)

Lerch, W. C, "The Influence of Gypsum on the
Hydration and Properties of Portland Cement Pastes,"
Proceedings, ASTM, Vol. 46, Philadelphia (1946)

Lerch, W. C. and Ford, C.L., "Long-Time Study of
Cement Performance in Concrete - Chapter 3. Chemical
and Physical Tests of Cements," *ACI Journal*, Proc.
Vol.44 (1948)

Lerch, W. C., "Studies of Some Methods of Avoiding
Expansion and Pattern Cracking Associated with the
Alkali-Aggregate Reaction," *Symposium on Use of
Pozzolanic Materials for Mortars and Concrete*, ASTM
STP 99, Philadelphia (1950)

Lerch, W. C., "Concrete Aggregates -- Chemical
Reactions", *Significance of Test and Properties of
Concrete and Concrete Aggregates*, Philadelphia (1956)

Lerch, W. C., "Plastic Shrinkage," *ACI Journal*, Proc.
Vol.53 (1957)

Leviant, I.,   "A Graphical Method for Concrete
Proportioning," *Civil Engineering & Public Works
Review*, London (1966)

Lewis, D. W., Dolch, W. L., and Woods, K. B.,
"Porosity Determinations and the Significance of Pore
Characteristics of Aggregates," *Proceedings, ASTM*,
Vol. 53 (1953)

Lewis, D. W.,   "Lightweight Concrete and Aggregates,"
*Significance of Tests and Properties of Concrete and
Concrete Making Materials*, ASTM STP 169A,
Philadelphia (1966)

Lezy, R., and Paillere, A., "Betons, mortiers et coulis. Amelioration par addition de resine," (Concretes, Mortars and Grouts. Improvement by Addition of a Resin) *Synthetic Resin in Building Construction*, Vol. 1, RILEM Symposium, Paris (1967)

Lhopitallier, P., "Calcium Aluminates and High-Alumina Cements," *Chemistry of Cement*, Proccedings of the Fourth International Symposium, *NBS Monograph* 43, Vol. II, Washington, D.C. (1960)

Li, S. T., "Wear-Resistant Concrete Construction," *ACI Journal*, Proc. Vol. 55 (1959)

Li, S. T., "Expansive Cement Concretes — A Review," *ACI Journal*, Proc. Vol. 62 (1965)

Li, S. T., "Selected Bibliography on Gap Grading and Gap-Graded Concrete," *ACI Journal*, Proc. Vol. 67, No. 7 (1970)

Li, S., and Ramakrishnan, V., "Gap-Graded Aggregates for High-Strength Quality Concrete," *Highway Research Record No. 441, Grading of Concrete Aggregates*, Highway Research Board, Washington, D.C. (1973)

Lin, C. H., Walker, R. D., and Payne, W. W., "Effects of Cooling Rates on the Durability of Concrete," *Transportation Research Record* No. 539, Transportation Research Board, Washington, D.C. (1975)

Lin, C. H., Walker, R. D., and Payne, W. W., "Chert-Aggregate Concrete Durability After Antifreeze Treatment," *Living with Marginal Aggregates*, *ASTM STP* 597, Philadelphia (1976)

Litvan, G. G., "Further Study of Particulate Admixtures for Enhanced Freeze-Thaw Resistance of Concrete," *ACI Journal*, Proc. Vol. 82, No. 5 (1985)

Locher, F. W., and Richartz, W., "Study on the Hydration Mechanism of Cement," A Principal Paper, *The VIth International Congress on the Chemistry of Cement*, Moscow (1974a)

Locher, F. W., and Sprung, S., "Ursache und Wirkungsweise der Alkalireaktion" (Cause and Mechanism of Alkali-Aggregate Reaction), *Betontechnische Berichte 1973*, Beton-Verlag, Dusseldorf (1974b)

Locher, F. W., and Sprung, S., "Einflusse auf die Alkali-Kielselsaure-Reaktion im Beton" (Influences on the Alkali-Silica Reaction in Concrete), *Zement-Kalk-Gips*, Vol. 28, No. 4 (1975)

Locher, F. W., "Die Festigkeit des Zements" (Strength of Cement), *beton*, Vol. 26, Nos. 7 and 8 (1976a)

Locher,F. W., "Chemie des Zements und der Hydrations-Produkte" (Chemistry of Cements and Hydration Products), *Zement-Taschenbuch 1976-77*, Bauverlag, Wiesbaden, Berlin (1976b)

Locher, F. W., and Wischers, G., "Aufbau und Eigenshaften des Zementsteins" (Structure and Properties of the Hardened Cement Paste), *Zement-Taschenbuch 1976-1977*, Bauverlag, Wiesbaden, Berlin (1976c)

Locher, F. W., "Die Festigkeit des Zements" (Strength of Cement), *Betontechnische Berichte 1976*, Beton-Verlag, Dusseldorf (1977)

Locher,F. W., "Geschichtliche Entwicklung und heutige Zementarten" (Historical Development and Present Cement Types), *Zement Taschenbuch*, 48th Edition, Bauverlag GmbH, Wiesbaden-Berlin (1984a)

Locher, F. W. and Sillem, H., "Zementherstellung," (Cement Manufacturing), *Zement Taschenbuch*, 48th Edition, Bauverlag GmbH, Wiesbaden-Berlin (1984b)

Loudon, A. G., "Computation of Permeability from Simple Soil Tests," *Geotechnique*, Vol. 3 (1952-1953)

Loveland, R. P. and Trivelli, A. P. H., "Mathematical Methods of Frequency Analysis of Size of Particles" *Journal of the Franklin Inst.* (1927)

Lovewell, C. E., and Hyland, E. J., "Effects of Combining Two or More Admixtures in Concrete," *Highway Research Record 370, Concrete*, Highway Research Board, Washington, D .C. (1971)

Ludwig, N. C. and Pence, S.A., "Conduction Calorimeter for Measuring Heat of Hydration of Portland Cement at Elevated Temperatures and Pressures," *ACI Journal*, Proc.Vol. 53 (1956)

Ludwig, U., and Schwiete, H. E., "Researches on the Hydration of Trass Cements," *Chemistry of Cement*, Proceedings of the Fourth International Symposium, *NBS Monograph* 43, Vol.II, Washington, D.C. (1960)

Ludwig, U., "Investigations on the Hydration Mechanism of Clinker Minerals," A Principal Paper, *VIth International Congress on the Chemistry of Cement*, Moscow (1974)

Luhr, H. P., and Friede, H., "Beitrag zur Umrechnung von Rundloch und Quadratlochsieblinien" (Contribution to the Conversion from Round to Square Openings), *Beton*, Vol. 20, No. 12 (1971)

Lyndon, F. D., "Current Practice Sheet No. 26 - Artificial Aggregates for Concrete," *Concrete*, Vol. 9, no. 9, London (1975)

Lyon, E. V., and Butzine, J., "Expansive Cements," *Special Bibliography* 157, PCA Research and Development Division Library (1965)

Lyse, I., "Tests on Consistency and Strength of Concrete Having Constant Water Content," *ASTM Proceedings*, Vol. 32, Part II (1932)

MacInnis, C., and Lau, E. C., "Maximum Aggregate Size Effect on Frost Resistance of Concrete," *ACI Journal*, Proc. Vol. 68, No. 2 (1971)

Mackenzie, R. C., "Differential Thermal Analysis," *The Chemistry of Cements*, ed. H. F., W. Taylor, Vol. II, Chap. 22, Academic Press, London and New York (1964)

MacLean, D. J., and Shergold, F. A., "The Polishing of Roadstones in Relation to Their Selection for Use in Road Surfacings," *Journal of the Institution of Highway Engineers* Vol. 6, No. 3, London (1959)

Majumdar, A. J., and Chatterji, S.,"Studies of the
Early Stages of Paste Hydration of High Alumina
Cements-2: Hydration of Iron and Silica Bearing
Compounds," *Indian Concrete Journal*, Vol. 40, No. 4
(1966)

Malhotra, V. M., "Correlation Between Particle Shape
and Surface Texture of Fine Aggregates and Their
Water Requirement," *Materials Research & Standards*,
Vol. 4, No. 12 (1964)

Malhotra, V. M., and Malanka, D., "Performance of
Superplasticizers in Concrete: Laboratory
Investigation - Part I," *Superplasticizers in
Concrete*, ACI Publication SP-62, American Concrete
Institute, Detroit (1979)

Malhotra, V. M., and Carette, G. G., "In-Situ
Testing- A Review," *Progress in Concrete Technology*,
V. M. Malhotra, Editor, Energy, Mines and Resources,
Ottawa, Canada, MRP/MSL 80-89 TR (1980)

Malhotra, V. M., "Superplasticizers: Their Effect on
Fresh and Hardened Concrete," *Concrete
International: Design & Construction*, Vol. 3, No. 5
(1981)

Malhotra, V. M. (Editor), *Proceeding of the
CANMET/ACI First International Conference on the Use
of Fly Ash, Silica Fume, Slag and Other Mineral
By-Products in Concrete*, Vol. 1, ACI Publication
SP-79, American Concrete Institute (1983a)

Malhotra, V. M. (Editor), *Proceedings of the
CANMET/ACI First International Conference on the Use
of Fly Ash, Silica Fume, Slag and Other Mineral
By-Products in Concrete*, Vol. 2, ACI Publication
SP-79, American Concrete Institute (1983b)

Malhotra, V. M., (Editor), *Supplemetary Cementing
Materials for Concrete*, CANMET SP 86-8E, Ottawa,
Canada (1987)

Malhotra, V. M. (Ed.) *Fly Ash, Silica Fume, Slag and
Natural Pozzolans in Concrete*, Vol. 1, *Proceeding,
Third International Conference*, Trondheim, Norway,
1989, ACI Publication SP-114, American Concrete
Institute (1989a)

Malhotra, V. M. (Ed.), *Fly Ash, Silica Fume, Slag, and Natural Pozzolans in Concrete*, Vol. 2, *Proceedings, Third International Conference*, Trondheim, Norway, 1989, *ACI Publication SP*-114, American Concrete Institute (1989b)

Malquori, G., "Portland-Pozzolan Cement," *Chemistry of Cement*, Proceedings of the Fourth International Sympoium, *NBS Monograph* 43, Vol.II, Washington, D.C. (1960)

Manns, W., "Leichtzuschlag" (Lightweight Aggregate), *Zement-Taschenbuch 1976-77* (1976)

Marek, C. R., Herrin, M., Kesler, C. E., and Barenberg, E. J., "Promising Replacements for Conventional Aggregates for Highway Use," *NCHRP Report* 135, Highway Research Board, Washington, D.C. (1972)

Markestad, A. and Rudjord, A., "Investigation into Cement Testing Methods and into the Correlation Between the Strength of Cement and that of Concrete," *Rilem Bulletin*, New Series No. 26 (1965)

Massazza, F., "Chemistry of Pozzolanic Additions and Mixed Cements," A Principal paper, *VIth International Congress on the Chemistry of Cement*, Moscow (1974)

Mather, B., "Tests of Fine Aggregate for Organic Impurities and Compressive Strength in Mortars," *ASTM Bulletin* No. 178 (1951)

Mather, B., "Laboratory Tests of Portland Blast-Furnace Slag Cements," *ACI Journal*, Proc. Vol. 54, No.3 (1957)

Mather, B., "The Partial Replacement of Portland Cement in Concrete," *Cement and Concrete*, *ASTM STP* 205, Philadelphia (1958)

Mather, B., "Investigation of Expanding Cements, Report 1. Summary of Information Available as of 1 July, 1963," *Technical Report* 6-691, USAE, Engineer Waterways Experiment Station, Vicksburg, Missi. (1965)

Mather, B., "Shape, Surface Texture, and Coatings," *Significance of Tests and Properties of Concrete and Concrete Making Materials, ASTM STP 169-A,* Philadelphia (1966)

Mather, B., "Cement Performance in Concrete," *Technical Report* 6-787, USAE Waterways Experiment Station, Vicksburg, Missi. (1967)

Mather, B., "New Concern Over Alkali-Aggregate Reaction", *NSGA Circular No. 122, NRMCA Publication No. 149,* Silver Spring, Maryland (1975)

Mather, K., "High Strength, High Density Concrete," *ACI Journal,* Proc. Vol. 62, No. 8 (1965)

Mather, K., "Petrographic Examination," *Significance of Tests and Properties of Concrete and Concrete Making Materials, ASTM STP* 169-A, Philadelphia (1966)

Matouschek, F. "Der Kornaufbau der Zement" (Grading of Cement),*Schweizer Archiv,* No. 2-3 (1947)

Mayfield, B., and Bettison, J., "Ultrasonic Pulse Testing of High Alumina Cement Concrete - In the Laboratory," *Concrete,* Vol. 8,No. 9, London (1974)

McCall, J. T., and Claus, R. J., "Effect of Pellet and Flake Forms of Calcium Chloride in Concrete," *Highway Research Board Bulletin* 75, Washington, D.C. (1953)

McCoy, W. J., "Mixing and Curing Water for Concrete," *Significance of Tests and Properties of Concrete and Concrete Making Materials, ASTM STP 169-A,* Philadelphia (1966)

McCurrich, L. H., Hardman, M. P. and Lammiman, S. A., "Chloride-Free Accelerators," *Concrete,* Vol. 13, No. 3, London (1979)

McGeary, R. K., "Mechanical Packing of Spherical Particles," *Journal of the American Ceramic Society,* Vol. 44, No. 10 (1961)

Mchedlov-Petrosyan, 0. P., and Babushkin, V. I.,
"Thermodynamics and Thermochemistry of Cement," A
Principal Paper, *VIth International Congress on the
Chemistry of Cement*, Moscow (1974)

McIntosh, J. D., Jordan, J. P. R., and O'Callaghan,
W., "The Effect on Some Properties of Concrete of
Partially Replacing Portland Cement by Pulverized
Fuel Ash," *Technical Report TRA/324*, Cement and
Concrete Association, London (1960)

McLaughlin, J. F., Woods, K. B., Mielenz, R. C., and
Rockwood, N. C., "Distribution, Production and
Engineering Characteristics of Aggegates," *Highway
Engineering Handbook*, ed. K.B.Woods, sec.16,
McGraw-Hill (1960)

McMahon, C. J., Jr., "Hardness," *Encyclopedia of
Chemical Technology*, Kirk-Othmer (Ed.), Second
Edition, Volume 10, Interscience Publishers, New
York, London, Sydney (1966)

McMillan, F. R., Stanton, T.E., Tyler, I.L. and
Hansen, W.C., "Concrete Exposed to Sulfate Soils -
Chapter 5, Long-Time Study of Cement Performance in
Concrete," *ACI Special Publication* (1949)

Mehta, P. K., "Retrogression in the Hydraulic
Strength of Calcium Aluminate Cement Structures,"
Mineral Processing, Vol. 5, No. 11 (1964)

Mehta, P. K., and Klein, A., "Investigations on the
Hydration Products in the System 4CaO •3Al2O3•
SO3-CaSO4-CaO-H2O," *Symposium on Structure of
Portland Cement Paste and Concrete, Special Report
90*, Highway Research Board, Washington, D.C. (1966)

Mehta, P. K., and Klein, A., "Formation of Ettringite
in Pastes Containing Calcium Aluminoferrites and
Gypsum," *Highway Research Record 192*, Highway
Research Board, Washington, D.C. (1967)

Mehta, P. K., and Polivka, M., "Expansive Cements," A
Principal Paper, *VIth International Congress on the
Chemistry of Cement*, Moscow (1974)

Mehta, P. K., and Polivka, M., "Use of Highly Active Pozzolans for Reducing Expansion in Concretes Containing Reactive Aggregates," *Living with Marginal Aggregates*, ASTM STP 597, Philadelphia (1976)

Mehta, P. K.,(Editor), *Cement Standards, Evolutionand Trends*, ASTM STP 663, Philadelphia (1978)

Mehta, P. K., "Influence of Different Crystalline Forms of C3A on Sulfate Resistance of Portland Cement," *7th International Congress on the Chemistry of Cement*, Vol. IV., Editions Septima, Paris (1980a)

Mehta, P. K., ank Polivka, M., "Expansive Cements and Their Application," *Progress in Concrete Technology*, V. M. Halhotra, Editor. Energy, Mines and Resources, Ottawa, Canada, MRP/MSL 80-89 TR (1980b)

Mehta, P. K., *CONCRETE - Structure, Properties, and Materials*, Prentice-Hall, Inc., Englewood Cliffs, N.J. (1986)

Meininger, R. C., "Influence of Aggregate Pore Properties on the Resistance of Concrete to Freezing Damage," Thesis submitted to the Faculty of the University of MaryLand in partial fulfilment of the requirements for the degree of Master of Science (1964)

Meininger, R. C., "Recycling Mixer Wash Water - Its Effect on Ready Mixed Concrete," Presentation at Annual Convention of NRMCA, Chicago (1972)

Meyer, A., "Normen fur die Festigkeitsprufung von Zement" (Standards for Testing the Strength of Cement), *Betonstein-Zeitung*, Vol.32, No.2 (1966)

Meyer, A., "Experience in the Use of Superplasticizers in Germany," *Superplasticizers in Concrete, ACI Publication SP-62*, American Concrete Institute, Detroit (1979a)

Meyer, L. M., and Perenchio, W. F., "Theory of Concrete Slump Loss as Related to the Use of Chemical Admixture," *Concrete International: Design & Construction*, Vol 1, No. 1 (1979b)

Meyers, B. L., "A Review of Literature Pertaining to Creep and Shrinkage of Concrete," *Engineering Bulletin Series* No. 56, Engineering Experiment Station, University of Missuori, Columbia, Mo. (1963)

Michaelis, W., Sr., The Hardening Process of Hydraulic Cements (in German), *Chemiker Zeitung*, Vol. 17 (1893) (English translation by W. Michaelis, Jr., *Cement and Engineering News*, Chicago (1907)

Midgley, H. G., "The Staining of Concrete by Pyrite," *Magazine of Concrete Research*, Vol. 10, No. 29, London (1958)

Midgley, H. G. and Taylor, H.F.W., "Optical Microscopy," *The Chemistry of Cements*, ed. H. F. W. Taylor, Vol. II, Chap. 20, Academy Press, London and New York (1964a)

Midgley, H. G., "The Formation and Phase Composition of Portland Cement Clinker," *The Chemistry of Cements*, ed. H.F.W. Taylor, Vol.I, Chap.3, Academic Press, London and New York (1964b)

Midgley, H. G., "The Effect of Lead Compounds in Aggregate Upon the Setting of Portland Cement," *Magazine of Concrete Research*, Vol. 22, No. 70, London (1970)

Midgley, H. G., "Electron Microscopy of Set Portland Cement," *Structure, Solid Mechanics and Engineering Design*, Proceedings of the Southampton 1969 Civil Engineering Materials Conference, Part I, Wiley-lnterscience, London, New York (1971)

Midgley, H. G., and Midgley, A., "The Conversion of High Alumina Cement," *Magazine of Concrete Research*, Vol. 27, No. 91, London (1975)

Mielenz, R. C., and Witte, L. P., " Tests Used by the Bureau of Reclamation fo Identifying Reactive Concrete Aggregates," *ASTM Proceeding*, ASTEA, Vol. 48, Philadelphia (1948)

Mielenz, R. C., Witte, L. P., and Glantz, 0. J., "Effect of Calcination on Natural Pozzolans," *Symposium on Use of Pozzolanic Materials in Mortars and Concrete*, ASTM STP 99, Philadelphia (1950)

Mielenz, R. C., Wolkodoff, V. E., Backstrom, J. E., Burrown, R. W., and Flack, H. L., "Origin, Evolution and Effects ot the Air Void System in Concrete," Part 1, *ACI Journal*, Proc. Vol. 55 (1958a)

Mielenz, R. C., Wolkodoff, V. E., Backstrom, J. E. and Burrows, R. W., "Origin, Evolution, and Effects of the Air Void System in Concrete," Part 4, *ACI Journal* (1958b)

Mielenz, R. C. and Benton, E. J., "Evaluation of the Quick Chemical Test for Alkali Reactivity of Concrete Aggregate," *Highway Research Board Bulletin* No. 171 (1958c)

Mielenz, R. C., "Water-Reducing Admixtures and Set-Retarding Admixtures for Concrete: Uses; Specifications; Research Objectives," *Symposium on Effect of Water-Reducing Admixtures and Set-Retarding Admixtures on Properties of Concrete*, ASTM STP 266, Philadelphia (1960)

Mielenz, R. C., "Reactions of Aggregates Involving Solubility, Oxidation, Sulfates, or Sulfides," *Highway Research Record* No. 43, *Properties of Concrete*, Highway Research Board, Washington, D.C. (1963)

Mielenz, R. C., "Concrete, as a Modern Material," *Modern Materials*, Vol. 5, ed. B. W. Gonser, Academic Press, New York and London (1965)

Mielenz, R. C., "Petrographic Examination," *Significance of Tests and Properties of Concrete and Concrete Making Materials*, ASTM STP 169-A, Philadelphia (1966)

Mielenz, R. C., "Use of Surface-Active Agents in Concrete," *Proceedings of the Fifth International Symposium on Chemistry of Cement*, Part IV, *Admixtures and Special Cements*, Tokyo (1969)

Mielenz, R. C., "Mineral Admixtures - History and Background," *Concrete International: Design and Construction*, Vol. 5, No. 8 (1983)

Mielenz, R. C., "History of Chemical Admixtures for Concrete," *Concrete International: Design & Construction*, Vol. 6, No. 4 (1984)

Mikhailov, V. V., "Stressing Cement and the Mechanism of Self-Stressing Concrete Regulation," *Chemistry of Cement*, Proceedings of the Fourth International Symposium, *NBS Monograph* 43, Vol. II, Washington, D.C. (1960)

Miller-Warden Associates, *Effects of Different Methods of Stockpiling Aggregates, Interim Report, NCHRP Report* 5, Highway Research Board, Washington, D.C. (1964)

Miller-Warden Associates, *Development of Guidelines for Practical and Realistic Construction Specifications, NCHRP Report* 17, Highway Research Board, Washington, D.C. (1965)

Miller-Warden Associates, *Evaluation of Construction Control Procedures, Interim Report, NCHRP Report* 34, Highway Research Board, Washington, D.C. (1967a)

Miller-Warden Associates, *Effects of Different Methods of Stockpiling and Handling Aggregates, NCHRP Report* 46, Highway Research Board, Washington, D.C. (1967b)

Mills, R. H., "Influence of Water in Areas of Restricted Adsorption on Properties of Concrete," *Materials and Structures - Research and Testing*, Vol. 1, No. 6 (1968)

Mills, W. H., and Fletcher, O. S., "Control and Acceptance of Aggregate Gradation by Statistical Methods," *Highway Research Record* No. 290, Highway Research Board, Washington, D.C. (1969)

Minnick, L. J. and Corson, W. H., "Lightweight Concrete Aggregate from Sintered Fly Ash," *Highway Research Record* No. 307, Highway Research Board, Washington, D.C. (1970)

Mohan, L., and Ghosh, S. N., "Magnesia in Cement," *Il Cemento*, Vol. 85 (1988)

Monfore, C. E., "Properties of Expansive Cement Made with Portland Cement, Gypsum and Calcium Aluminate Cement," *Journal of the PCA Research and Development Laboratories*, Vol. 6, No. 2 (1964)

Moore, A. E., "Comparison of the Results Obtained for the Compound Composition of Portland Cements by X-Ray Diffraction, Microscopy and Wet Chemical Methods," *SCI Monograph* 18, Society of Chemical Industry, London (1977)

Morgan, D. R., "Possible Mechanism of Influence of Admixtures on Drying Shrinkage and Creep in Cement Paste and Concrete," *Materials and Structures - Research and Testing*, Vol. 7, No. 40, Paris (1974)

Mortureux, B., Hornain, H. and Regourd, M., "Role of C3A in the Attack on Cements by Sulfate Solutions," *7th International Congress on the Chemistry of Cement*, Vol. IV., Editions Septima, Paris (1980)

Moyer, R. A., "California Skid Tests with Butyl Rubber Tires and Report of Visit to Road Research Laboratories in Europe Engaged in Skid Prevention Research," *Highway Research Record* No. 28, *Roadway Surface Properties*, Washington, D.C. (1963)

Murdock, L. J., "The Workability of Concrete," *Magazine of Concrete Research*, Vol. 12, No. 36, London (1960)

Murphy, G., *Similitude in Engineering*, The Ronald Press, New York (1950)

Nadai, A., *Theory of Flow and Fracture of Solids*, 2nd ed., Vol. 1, McGraw-Hill, New York (1950)

Nagai, S., "Special Masonry Cement Having a High Slag Content," *Chemistry of Cement*, Proceedings of the Fourth International Symposium, *NBS Monograph* 43, Vol. II , Washington D.C. (1960)

Narver, D. L., Jr., "Proportioning Mixes for Steel Coarse Aggregate and Limonite and Magnetite Matrix Heavy Concretes," *ACI Journal*, Proc. Vol. 52 (1962)

Nawy, E. G., Ukadike, M. M., and Sauer, J. A., "Optimum Polymer Content in Concrete Modified by Liquid Epoxy Resins," *Polymers in Concrete*, ACI *Publication* SP-58, American Concrete Institute, (1978)

Nelson, G. H., and Frei, O. C., "Lightweight Structural Concrete Proportioning and Control," *ACI Journal*, Proc. Vol. 54 (1958)

Neumann, D. L., "Mathematical Method for Blending Aggregates," *Journal of the Construction Division*, Proc. ASCE, CO 2 (1964)

Neville, A. M., *Properties of Concrete*, John Wiley and Sons, New York (1963)

Neville, A. M., *High Alumina Cement Concrete*, John Wiley and Sons, New York (1975)

Neville, A., "High Alumina Cement - Its Properties, Applications, and Limitations," *Progress in Concrete Technology*, V. M. Malhotra, Editor, Energy, Mines and Resources, Ottawa, Canada (1980)

Neville, A. M., *Properties of Concrete*, Third Edition, John Wiley and Sons, New York (1981)

Newlon, H. H., Jr. and Sherwood, W. C., "An Occurance of Alkali-Reactive Carbonate Rock in Virginia," *Highway Research Board Bulletin* 355, Washington, D.C. (1962)

Newlon, H. H., Jr., Discussion of article by L. M. Meyer and W. F. Perenchio "Theory of Concrete Slump Loss as Related to the Use of Chemical Admixtures," *Concrete International: Design and Construction*, Vol. 1, No. 7 (1979)

Newman, A. J., and Teychenne, D. C., "A Classification of Natural Sands and its Use in Concrete Mix Design" *Symposium on Mix Design and Quality Control of Concrete*, Cement and Concrete Association, London (1954)

Nicholls, R., *Composite Construction Mateerials Handbook*, Prentice-Hall, Inc. Englewood Cliffs, N.J. (1976)

Nilsson, L.-O. (Ed.), *Durability of concrete. Aspects of admixtures and industrial by-products*, Swedish Council for Building research, Stokholm (1988)

North, O. S., "Expansive Cements. Properties - Manufacture - Research - Development - Formulas," *Minerals Processing*, Vol. 6, No. 1 (1965a)

North, O. S., "White Cements," *Minerals Processing*, Vol. 6, No. 2 (1965b)

Novgorodsky, M., *Testing of Building Materials and Structures*, Mir Publishers, Moscow (1973)

Nurse, R. W., "Slag Cements," *The Chemistry of Cements*, ed. H.F.W.Taylor, Vol. II, Chap. 13, Academic Press, London and New York (1964)

Nurse, R. W., Midgeley, H.G., Gutt,W. and Fletcher, K., "Effect of Polymorphism of Tricalcium Silicate on Its Reactivity," *Symposium on Structure of Portland Cement Paste and Concrete, Special Report 90*, Highway Research Board, Washington, D.C. (1966)

Nurse, R. W., "Phase Equilibria and Formation of Portland Cement Minerals," *Proceedings of the Fifth International Symposium on the Chemistry of Cement, Part I, Chemistry of Cement Clinker*, Tokyo (1969)

Orchard, D. F., *Concrete Technology*, Vol. 1, *Properties of Materials*, Third Ed., John Wiley and Sons, Inc., New York, etc. (1973a)

Orchard, D. F., *Concrete Technology*, Vol.2, *Practice*, 3rd ed., Contractors Record, London, Applied Science (1973b)

Orchard, D. F., *Concrete Technology*, 3rd edition, Vol. 3, *Properties and Testing of Aggregates*, John Wiley and Sons, New York, etc. (1976)

Ozol, M. A., "The Portland Cement Aggregate Bond: Influence of Surface Area of the Coarse Aggregate as a Function of Lithology," *VHRC 71-R40*, Virginia Highway Council, Charlottsville, Va. (1972)

Palotas, L., The Pratical Significance of the
Fineness Modulus of Abrams (In Hungarian),
Anyagvizsgalok Kozlonye, Vol. 11, No. 9 - 10,
Budapest (1933)

Palotas, L., Optimum Grading of Concrete Aggregates
(In Hungarian), Anyagvizsgalok Kozlonye, Vol. 12. No.
9 - 10, Budapest (1934)

Palotas, L., Verfahren zur Verbesserung der
Betonzuschlagstoffe (Method for Improving Concrete
Aggregates), Zement, Vol. 25, No. 18 (1936)

Palotas, L., and Kilian, J., "Durability of Expansive
Aluminous Cement Mortars Produced in Hungary,"
Proceedings of the Technical University of Building
and Transport Engineering, Vol. XII, No. 6, Budapest
(1967)

Peray, K. E., Cement Manufacturer's Handbook,
Chemical Publishing Co., New, York (1979)

Peter, E., "Die quantitative Bestimmung der
Klinkenmineralien mit dem Diffraktometer"
(Quantitative Determination of Clinkerminerals with
Diffraktometer), Cement and Concrete Research, Vol.
1, No. 1 (1971)

Peterman, M. B. and Carrasquillo, R.L., Production of
High Strength Concrete, Noyes Publications, Park
Ridge, N J. (1986)

Petersen, P. H., "Lightweight Concrete," Concrete
Construction Handbook, 2nd Edition, Editor J. J.
Waddell, McGraw-Hill, New York (1974)

Pfeiffer, D. W., "Fly Ash Aggregate Lightweight
Concrete," ACI Journal, Proc. Vol. 68, No. 3 (1971)

Philleo, R. E., "Freezing and Thawing Resistance of
High-Strength Concrete," National Cooperative
Highway Research Program Synthesis of Highway
Practice, 129, Transportation Research Board,
Washington, D.C. (1986)

Pike, R. G., Hubbard, D., and Newman, E. S., "Binary
Silicate Glasses in the Study of Alkali-Aggregate
Reaction," Highway Research Board Bulletin, No. 275,
Washington, D.C. (1960)

Pilny, F., "Natursteine," *HUTTE, Taschenbucher der Technik*, 29th ed., *Bautechnik 1*, Springer Verlag, Berlin, Heidelberg, New York (1974)

Pistilli, M. F., Peterson, C. F., and Shah, S. P., "Properties and Possible Recycling of Solid Waste from Ready-Mix Concrete," *Cement and Concrete Research*, Vol. 5, No. 3 (1975)

Plagemann, W., "Zur Kennzeichnung der Kornform von Betonzuschlagstoffen" (Characterization of the Particle Shape of Concrete Aggregates), *Baustoffindustrie*, Vol. 7, No. 4 (1964a)

Plagemann, W., "Zur Ermittlung der Oberflache eines Zuschlagstoffgemenges" (Determination of the Surface Area of Aggregate), Parts I and II, *Baustoffindustrie*, Vol. 7, Nos. 8 and 9 (1964b)

Plum, N. M., "The Predetermination of Water Requirement and Optimum Grading of Concrete," *Building Research Studies 3*, Danish National Institiute of Building Research, Copenhagen (1950)

Poijarvi, H., On the Effects of the Finest Part of Aggregate on the Properties of Concrete, (In Finnish), *Julkaisu Publication* 110, State Institute for Technical Research, Finland, Helsinki (1966)

Polivka, M. and Klein, A., "Effect of Water-Reducing Admixtures and Set-Retarding Admixtures as Influenced by Portland Cement Composition," *Symposium on Effect of Water-Reducing Admixtures on Properties of Concrete*, ASTM STP 266, Philadelphia (1960)

Polivka, M., "Radiation Effects and Shielding," *Significance of Tests and Properties of Concrete and Concrete Making Materials*, ASTM STP 169-A (1966)

Polivka, M., and Mehta, P. K., "Use of Aggregates Producing High Shrinkage with Shrinkage-Compensating Cements," *Living with Marginal Aggregates*, ASTM STP 597, Philadelphia (1976)

Pollitt, H. W. W., "Raw Materials and Processes for Portland Cement Manufacture," *The Chemistry of Cements*, ed. H.F.W. Taylor, Vol.1, Chap.1, Academic Press, London and New York (1964)

Popovics, S., Evaluation of Grading of Concrete
Aggregate (In Hungarian), *Epitestudomanyi
Kozlemenyek*, No. 3 - 4, Budapest (1950)

Popovics, S., On the Numerical Characteristics of the
Grading of Aggregates (In Hungarian), *Muszaki Tud.
Osztaly Kozlemenyei*, Magyar Tudomanyos Akademia, Vol.
7, No 1 - 3 Budapest (1952)

Popovics, S. and Ujhelyi, J., Results of the Cement
Tests Conducted in 1952 (In Hungarian), *Epitoanyag*,
Vol.5, No.10, Budapest (1953a)

Popovics, S., and Ujhelyi, J., Statistical Evaluation
of the Strenght Results of Concretes Made in the
Years 1952-53 (In Hungarian), *Magyar Epitoipar*, Vol.
2, No. 12, Budapest (1953b)

Popovics, S. and Ujhelyi, J., Comparison of the
Standardized Methods of Hungary for Testing Cement
Strenght (In Hungarian), *ETI Tudomanyos Kozlemenyek*
No 7, Budapest (1955a)

Popovics, S., *Epitoanyagok (Building Materials)*, 2nd
ed., Epitesugyi Kiado, Budapest (1955b)

Popovics, S., "Zahlenmassige Bewertung der
Kornzusammensetzung von Betonzuschlagstoffen"
(Numerical Evaluation of Grading of Concrete
Aggregate), *Acta Technica*, Acad. Sc. Hung., Tom.
XIII., Fasc. 1 - 2, Budapest (1955c)

Popovics, S., "How to Improve the Flexural Strength
of Concrete Made with Alabama Aggregates," *Research
Report* for the Alabama State Highway Department,
Auburn University (1961a)

Popovics, S., "A Method of Evaluating Gradings of
Concrete Aggregates", Thesis submitted to the Faculty
of Purdue University in partial fulfillment of the
requirements for the degree of Doctor of Philosophy
(1961b)

Popovics, S., "Formulas on Fineness Modulus and
Specific Surface Area," *RILEM Bulletin*, No. 16, Paris
(1962a)

Popovics, S., "Comparison of Several Methods of Evaluating Aggregate Grading," *RILEM Bulletin*, No. 17 (1962b)

Popovics, S., "General Relation Between Mix Proportion and Cement Content of Concrete," *Magazine of Concrete Research*, Vol. 14, No. 42, London (1962c)

Popovics, S., "Contribution to the Prediction of Consistency," *RILEM Bulletin*, No. 20, Paris (1963)

Popovics, S., "Tables for Concrete Mix Proportioning," *ACI Journal*, Proc. Vol. 61, No. 1 (1964a)

Popovics, S., "Theory and Application of Triangular Diagrams," *RILEM Bulletin*, New Series, No. 22, Paris (1964b)

Popovics, S., "Investigation of the Grading Requirements for Mineral Aggregates of Concrete," (Final Report) *Alabama Highway Research*, HPR Report No. 6, Montomery (1964c)

Popovics, S., "An Investigation of the Unit Weight of Concrete," *Magazine of Concrete Research*, Vol. 16, No. 49, London (1964d)

Popovics, S., "Uber den Einfluss des Wassergehalts auf die Konsistenz" (Influence of Water Content on the Consistency), *Betonstein-Zeitung*, Vol. 32, No. 12 (1966a)

Popovics, S., "The Use of Fineness Modulus for the Grading Evaluation of Aggregates for Concrete," *Magazine of Concrete Research*, Vol. 18, No. 56, London (1966b)

Popovics, S., "A Method for Evaluating How Well Observed Data Fit the Line Y = X," *Materials, Research and Standards*, Vol.7, No.5 (1967a)

Popovics, S., "A Model for the Kinetics of the Hardening of Portland Cement," *Highway Research Record*, No. 192, Highway Research Board, Washington, D.C. (1967b)

Popovics, S., "Factors Affecting the Relationship Between Strength and Water-Cement Ratio," *Materials, Research and Standards*, MTRSA, Vol. 7, No. 12 (1967c)

Popovics, S., "Analysis of the Influence of Water Content on Consistency," *Highway Research Record*, No. 218, Highway Research Board, Washington D.C. (1968a)

Popovics, S., "Berechnung der Festigkeits-entwicklung von Morteln und Betonen unter Berucksichtigung der Klinkerphasen des verwendeten Portlandzementes," Referate, (Calculation of the Strength Development of Mortars and Concretes from the Compound Composition of the Portland Cement Used), *Betonstein-Zeitung*, Vol.34, No.11 (1968b)

Popovics, S., Discussion of Ref.(Dise 1968), *Cement, Comparison of Standards and Significance of Particular Tests*, ASTM STP 441, Philadelphia (1968c)

Popovics, S., "Examples for the Application of Mathematics in Concrete Technology," *Reports*, IV. International Congress on the Application of Mathematics in Engineering, Weimar, DDR, 1967, Vol.I, VEB Verlag fur Bauwesen, Berlin (1968d)

Popovics, S., "What Should an Engineer Know About the Nature of Admixtures," *Concrete*, Vol.2 No.7, London (1968e)

Popovics, S., "Effects of Admixtures on Concrete Strengths," *Materials Technology - An Interamerican Approach*, Inter-American Conference on Materials Technology, San Antonio, Texas (1968f)

Popovics, S., "Efectos del Arido Sobre Ciertas Propiedades del Hormigon de Cemento Portland" (Effects of Aggregate on Certain Properties of Portland Cement Concrete), *Informes de la Construccion*, No. 197, Instituto Eduardo Torroja, Madrid (1968g)

Popovics, S., "Estimating Proportions for Structural Concrete Mixtures," *ACI Journal*, Proc. Vol. 65, No.2 (1968h)

Popovics, S., "Comparison of Various Measurements Concerning the Kinetics of Hydration of Portland Cement," *Proceedings of the Fifth International Symposium on the Chemistry of Cement*, Part III., *Properties of Cement Paste and Concrete*, Tokyo (1969a)

Popovics, S., What Should an Engineer Know About the Nature of Admixtures? (In Finnish) *Betonituote*, No. 2, Helsinki (1969b)

Popovics, S., "Structural Model Approach to Two-Phase Composite Materials: State of the Art," *American Ceramic Society Bulletin*, Vol. 48, No. 11 (1969c)

Popovics, S., "Que Debe Saber un Ingeniero Respecto a la Naturaleza de los Aditivos?" (What Should an Engineer Know About the Nature of Admixtures?), *Revista IMCYC*, Vol. 8, No. 48, Mexico City (1971a)

Popovics, S., "Physical Aspects of the Setting of Portland Cement," *Journal of Materials*, JMLSA, Vol.6, No.1 (1971b)

Popovics, S., Effect of Kinetics on the Ultimate Strength of Portland Cement Pastes (in Russian), *Beton i Zhelezbeton*, Moscow (1972)

Popovics, S., "Segregation and Bleeding" (General Report), *Fresh Concrete: Important Properties and Their Measurement*, RILEM Seminar Held at the University of Leeds, Vol.2, Paper 6.1 (1973a)

Popovics, S., "Aggregate Grading and the Internal Structure of Concrete," *Highway Research Record* No. 441, *Grading of Concrete Aggregates*, Highway Research Board, Washington, D.C. (1973b)

Popovics, S., " Methods for the Determination of Required Blending Proportions for Aggregates," *Highway Research Record Number 441, Grading of Concrete Aggregates*, Highway Research Board (1973c)

Popovics, S., "Strength Development of Portland Cement Paste," *The VI. International Congress on the Chemistry of Cement*, Supplementary Paper, Section II, Moscow (1974a)

Popovics, S., "Polymer Cement Concretes for Field Construction," *Journal of the Construction Division,* ASCE, Vol. 100, No. CO3, Proc. paper 10806 (1974b)

Popovics, S., "Proportioning Concrete for a Specified Cement Content and/or for a Specified Weight," *Proportioning Concrete Mixes,* Publications SP-46, American Concrete Institiute (1974c)

Popovics, S., "Polymer and Portland Cement Concrete Combinations," *Proceedings of the 23rd Annual Arizona Conference on Roads and Streets,* The University of Arizona (1974d)

Popovics, S., "Phenomenological Approach to the Role of C3A in the Hardening of Portland Cement Pastes," *Cement and Concrete Research,* Vol. 6, No. 3 (1976a)

Popovics, S., "Cement-Mortar Experiments Concerning the Addition of Water-Dispersible Epoxy or Furfuryl Alchohol Systems," *Polymers in Concretes,* Paper 2.7, London, May 5-7, 1975, The Construction Press, Ltd., London (1976b)

Popovics, S., "Frost- und Witterungsbestandigkeit von Beton in Abhangigkeit von den Eigenschaften des Zuschlags" (Frost Resistance and Durability of Concrete as Influenced by Aggregate Properties), *Beton,* Vol. 26, No. 6 (1976c)

Popovics, S., "Reef Shell - Beach Sand Concrete," Living with Marginal Aggregates, *ASTM STP* 597 (1976d)

Popovics, S., "Heat and Noise Protection of Buildings," *Journal of the Construction Division* ASCE, Vol. 103, No. CO3 (1977)

Popovics, S., *Concrete Making Materials,* McGraw-Hill Book Company, New York, etc. and Hemisphere Publishing Corporation, Washington, etc. (1979)

Popovics, S., "Possibility of a Catalytic Role of C3A in the Hardening of Portland Cements," *7th International Congress on the Chemistry of Cement,* Vol. IV, Editions Septima (1980)

Popovics, S., "Extended Model for Estimating the
Strength Developing Capacity of Portland Cement,"
Magazine of Concrete Research, Vol. 33, No. 116,
London (1981a)

Popovics, S., "Generalization of the Abram's Law -
Prediction of Strength Development of Concrete from
Cement Properities," *ACI Journal*, Proc. Vol. 78, No.
2 (1981b)

Popovics, S., *Fundamentals of Portland Cement
Concrete: A Quantitative Approach*, Vol.1: *Fresh
Concrete*, John Wiley & Sons, New York,etc. (1982a)

Popovics, S., "Strength Relationships for Fly Ash
Concrete," *ACI Journal*, Proc.Vol. 79, No. 1 (1982b)

Popovics, S., "Production Schedule of Concrete for
Maximum Profit," *Materials and Structures - Research
and Testing*, Vol. 15, No. 86 (1982c)

Popovics, S., "Accelerator Additive for Cementitious
Compositions," *U.S, Patent* No. 4,419,138 (1983)

Popovics, S., "New Results with Epoxy Modification of
Portland Cement Concrete," *Polymers in Concrete*, H.
Schulz Editor, Fourth International Congress,
Institut fur Spanende Technologie und Werkzeug-
maschinen, Technische Hochschule Darmstadt (1984a)

Popovics, S., Modification of Mortars and Concretes
with the Addition of Polymer, (In Romanian),
Materiale Plastice, Vol. XXI, No. 2, Institutul
Central de Chimie, Bucuresti (1984b)

Popovics, S., "Effects of Aggregate Properties on the
Frost Resistance of Concrete," *VTT Symposium 50*,
Third International Conference on the Durability of
Building Materials and Components, Vol. 3, Technical
Research Center of Finland, Espoo (1984c)

Popovics, S., "Modification of Portland Cement
Concrete with Epoxy as Admixture," *Polymer Concrete
Uses, Materials and Properties*, ACI Publication
SP-89, J. T. Dikeou and D. W. Fowler, Editors,
American Concrete Institute (1985a)

Popovics, S., "Improved Utilization of Fly Ash in Concrete Through a Chloride-Free Accelerator," *ASTM Cement, Concrete and Aggregates*, CCAGDP, Vol. 7, No. 1 (1985b)

Popovics, S., "Effect of Curing Method and Final Moisture Condition on Compressive Strength of Concrete," *ACI Journal*, Proc. Vol. 83, No. 4 (1986a)

Popovics, S., "Stato attuale della determinazione della resistenza del calcestruzzo mediante la velocita degli impulsi in America" (Present State of the Determination of Concrete Strength by Pulse Velocity in America), *Il Cemento*, Anno 83, No. 3 (1986b)

Popovics, S., "Model for the Quantitative Description of the Kinetics of Hardening of Portland Cements," *Cement and Concrete Research*, Vol. 17, No.5 (1987a)

Popovics, S., "Strength Losses of Polymer-Modified Concretes under Wet Conditions," *Polymer Modified Concrete*, ACI Publication SP-99, American Concrete Institute, Detroit (1987b)

Popovics, S., "Strength-Increasing Effects of a Chloride-Free Accelerator," *Corrosion, Concrete, and Chlorides*, F. W. Gibson, Editor, ACI Publication SP-102, American Concrete Institute (1987c)

Popovics, S., "A Classification of the Deterioration of Concrete Based on Mechanism," *Concrete Durability, Katharine and Bryant Mather International Conference*, J. M. Scalon (Ed.), Vol. 1, ACI SP-100, American Concrete Institute (1987d)

Popovics, S., Rajendran, N., and Penko, M., "Rapid Hardening Cements for Repair of Concrete," *ACI Materials Journal*, Vol. 84, No. 1 (1987e)

Popovics, S., and Rajendran, N., "Early Age Properties of Magnesium Phospate-Based Cements Under Various Temperature Conditions," *Transportation Research Record* 1110, *Concrete and Concrete Construction*, Transportation Research Board, Washington, D.C. (1987f)

Popovics, S., "Attempts to improve the bond between cement paste and aggregate," *Materials and Structures - Research and Testing*, Vol. 20, No. 115 (1987g)

Popovics, S., "Effects of a chloride-free accelerator on concrete strength," *Second International Colloquium on Concrete in Developing Countries*, Session 8: Mineral and Fibre Admixtures to Concrete, Bombay, India (1988a)

Popovics, S., and Rajendran, N., "Test Methods for Rapid-Hardening Magnesium Phosphate-Based Cements," *ASTM Cement, Concrete, and Aggregates*, CCAGDP, Vol. 10, No. 1 (1988b)

Popovics, S., and McDonald, W. E., "Inspection of the Engineering Condition of Underwater Concrete Structures," *Technical Report*, REMR-CS-9, USAE Waterways Experiment Station, Vicksburg, Missi. (1989)

Popovics, S., "Stabilization of a Bauxite Residue for Construction Purposes," *The Journal of Resource Management and Technology*, Vol. 18, No. 1 (1990a)

Popovics, S., "Analysis of the Concrete Strength vs. Water-Cement Ratio Relationship," *ACI Materials Journal* (1990b)

Popovics, S., "A Study on the Use of a Chloride-Free Accelerator," *Admixture for Concrete - Improvement of Properties*, (E. Vazquez, Editor), Proceedings of the International RILEM Symposium, Chapman and Hall, London, etc. (1990c)

Popovics, S., "Test Methods for the Evaluation of Treated Sludges for Construction Purposes," *Hazardous and Industrial Waste*, J. P. Martin, S.-C. Cheng, M. A. Susavidge, Editors, *Proceeding of the Twenty-Second Mid-Atlantic Industrial Waste Conference*, Drexel University, Philadelphia (1990d)

Popovics, S., "A Model for Estimation of the Contribution of Fly Ash to Concrete Strength," *Blended Cements in Construction*, R. N. Swamy, Editor, Elsevier Applied Science, London and New York (1991)

Portland Cement Association, *Concrete Information*, (1965-1972)

Portland Cement Association, "Design and Control of Concrete Mixtures," *Engineering Bulletin*, Eleventh edition, PCA, Skokie, IL. (1968)

Portland Cement Association, *Special Concretes, Mortars and Products*, Wiley, New York (1975a)

Portland Cement Association, *Principles of Quality Concrete*, Wiley, New York (1975b)

Powers, T. C., "Studies of Workability of Concrete," *ACI Journal*, Proc. Vol. 28 (1932)

Powers, T. C., "The Bleeding of Portland Cement Paste, Mortar and Concrete," *ACI Journal*, Proc. Vol.35 (1939)

Powers, T. C., "A Working Hypothesis for Further Studies of Frost Resistance of Concrete," *ACI Journal*, Proc. Vol. 41 (1945a)

Powers, T. C., "Should Portland Cement Be Dispersed?," *ACI Journal*, Proc. Vol. 42 (1945b)

Powers, T. C., and Brownyard, T. L., "Studies of the Physical Properties of Hardened Portland Cement Paste-Part 2. Studies of Water Fixation," *ACI Journal*, Proc. Vol. 43 (1946)

Powers, T. C., and Brownyard, T. L., "Studies of the Physical Properties of Hardened Portland Cement Paste," Part 1-Part 9, *ACI Journal*, Proc. Vol. 43 (1947-1948)

Powers, T. C., "The Nonevaporable Water Content of Hardened Portland Cement Paste - Its Significance for Concrete Research and Its Method of Determination," *ASTM Bulletin*, No. 158 (1949a)

Powers, T. C., "The Air Requirement of Frost-Resistant Concrete," *Proceedings, Highway Research Board*, Vol. 29 (1949b)

Powers, T. C., and Helmuth, R. A., "Theory of Volume Changes in Hardened Portland Cement Paste During Freezing," *Proceedings, Highway Research Board*, Vol. 32 (1953a)

Powers, T. C., A Discussion of C. A. G. Weymouth's Theory of Parictle Interference, *Research Laboratories of the PCA*, Chicago (1953b)

Powers, T. C., "Void Spacing as a Basis for Producing Air-Entrained Concrete," *ACI Journal*,   Proc. Vol. 50 (1954a)

Powers, T. C., Copeland, L. E., Hayes, J. C., and Mann, H. M., "Permeability of Portland Cement Paste," *ACI Journal*, Proc. Vol. 51   (1954b)

Powers, T. C., "Basic Considerations Pertaining to Freezing and Thawing Tests," *Proceedings, ASTM*, Vol. 55 (1955a)

Powers, T. C., and Steinour. H. H., "An Interpretation of Published Researches on the Alkali-Aggregate Reaction," Parts I and II, *ACI Journal*, Vol. 51 (1955b)

Powers, T. C., "The Physical Structure and Engineering Properties of Concrete," *Cement Lime and Gravel*, No. 10, London (1956)

Powers, T. C., " The Physical Structures and Engineering Properties of Concrete," *Bulletin* 90, Research and Development Laboratories, Portland Cement Association, Skokie, IL. (1958a)

Powers, T. C., "Structure and Physical Properties of Hardened Portland Cement Paste," *Journal of the American Ceramic Society*, Vol.41, No. 1 (1958b)

Powers, T. C., Copeland, L. E., and Mann, H. M., "Capillary Continuity or Discontinuity in Cement Pastes," *Journal of the PCA Research and Development Laboratories*, Vol. 1, No. 2 (1959)

Powers, T. C., "Physical Properties of Cement Paste," *Chemistry of Cement*, Proceedings of the Fourth International Symposium, *NBS Monograph* 43, Vol. II, Washington, D.C. (1960)

Powers, T. C. "Some Physical Aspects of the Hydration of of Portland Cement," *Journal of the PCA Research and Development Labortories*, Vol. 3, No. 1 (1961)

Powers, T. C., "A Hypothesis on Carbonation Shrinkage," *Journal of the PCA Research and Development Laboratories*, Vol.4, No.2 (1962)

Powers, T. C., "The Physical Structure of Portland Cement Paste," *The Chemistry of Cements*, ed. H. F. W. Taylor, Vol. 1, Chap. 10, Academic Press, London and New York (1964)

Powers, T. C., "The Nature of Concrete," *Signifcance of Tests and Properties of Concrete and Concrete Making Materials*, ASTM STP 169-A, Philadelphia (1966)

Powers, T. C., *The Properties of Fresh Concrete*, John Wiley and Sons, New York, etc.(1968)

Price, W. H., "Erosion of Concrete by Cavitation and Solids in Flowing Water," *ACI Journal*, Proc. Vol. 19 (1947)

Price, W. H., "Factors Influencing Concrete Strength," *ACI Journal*, Proc. Vol. 47 (1951)

Price, W. H., "The Practical Qualities of Cement," *ACI Journal*, Proc. Vol. 71 (1974)

Price, W. H, "Pozzolans - A Review," *ACI Journal*, Proc. Vol. 72, No. 5 (1975)

Price, W. H., "Admixtures and How They Developed," *Concrete Construction*, Vol. 21, No. 4 (1976)

Prior, M. R., and Adams, A. B., "Introduction to Producers' Papers on Water-Reducing Admixtures and Set-Retarding Admixtures for Concrete," *Symposium on Effect of Water-Reducing Admixtures and Set-Retarding Admixtures on Properties of Concrete*, ASTM STP 266, Philadelphia (1960)

Proudley, C. E., "Sampling of Mineral Aggregates," *Symposium on Mineral Aggregates*, ASTM STP 83, Philadelphia (1948)

Quon, D. H. H., and Malhotra, V. M., "Effect of Superplasticizers on Slump, Strength and Degree of Conversion of High-Alumina Concrete," *Developments in the Use of Superplasticizers*, ACI Publication SP 68, American Concrete Institute, Detroit (1981)

Quon, D. H. H., and Malhotra, V. M., "Performance of High Alumina Cement Concrete Stored in Water and Dry Heat at 25, 35, and 50 C," *ACI Journal* , Proc. Vol. 79, No. 3 (1982)

Ramachandran, V. S., "Possible States of Chloride in the Hydration of Tricalcium Silicate in the Presence of Calcium Chloride," *Materials and Structures - Research and Testing*, Vol. 4, No. 19, Paris (1971)

Ramachandran, V. S., *Calcium Chloride in Concrete: Science and Technology*, Applied Science Publishers, London (1976)

Ramachandran, V. S., "Role of Calcium Chloride in Concrete," *Progress in Concrete Technology*, V. M. Halhotra, Editor, Energy, Mines and Resources, Ottawa, Canada, MRP/MSL TR (1980)

Ramachandran, V. S., Feldman, R. F., and Beaudoin, J.J., *Concrete Science*, Heyden and Son, Ltd., London - Philadelphia - Rheine (1981)

Ramachandran, V. S. (Ed.) *Concrete Admixtures Handbook*, Noyes Publications, Park Ridge, N.J. (1984)

Ramirez, J. L., Barcena, J. M., Urreta, J. I., "Sables calcaires a fines calcaires et argileuses: influence et nocivite dans les mortiers de ciment" (Calcareous sand with calcareous and clay fines: effect and noxiousness in cement mortars), *Materials and Structures - Research and Testing*, RILEM, Vol. 20, 117 Paris (1987)

Rauen, A., "Zum Wirkungsmechanismus von Betonverflussigern auf der Basis von wasserloslichen Melaminharzen (Working Mechanism of Water-Soluble Melamine Resin Plasticizers), *Cement and Concrete Research*, Vol. 6, No. 1 (1976)

Regourd, M. and Guinier, A., "The Crystal Chemistry of the Constituents of Portland Cement Clinker," A Principal Paper, *VIth International Congress on the Chemistry of Cement*, Moscow (1974)

Reichard, T. W., "Creep and Drying Shrinkage of Lightweight and Noprmal-Weight Concretes," *National Bureau of Standards Monograph* 74, Washington, D.C. (1964)

Reilly, W. E., "Hydrothermal and Vacuum Saturated Lightweight Aggregate for Pumped Structural Concrete," *ACI Journal,* Proc. Vol. 69, No. 6 (1972)

Reinsdorf, S., "Einfluss von Mahlfeinheit, Lagerdauer und Lagerungsart des Zements auf seine Fruhfestigkeiten," (Influence of Fineness, Duration and Type of Storage of Cement on the Early Strength), *Silikattechnik*, Vol.13, No.1 (1962)

Revay, M., "The Pre-Estimation of the Expected Decrease of Strength of High Alumina Cement Concretes," Supplementary Paper III-4, *VIth International Congress on the Chemistry of Cement,* Moscow (1974)

Rex, H. M., and Peck, R. A., "A Laboratory Test to Evaluate the Shape and Surface Texture of Fine Aggregate Particles," *Public Roads*, Vol. 29, No. 5 (1956)

Rhoades, R., and Mielenz, R. C., "Petrography of Concrete Aggregate," *ACI Journal*, Vol. 42 (1946)

Richartz, W., "Electron Microscopic Investigations About the Relations Between Structure and Strength of Hardened Cement," *Proceedings of the Fifth International Symposium on the Chemistry of Cement, Part III, Properties of Cement Paste and Concrete,* Tokyo (1969)

RILEM, "Resin Concretes," Symposium by Correspondence, *RILEM Bulletin*, New Series No. 28, Paris (1965)

RILEM-ABEM, "International Symposium on Admixtures for Mortar and Concrete, Brussels, 1967," *Materials and Structures - Research and Testing* , Vol. 1, No. 2 (1968)

RILEM IS-PM Technical Committee, "Determination of Pore Properties of Constructional and Other Materials," *Materials and Structures—Research and Testing*, Vol. 6, No.33 (1973)

RILEM Committee 11a, "Rapport final du Groupe de Travail Adjuvant" (Final Report of Working Group Admixtures), *Materials and Structures - Research and Testing*, Vol.8, No. 48 (1975)

RILEM TC FAB-67, "Test Methods for Determining the Properties of Fly Ash for Use in Building Materials," *Materials and Structures - Research and Testing*, Vol. 22, No. 130 (1989)

Rixom, M. R. (Editor), *Concrete admixtures: use and application*, Construction Press, London (1977)

Rixom, M. R., *Chemical Admixtures for Concrete*, E. & F. N. Spon, Ltd. (1978)

Road Research Laboratory, "Design of Concrete Mixes," *Road Note* No. 4, Dept. of Scientific and Industrial Research, H.M.S.D. London (1950)

Road Research Laboratory, *Concrete Roads*, Her Majesty's Stationery Office, London (1955)

Robson, T. D., "Characteristics and Applications of Mixtures of Portland and High Alumina Cements," *Chemistry and Industry*, No. 1 (1952)

Robson, T. D., *High-Alumina Cements and Concretes*, Contractors Record, London, John Wiley and Sons, New York, etc. (1962)

Robson, T. D., "Aluminous Cement and Refractory Castables, *The Chemistry of Cements*, ed. H. F. W. Taylor, Vol. II, Chap. 12, Acadecmic Press, London and New York (1964)

Robson, T. D., "The Chemistry of Calcium Aluminates and Their Relating Compounds," *Proceedings of the Fifth International Symposium on the Chemistry of Cement*, Part 1, *Chemistry of Cement Clinker*, Tokyo (1969)

Rockwood, N. C., "Production and Manufacture of Fine and Coarse Aggregates," *Symposium on Mineral Aggregates, ASTM STP* 83, Philadelphia (1948)

Ros, M., "Die materialtechnischen Grundlagen und Probleme des Eisenbetons im Hinblickauf die zukunftige Gestaltung der Stahlbeton-Bauweise" (Material-Technological Foundation and Problems of Reinforced Concrete.), *Bericht, no. 162*, Eidgenossische Materialprufungs- und Versuchsanstalt fur Industrie, Bauwesen und Gewerbe, Zurich (1950)

Rose, J. G. and Ledbetter, W. B. " *Summary of Surface Factors Influencing the Friction Properties of Concrete Pavements*" Highway Research Record, Number 357, *Quality Assurance Concrete, Construction, Bridge Deck Coating, and Joint seals*, Highway Research Board, Washington D.C. (1971)

Rosenberg, A. M., "Study of the Mechanism Through Which Calcium Chloride Accelerates the Set of Portland Cement, *ACI Journal*, Proc. Vol. 61 (1964)

Rosenberg, A. M., Gaidis, J. M., Kossivas, T. G. and Previte, R. W., "A Corrosion Inhibitor Formulated with Calcium Nitrite for Use in Reinforced Concrete," *Chloride Corrosion of Steel in Concrete*, STP 629, American Society for Testing and Materials, Philadelphia (1977)

Rosin, P., "Gesetzmassigkeit in der Kornzusammensetzung des Zements" (A Pattern in the Particle Size Distribution of Cement), *Zement,* Vol.22 (1933)

Rothfuchs, G., "Umrechung der Lochwerte von Rundlock und Maschensieben" (Conversion from Round Openings to Square Openings of Sieves), *Zement* (1934)

Rothfuchs, G., "Kornzusammensetzung von Betonzuschlagen zur Erzielung grosster Dichte" (Grading of Concrete Aggregate to Obtain Greatest Denseness), *Zement*, Vol. 24, No. 1 (1935)

Rothfuchs, G., *Betonfibel* Band 1. (Concrete Primer, Vol. 1.), Bauverlag GMBH. Weisbaden-Berlin (1962)

Rothfuchs, G., *Betonfibel*, Band II (*Concrete Primer,* Vol. 2), Bauverlag, Wiesbaden-Berlin (1964)

Sakurai, T., Sato, T., and Yoshinaga, A., "The Effect of Minor Components on the Early Hydraulic Activity of the Major Phases of Portland Cement Clinker," *Proceedings of the Fifth International Symposium on the Chemistry of Cement,* Part I, *Chemistry of Cement Clinker,* Tokyo (1969)

Sargent, Ch., "Economic Combinations of Aggregates for Various Types of Concrete," *Highway Research Board Bullentin 275* (1960)

Sasse, H. R., "Water-Soluble Plastics as Concrete Admixtures," *First International Congress on Polymer Concretes,* Session E, Paper 2, London, May 5-7 (1975)

Satarin, V. I., "Slag Portland Cement," A Principal Paper, *VIth International Congress on the Chemistry of Cement,* Moscow (1974)

Schaffler, H., "Beurteilung der Kornzusammensetzung nach einer Kornungsziffer" (Assessment of the Grading with the Aid of a Grading Coefficient), *Betonstein Zeitung 22,* Heft 1 (1956)

Scheidegger, A. E., *The Physics of Flow Through Porous Media,* Macmillan, New York (1957)

Schiel, F., "Bestimmung der Kornform" (Determination of the Particle Shape), *Abhandlungen uber Bodenmechanik und Grundbau,* Die Forschungsgesellschaft d. Strassenwesen (1948)

Schramli, W., "An Attempt to Assess Beneficial and Detrimental Effects of Aluminate in the Cement on Concrete Performance," Parts 1 and 2, *World Cement Technology,* Vol. 9, Nos. 3 and 4 (1978)

Schroder, F., "Slags and Slag Cement," *Proceedings of the Fifth International Symposium on the Chemistry of Cement,* Part IV, *Admixtures and Special Cements,* Tokyo (1969)

Schulze, W., and Altner, W., "The Signifcance of the Saturation Degree for Estimating the Frost-Resistance of Concrete," *RILEM Durability of Concrete, Final Report of the International Symposium,* Praha (1962)

Schuster, R. L., and McLaughlin, J. F., "A Study of Chert and Shale Gravel in Concrete," *Highway Research Board Buletin*, No. 305 (1961)

Schutz, R. J., "On New ASTM Standards - Epoxy Resins," *Concrete International: Design & Counstruction*, Vol. 4, No. 1 (1982)

Schwanda, F., "Der Bestwert der Kornzusammensetzung von Betonzuschlagstoffen" (Optimum Grading for Concrete Aggregates), *Der Bauingenieur* (1956)

Schwartz, D. R., "D-Cracking of Concrete Pavements," *NCHRP Synthesis of Highway Practice*, 134, Transportation Research Board, Washington, D. C. (1987)

Schwiete, H. E., and Ludwig, U., "Crystal Structures and Properties of Cement Hydration Products (Hydrated Calcium Aluminates and Ferrites)," *Proceedings of the Fifth International Symposium on the Chemistry of Cement*, Part II, *Hydration of Cements*, Tokyo (1969)

Selig, E. T. and Roner, C. J., "Effects of Particle Characteristics on Behavior of Granular Material," *Transportation Research Record* 1131, Washington, D.C. (1987)

Seligmann, P., and Greening, N. R., "Studies of Early Hydration Reactions of Portland Cement by X-Ray Diffraction," *Highway Research Record*, No. 62, Highway Research Board, Washington, D.C. (1963)

Seligmann, P., and Greening, N. R., "Phase Equilibria of Cement-Water," *Proceedings of the Fifth International Symposium on the Chemistry of Cement*, Part II, *Hydration of a Cements*, Tokyo (1969)

Shacklock, B. W., and Walker, W. R., "The Specific Surface of Concrete Aggregates and Its Relation to the Workability of Concrete," *Technical Report* TRA/241, Cement and Concrete Association, London (1957)

Shacklock, B. W., *Concrete Constituents and Mix Proportions*, Cement and Concrete Association, London, (1974)

Shaefer, T. E., "Application of Trilinear Charts to Aggregate Grading Control," *Concrete* (1930)

Shakoor, A. and Scholer C. F., "Comparison of Aggregate Pore Characteristics as Measured by Mercury Intrusion Porosimeter and Iowa Pore Index Tests," *ACI Journal*, Proc. Vol. 82, No. 4 (1985)

Shalon, R., and Raphael, M., "Influence of Sea Water on Corrosion of Reinforcement," *ACI Journal*, Proc. Vol. 55 (1959)

Shelburn, T. E., and Dillard, J. H., "An Appraisal of Skid Prevention," *Proceedings*, Ninth Pan American Highway Congress, Chapter VI, Topic 15, Doc. no. 90, Organization of American States (1963)

Shergold, F. A., "The Percentage Voids in Compacted Gravel as a Measure of its Angularity," *Magazine of Concrete Research* (London), Vol. 5, no. 13 (1953)

Shideler, J. J., "Calcium Chloride in Concrete," *ACI Journal*, Proc. Vol. 48 (1952)

Shideler, J. J., "Lightweight Aggregate Concrete for Structural Use," *ACI Journal*, Proc. Vol. 54, No. 4 (1957)

Short, A. and Kinniburgh, W., *Lightweight Concrete*, C. R. Books - Wiley (1963)

Shupe, J. W., and Lounsbury, R. W., "Polishing Characteristics of Universal Aggregates," *Proceedings*, First International Skid Prevention Conference, Part II, Virginia Council of Highway Investigation, Charlottesville, Va. (1959)

Shupe, J. W., "Pavement Slipperiness," *Highway Engineering Handbook*, ed. K. B. Woods, Section 20, McGraw-Hill, New York-Toronto-London (1960)

Singh, B. G., "Effect of the Specific Surface of Aggregates on Consistency of Concrete" *ACI Journal*, Proc. Vol. 53 (1957)

Skalny, J., and Odler, I., "The Effect of Chlorides upon the Hydration of Portland Cement and upon Some Clinker Minerals," *Magazine of Concrete Research*, Vol. 19, No. 61, London (1967)

Skalny, J., Jawed, I. and Taylor,H.F.W., "Studies of
Hydration of Cement - Recent Developments," *World
Cement Technology*, Vol. 9., No.6 (1978)

Skalny, J., "Cement Hydration: An Overview," *Advances
in Cement-Matrix Composites*, Editors: D.M. Roy, A.J.
Majumdar, S.P. Shah and J.A. Manson, Proceedings,
Symposium L, Materials Research Society, Boston
(1980)

Skeist, I., and Sommerville, G. R., *Epoxy Resins*,
Reinhold Publishing Corporation, New York (1962)

Skvara, F., Effect of the Granulometric Composition
of the Solid Phase on the Rheological Properties of
Cement Pastes, (In Slovakian), *Silikaty*, Vol. 20,
No.1, Prague (1976)

Smith, F. L., "The Effect of Aggregate Quality on
Resistance of Concrete to Abrasion Cement and
Concrete," *ASTM STP* 205, Philadelphia (1958)

Sneck, T., and Oinonen, H., "Measurement of Pore
Size Distribution of Porous Materials," *Julkaisu*
155, State Institute for Technical Research, Finland,
Helsinki (1970)

Soles, J. A., "Petrography in the Evaluation of
Aggregates and Concrete," *Progress in Concrete
Technology*, V. M. Malhotra, Editor, Energy, Mines
and Resources, Ottawa, Canada, MRP/MSL 80-89 TR
(1980)

Solomatov, V. I., *Polymer-Cement Concretes and
Polymer-Concretes*, Translation Series AEC-tr-7147,
Oak Ridge, Ten. (1970)

Solvey, 0. R., *Neue rationelle Betonerzeugung* (New
Rational Concrete Production), Springer-Verlag, Wien
(1949)

Soroka, I., and Sereda, P. J., "The Interrelation of
Hardness, Modulus of Elasticity and Porosity in
Various Gypsum Systems," *Journal of the American
Ceramic Society*, Vol. 51 (1968)

Soroka, I., and Sereda, P. J., "The Structure of Cement-Stone and the Use of Compacts as Structural Models," *Proceedings of the Fifth International Symposium on the Chemistry of Cement*, Part III, *Properties of Cement Paste and Concrete*, Tokyo (1969)

Soroka, I., *Portland Cement Paste and Concrete*, The McMillan Press, Ltd., London, etc. (1979)

Spiegel, M. R., *Theory and Problems of Statistics*, Schaum, New York (1961)

Spohn, E., Woermann, E. and Knoefel,D., "A Refinement of the Lime Standard Formula," *Proceedings of the Fifth International Symposium on the Chemistry of Cement*, Part I, *Chemistry of Cement Clinker*, Tokyo (1969)

Stanton, T. E., "Expansion of Concrete Through Reaction Between Cement and Aggregate," *Proc. ASCE*, Vol. 66 (1940a)

Stanton, T. E., "Influence of Cement and Aggregate on Concrete Expansion," *Engineering News Record*, Vol. 124 (1940b)

Stark, D., "Field and Laboratory Studies on the Effect of Subbase Type on the Development of D-Cracking," *Highway Research Record* No. 342, *Environmental Effects on Concrete*, Highway Research Board, Wasnington, D.C. (1970)

Stark, D., and Klieger, P., "Effect of Maximum Size of Coarse Aggregate on D-Cracking in Concrete Pavements," *Highway Research Record* No 441, *Grading of Concrete Aggregates*, Highway Research Board, Washington, D.C. (1973)

Stark, D., "Characteristics and Utilization of Coarse Aggregates Associated with D-Cracking," *Living with Marginal Aggregates*, ASTM STP 597, Philadelphia (1976)

Steinberg, M., Dikeou, J. T., et. al., *Concrete-Polymer Materials*, First Topical Report, *BNL 50134 (T-509)* and *USBR Gen. Rep. 41* (1968)

Steinour, H. H., "Further Studies of the Bleeding of Portland Cement Paste," *Bulletin* 4, Research Laboratory of the Portland Cement Association, London (1945)

Steinour, H. H., "The Setting of Portland Cement," *Research Department Bulletin* 98, Portland Cement Association (1958)

Steinour, H. H., "Concrete Mix Water - How Impure Can It Be?," *Journal of the PCA Research and Development Laboratories*, Vol. 2, No. 3 (1960a)

Steinour, H. H., "Who Invented Portland Cement?," *Journal of the PCA Research and Development Laboratories*, Vol.2, No.2 (1960b)

Stern, O., "Mittelbare Bewertung der Kornform von Zuschlagen" (Indirect Determination of the Shape ofAggregate Particles), *Zement*, Vol. 26, Nos. 42 and 43 (1937)

Stoll, U. W., "Hydrophobic Cement," *Cement and Concrete*, ASTM STP 205, Philadelphia (1958)

Stroeven, P., "Structural Investigations of Concrete by Means of Sterological Techniques," *Seminar on Fresh Concrete*, Paper 5.20, Leeds (1973)

Strunge, J., Knoefel, D. and Dreizler, I., "Einfluss der Alkalien und des Schwefels auf die Zementeigenschaften, Teil II: Einfluss der Alkalien und des Sulfates unter Berucksichtigung des Silicatmoduls auf den Zementeigenschaften" (Influence of Alkalies and Sulphur on the Properties of Cement. Part II: Influence of Alkalies and Sulphate on the Properties of Cement, Taking Account of the Silica Ratio), *Zement-Kalk-Gips*, Vol.38, No.8 (1985)

Sun, P. F., Nawy, E. G. and Sauer, J. A., "Properties of Epoxy-Cement Concrete System," *ACI Journal*, Proc. Vol. 72, No. 11 (1975)

Swayze, M. A., and Gruenwald, E., "Concrete Mix Design - A Modification of the Fineness Modulus Method" *ACI Journal*, Proc. Vol.43 (1947)

Swayze, M. A., "Early Concrete Volume Changes and Their Control," *ACI Journal*, Proc.Vol.38 (1942)

Sweet, H. S., and Woods, K. B., "A Study of Chert as a Deleterious Constituent in Aggregates," *Engineering Bulletin* (Purdue University), Vol. XXVI, no. 5 (1942)

Sweet, H. S., "Physical and Chemical Tests of Mineral Aggregates and Their Significance," *Symposium on Mineral Aggregates, ASTM STP* 83, Philadelphia (1948a)

Sweet, H. S., "Research on Concrete Durability as Affected by Coarse Aggregate," *Proceedings, ASTM*, Vol 48 (1948b)

Swenson, E. G., "A Reactive Aggregate Undetected by ASTM Tests," *ASTM Bulletin* 226, Philadelphia (1957)

Swenson, E. G. and Gillott, J. E., "Characteristics of Kingston Carbonate Rock Reaction," *Highway Research Board Bulletin* 275, Washington, D.C. (1960)

Swenson, E. G. and Gillott, J. E., "Alkali-Carbonate Rock Reaction" *Symposium on Alkali-Carbonate Rock Reactons, Highway Research Record* No. 45, Washington, D. C. (1964)

Swenson, E. G., and Gillott, J. E., "Alkali Reactivity of Dolimitic Limestone Aggregate," *Magazine of Concrete Research*, Vol. 19, No. 59, London (1967)

Swenson, E. G. (ed), *Performance of Concrete, Resistance of Concrete to Sulfate and Other Enviromental Conditions.* A Symposium in Honour of Thorbergur Thornvaldson. *Canadian Building Series* No.2, University of Toronto Press (1968)

Szeri, A. Z., "Introduction," *Tribology - Friction, Lubrication, and Wear*, A. Z. Szeri, (Ed.), Hemisphere Publishing Corporation, Washington D.C. etc. (1980)

Szuk, G. and Naray-Szabo, I., "Setting and Hardening of Hydraulic Materials, II. Conductometric Analysis of Setting of Cement Pastes Under Isometric Conditions," *Acta Technica*, Vol. 22, Hungarian Academy of Sciences, Budapest (1958)

Szuk, G., "Use of Special Instruments for Testing Setting Time of Cement Paste, Steam Curing of Concrete as Well as Hardening Process of Concrete," *RILEM Bulletin*, No.16 (1962)

Talaber, J., "Durabilite des ciments alumineux," (Durability of High-Alumina Cements), *RILEM Durability of Concrete*, Final Report of the International Symposium, Praha (1962)

Talaber, J., "High-Alumina Cements," A Principal Paper, *VIth International Congress on the Chemistry of Cement*, Moscow (1974)

Tamas, F. D., "Acceleration and Retardation of Portland Cement Hydration by Additives," *Symposium on Structure of Portland Cement Paste and Concrete*, Special Report 90, Highway Research Board Washington, D.C. (1966)

Tamas, F. D., and Fabry, M., "The Change in Reactivity of Silicate Anions During the Hydration of Calcium Silicates and Cement," *Cement and Concrete Research*, Vol. 3, No.6 (1973)

Tamas, F., and Varady, T., "Role of Poly-Reactions in the Hydration of Cement," *Hungarian Journal of Industrial Chemistry*, Vol. 3 (1975)

Tamas, F., Sankar, A. K., and Roy, D. M., "Effect of Variables upon the Silylation Products of Hydrated Cements," *Proceedings, Conference on Hydraulic Cement Pastes: Their Structure and Properties*, Sheffield, England (1976)

Taylor, H. F. W. (ed.), *The Chemistry of Cements*, Vols. I and II, Academic Press, London and New York (1964a)

Taylor, H. F. W., "Introduction," *The Chemistry of Cements*, ed. H.F.W. Taylor, Vol.I, Academic Press, London and New York (1964b)

Taylor, H. F. W., "The Steam Curing of Portland Cement Products," *The Chemistry of Cements*, ed. H. F. W. Taylor, Vol. 1, Chap. II, Academic Press, London and New York (1964c)

Taylor, H. F. W., "Tabulated Crystallographic Data,"
The Chemistry of Cements, ed. H.F.W. Taylor, Vol. II,
App. I, Academic Press, London and New York (1964d)

Taylor, H. F. W., "The Calcium Silicate Hydrates,"
Proceedings of the Fifth Internationa Symposium on
the Chemistry of Cement, Part II, Hydration of
Cements, Tokyo (1969)

Tenoutasse, N., "The Hydration Mechanism of C3A and
C3S in the Presence of Calcium Chloride and Calcium
Sulphate," Proceedings of the Fifth International
Symposium on Chemistry of Cement, Part II, Hydration
of Cements, Tokyo (1969)

Terrier, P., and Moreau, M., "Recherche sur le
mecanisme de l'action pouzzolanique des cendres
volantes dans le ciment" (Research Concerning the
Mechanism of the Pozzolanic Action of Fly Ash in
Cement), Revue des Materiaux de Construction, Ciments
et Betons , Nos. 613-614, Paris (1966)

Terrier, P., and Hornain, H., "Sur l'appli ation des
methods mineralogiques a l'industrie des liants
hydrauliques" (Application of Mineralogical Methods
in the Cement Industry), Revue des Materiaux de
Construction, "Ciments et Betons", March - June
(1967)

Teychenne, D. C., "Long-Term Research into the
Characteristics of High Alumina Cement Concrete,"
Magazine of Concrete Research, Vol. 27, No. 91,
London (1975)

Teychenne, D. C., "Long-Term Research Into the
Characteristics of High Aluminia Cement Concrete,"
Concrete, BRE Building Research Series, Vol. 1, The
Construction Press, Lancaster, London, New York
(1978)

Thomas, G. H., "A Comparative Assessment of the
Resistance of Super Sulphated, Sulphate Resistant
Portland, and Ordinary Portland Cements to Solutions
of Various Sulphates and Dilute Mineral Acids,"
Proceedings of the Fifth International Symposium on
the Chemistry of Cement, Part IV, Admixtures and
Special Cements, Tokyo (1969)

Thorenbohm, A. "Uber die Petrographie des Portland Zements" (The Petrography of Portland Cement), *Tonindustrie Zeitung*, Vol.21 (1897)

Tons, E., and Goetz, W. H., "Packing Volume Concept for Aggregates," *Highway Research, Record* No. 236, Highway Research Board, Washington, D.C. (1968)

Torrans, P. H., and Ivey, D. L., "Air Void Systems Affeccted by Chemical Admixtures and Mixing Methods," *Highway Research Record* No 226, Highway Research Board, Washington, D.C. (1968)

Transportation Research Board, *Cement-Aggregate Reactions*, Washington, D.C. (1974)

Transportation Research Board, *Superplasticizers in Concrete, Transportation Research Record* 720, Washington, D.C. (1979)

Tremper, B., and Spellman, D. L., "Tests for Freeze-Thaw Durability of Concrete Aggregates," *Highway Research Board Bulletin* No. 305 (1961)

Troxell, G. E., David, H.E. and Kelly, J.W., *Composition and Properties of Concrete*, 2nd ed., McGraw-Hill, New York (1968)

Turriziani, R. and Rio, A., "High Chemical Resistance Pozzolanic Cements," *Chemistry of Cement*, Proceedings of the Fourth International Symposium, *NBS Monograph* 43, Vol. II, Washington, D.C. (1960)

Turriziani, R., "Aspects of the Chemistry of Pozzolanas," *The Chemistry of Cements*, ed. H. F. W. Taylor, Vol. II, Chap. 14, Academic Press, London and New York (1964a)

Turriziani, R., "The Calcium Aluminate Hydrates and Related Compounds," *The Chemistry of Cements*, ed. H. F. W. Taylor, Vol. I, Chap. 6, Academic Press, London and New York (1964b)

Tyler, I. L., "Long-Time Study of Cement Performance in Concrete - Chapter 12, Concrete Exposed to Sea Water and Fresh Water," *ACI Journal*, Proc. Vol. 57 (1960)

Tynes, W. O., "Effect of Fineness of Continuously Graded Coarse Aggregate on Properties of Concrete," *Technical Report* 6-819, USAE Waterways Experiment Station, Vicksburg, Miss. (1968)

Tynes, W. D., "Investigation of Proprietary Admixtures," *Technical Report* C-77-1, USAE Waterways Experiment Station, Vicksburg, Miss. (1977)

U.S. Bureau of Reclamation, *Concrete Manual*, Seventh Ed., Revised Reprint, U. S. Government Printing Office, Washington, D.C. (1966)

U.S. Department of the Interior, *Thermal Properties of Concrete, Boulder Canyon Project, Final Reports*, Parts Vll, *Cement and Concrete Investigations, Bulletin 1.*, Bureau of Reclamation, Denver (1940)

U.S. Department of the Interior, *Investigations of Portland Cements, Boulder Canyon Project*, Final Reports, Part VII, *Cement and Concrete Investigations*, Bulletin 2, Bureau of Reclamation, Denver (1949)

Uchikawa, H., and Tsukiyama, K., "The Hydration of Jet Cement at 20°C," *Cement and Concrete Research*, Vol. 3, No.3 (1973)

Uchikawa, H., Uchida, S., and Hanehara, S., "Flocculation structure of fresh cement paste determined by sample freezing - back scattered electron image method," *Il Cemento*, New Series Vol. 84 (1987)

Ujhelyi, J., *Perlite Concrete - Perlite Mortar* (In Hungarian), Muszaki Konyvkiado, Budapest (1963)

Ujhelyi, J., (ed), *Handbook of Technology of Concrete and Mortar* (in Hungarian), Muszaki Konyvkiado, Budapest (1973)

Ullmann, G. R., "Re-Use of Wash Water as Mixing Water," Study at NRMCA Research Plant in Branchville, Md., and the Joint Research Laboratory of NSGA and NRMCA at the University of Maryland (1972)

Valenta, O., and Kucera, E., "Quel parti tirer des proprietes des epoxyles utilisees comme adjuvant du beton," (What Properties of Epoxies Are Important When Used as Admixture to Concrete) *Synthetic Resins in Building Construction*, Vol. I., RILEM Symposium (1967)

Valenta, O., "From the 2nd RILEM Symposium-Durability of Concrete - in Prague," *Materials and Structures - Research and Testing*, Vol. 3, No. 17, Paris. (1970)

Vallette, R., *Manuel de composition des betons (Manual of Concrete Compostion)* Editions Eyrolles, Paris (1963)

Valore, R. C., Jr., "Pumpability Aids for Concrete," *Significance of Tests and Properties of Concrete and Concrete Making Materials*, ASTM STP 169B, Chapter 50, Philadelphia (1978)

Van Keulen, J., "Density of Porous Solids," *Materials and Structures—Research and Testing,* Vol. 6, No. 33 (1973)

Van Wallendael, M., and Mahieu, 1., "Etude de l'action des "sucres" sur la prise des ciments. Analyse des sucres" (Study of the Action of "Sugars" on the Setting of Cements. Analysis of Sugars), RR CRIC 36-f-1973, Centre National de Recherches Scientifiques et Techniques pour l'Industrie Cimentiere (1973)

Vavrin, F., "Effect of Chemical Additions on Hydration Processes and Hardening of Cement," A Principal Paper, *VIth International Congress on the Chemistry of Cement*, Moscow (1974)

Venuat, M., "Methodes d'essais utilisables en laboratoire pour caracteriser les adjuvants" (Methods of Test Applicable in Laboratory to Characterize Admixtures), *RILEM Colloque international sur les adjuvants des mortiers et betons*, Bruxelles, August 30 - September 1 (1967a)

Venuat, M., "Les adjuvants des betons," (Admixtures for Concrete), *Revue des Materiaux, Ciments & Betons*, Nos. 626-627, Paris (1967b)

Venuat, M., "Effect of Elevated Temperatures and Pressures on the Hydration and Hardening of Cement," A Principal Paper, *VIth International Congress on the Chemistry of Cement*, Moscow (1974)

Verbeck, G. J., "The Camera Lucida Method for Measuring Air Voids in Hardened Concrete," *ACI Journal*, Proc. Vol. 43 (1947)

Verbeck, G. J. and Foster, C.W., "Long-Time Study of Cement Performance in Concrete with Special Reference to Heats of Hydration - Chapter 6," *Proceedings*, ASTM, Vol.50 (1950)

Verbeck, G. J., and Hass, W. E., "Dilatometer Method for Determination of Thermal Coefficient of Expansion of Fine and Coarse Aggregate," *Proc. Highway Research Board*, Vol. 30, Washington, D. C. (1951)

Verbeck, G. J., and Gramlich, C., "Osmotic Studies and Hypothesis Concerning Alkali-Aggregate Reaction," *Proc. ASTM*, Vol. 55 (1955)

Verbeck, G. J., "Carbonation of Hydrated Portland Cement," *Cement and Concrete, ASTM STP 205*, Philadelphia (1958)

Verbeck, G., and Landgren, R., "Influence of Physical Characteristics of Aggregates on Frost Resistance of Concrete," *Proceedings*, ASTM, Vol 60, Philadelphia (1960)

Verbeck, G. J., "Cement Hydration Reactions at Early Ages," *Journal of the PCA Research and Development Laboratories*, Vol.7, No.3 (1965)

Verbeck, G. J., "Pore Structure," *Significance of Tests and Properties of Concrete and Concrete Making Materials, ASTM STP 169-A*, Philadelphia (1966)

Verbeck, G. J., and Helmuth, R. H., "Structures and Physical Properties of Cement Pastes," *Proceedings of the Fifth International Symposium on the Chemistry of Cement*, Part III, *Properties of Cement Paste and Concrete*, Tokyo (1969)

Verein Deutscher Zementwerke, "Vorbeugende Massmahmen gegen Alkalireaction im Beton" (Preventive Measures Against Alkali-Aggregate Reactions in Concrete), *Schriftenreihe der Zementindustrie* No. 40, Beton-Verlag, Dusseldorf (1973)

Vicat, L., "Recherches experimentales sur les chaux de construction" (Experimental Investigations on Construction Limes), *Les Betons et les Mortiers*, Paris (1818)

Vivian, H. E., "Studies in cement-aggregate reaction: X. The effect on mortar expansion of amount of reactive component", *Commonwealth Scientific and Industrial Research Organization Bulletin*, Melbourne (1950)

Vivian, H. E., "Some Chemical Additions and Admixtures in Cement Paste and Concrete," *Chemistry of Cement*, Proccedings of the Fourth International Symposium, *NBS Monograph* 43, Vol. II, Washington, D.C. (1960)

Vivian, H. E., "Effect of Particle Size on the Properties of Cement Paste," *Symposium on Structure of Portland Cement Paste and Concrete*, *Special Report* 90, Highway Research Board, Washington, D.C. (1966)

Voellmy, A., "High Concrete Quality in Cold Weather" (General Report), RILEM Symposium, *Winter Concreting Theory and Pratice*, Session D, Special Report, Danish National Institute of Building Research, Copenhagen (1956)

Von Euw, M. and Gourdin, P., "Estimation of the Strength of Portland Cements," *Materials and Structures - Research and Testing*, RILEM, Vol. 3, No. 17 (1970)

Waddell, J. J., *Practical Quality Control for Concrete*, McGraw-Hill, New York, etc. (1962)

Waddell, J. J., "Cement," *Concrete Construction Handbook*, 2nd ed., J.J. Waddell, Chap.1, McGraw-Hill, New York, etc.(1974)

Wagner, H. B., "Water-Retentive Hydraulic Cements,"
Industrial and Engineering Chemistry,   Vol. 52, No. 3
(1960)

Wakley, J. T., "Use of Cones in Clarifying Wash Water
from Sand and Gravel Plants," *NSGA Circular* 114
(1972)

Walker, R. D., and McLaughlin, J. F., "Effects of
Heavy Media Separation on Durability of Concrete Made
with Indiana Gravels," *Highway Research Board
Bulletin*, No. 143 (1956)

Walker, R. D., "Identification of Coarse Aggregates
That Undergo Destructive Volume Changes When Frozen
in Concrete," *Highway Research Record* No. 120,
Aggregate Characteristics and Examination,
Washington, D.C. (1966)

Walker, R. D., and Hsieh, T. C., "Relationship
Between Aggregate Pore Characteristics and Durability
of Concrete Exposed to Freezing and Thawing," *Highway
Research Record* No. 226, Highway Research Board,
Washington, D.C. (1968)

Walker, S., "Effect of Grading of Gravel and Sand on
Voids and Weights," *Circular* 8, National Sand and
Gravel Association, Washington, D.C. (1930)

Walker, S., and Bartel, F. F., Discussion of a paper
by M. A. Swayze and E. Gruenwald: "Concrete Mix
Design - A Modificationof the fineness Modulus
Mehtod" *ACI Journal*, Proc. Vol. 43, Part 2 (1947)

Walker, S., and Bloem, D. L., "Effect of Heavy Media
Processing on Quality of Gravel," *NSGA Circular* 55
(1953)

Walker, S., and Bloem, D. L., "Design and Control of
Air-Entraining Concrete," *NRMCA Publication* No. 60,
Washington, D.C. (1955)

Walker, S. and Bloem, D.L., "Variations in Portland
Cement," *Proceedings, ASTM*, Vol. 58 (1958)

Walker, S., and Bloem, D.L., "Variations in Portland
Cement - Part 2," *Proceedings, ASTM*, Vol. 61 (1961)

Waller, H. F., Jr., "Stockpiling of Aggregate for Gradation Uniformity," *Proc. of National Conference on Statistical Quality Control Methodology in Highway and Airfield Construction* (1966)

Walters, G. D., "What Are Latexes," *Concrete International: Design & Construction*, Vol. 9, No. 12 (1987)

Walz, K., "Die Bestimmung der Kornform der Zuschlagstoffe" (Determination of the Particle Shape of Aggregate), *Die Betonstrasse*, No. 2 (1936)

Walz, K., *Undurchlassiger Beton* (Impermeable Concrete), *Bautechnik-Archiv*, Heft 13, Berlin (1956)

Walz, K., "Beton- und Zementdruckfestigkeiten in den USA und die Umrechnung auf deutsche Prufwerte" (Strengths of Cement and Concrete in the USA and their Conversion to German Testing Values), *Betontechnische Berichte 1962*, Beton Verlag, Dusseldorf (1963)

Walz, K., Zur Beurteilung der Eigenschafen des Betons mit Ausfallkornungen (The Assessment of the Properties of Gap-Graded Concrete), Betontechnische Berichte 1974, pp. 163-188, Beton-Verlag, Dusseldorf (1974)

Walz, K., "Beurteilung der Betonzusatzmittel nach den Wirksamkeitsprufrichtlinien" (Evaluation of Concrete Admixtures According to the Guide Lines for Testing Effectiveness), *Beton Herstellung Verwendung*, No. 2, and No. 3, Beton-Verlag, Dusseldorf (1975)

Washburn, E. W., "Note on a Method of Determining the Distribution of Pore Sizes in a Porous Material," *Proceedings*, National Academy of Science, Vol. 7 (1921)

Wesche, K., *Baustoffe fur Tragende Bauteile*, Band 2, *Beton, Mauerwerk*. (Building Materials for Load-Bearing Elements, Vol. 2, Concrete, Masonry), Bauverlag, Wiesbaden und Berlin (1974)

Weymouth, C. A. G., Effect of Particle Interference in Mortars and Concrete, *Rock Products*, vol. 36, no. 2, pp. 26-30, 1933.

Whitehurst, E. A., "Soniscope Tests Concrete Structures" *Journal of the American Concrete Institute,* Proc. Vol. 47 (1951)

Wieker, W., "New Methods of Investigations of the Hydration Processes of Portland Cements," A Principal Paper, *VIth International Congress on the Chemistry of Cement,* Moscow (1974)

Williamson, R. B., and Tewari, R. P.,"Effects of Microstructure on Deformation and Fracture of Portland Cement Paste," *Electron Microscopy and Structure of Materials,* ed. G. Thomas, University of California Press, Berkeley, Los Angeles, London (1972)

Wills, M. H., Jr., Lepper, H. A., Jr., Gaynor, R. D., and Walker, S., "Volume Change as a Measure of Freezing-and-Thawing Resistance of Concrete Made with Different Aggregates," *Proceedings, ASTM,* Vol. 63 (1963)

Wills, M. H., Jr., "How Aggregate Particle Shape Influences Concrete Mixing Water Requirements and Strength," *Journal of Materials,* Vol. 2, No. 4 (1967)

Wills, M. H., Jr., "Lightweight Aggregate Particle Shape Effect on Structural Concrete," *ACI Journal,* Proc. Vol. 71, No. 3, (1974)

Wilson, F. C., Concrete Mix Proportioning for Pumping-Some Influences of Material Properties - A Graphical Approach to Optimum Gradation, *Proportioning Concrete Mixes,* Publication SP-46-3, American Concrete Institute (1974)

Winslow, D. N., and Diamond, S., "A Mercury Porosimetry Study of the Evolution of Porosity in Portland Cement," *Journal of Materials,* JMLSA, Vol. 5, No. 3 (1970)

Winslow, D. N., and Diamond, S., "The Specific Surface of Hydrated Portland Cement Paste as Measured by Low-Angle X-Ray Scattering," *Journal of Colloid and Interface Science,* Vol. 45, No. 2 (1973)

Winslow, D. N., and Diamond, S., "Specific Surface of Hardened Portland Cement Paste as Determined by Small-Angle X-Ray Scattering," *Journal of the American Ceramic Society*, Vol. 57, No. 5 (1974)

Wischers, G., "Zur Normung von Zement" (Notes on the Revised German Cement Standard), *Betontechnische Berichte 1971*, Beton-Verlag, Dusseldorf (1972)

Wischers, G., and Krumm, E., "Zur Wirksamkeit von Betondichtungsmitteln" (The Effectiveness of Water Repelling Concrete Admixtures), *Betontechnische Berichte 1975*, Beton-Verlag, Dusseldorf (1976a)

Wischers, G., "Bautechnische Eigenschaften des Zements" (Technical Properties of Cement), *Zement Taschenbuch 1976-77*, Bauverlag, Wiesbaden, Berlin (1976b)

Wise, M. E., "Dense Random Packing of Unequal Spheres," *Philips Research Report*, Vol. 7, No. 5 (1952)

Witt, J. C., *Portland Cement Technology*, Chemical Publishing Co., New York (1966)

Woods, H., Starke, H. R., and Steinour, H. H., "Effect of Cement Compositions on Mortar Strength," *Engineering News Record*, Vol. 109, No. 15 (1932)

Woods, H., "Rational Development of Cement Specifications," *Journal of the PCA Research Development Laboratories*, Vol.I, No.1 (1959)

Woods, H., "Durability of Concrete Construction," *American Concrete Institute Monograph* 4, American Concrete Institute and The Iowa State University Press (1968)

Woods, K. B., and McLaughlin, J. F., "The Role of Mineral Aggregates in the Design, Construction and Performance of Highway Pavements and Bridges," *Proceedings*, Ninth Pan American Highway Congress, Chapter IV, Topic 10, Doc. no. 49, Organization of American States (1963)

Woolf, D. O.,   "The Relation Between Los Angeles Abrasion Test Results and the Service Roads of Coarse Aggregates," *Proc. Highway Research Board*,   Vol. 17 (1937)

Woolf, D. O.,   "Toughness, Hardness, Abrasion, Strength, and Elastic Properties," *Significance of Tests and Properhes of Concrete and Concrete Maing Materials*,   ASTM STP 169-A, Philadelphia (1966)

Wray, F. M., and Lichtefeld, H. J.,   "The Influence of Test Methods on Moisture Absorption and Resistance of Coarse Aggregate to Freezing and Thawing," *Proceedings, ASTM*,   Vol. 40 (1940)

Wright, P. J. F.,   "A Method of Measuring the Surface Texture of Aggregate," *Magazine of Concrete Research*, Vol. 7, No. 21, London (1955)

Wuerpel, C. E., and Rexford, E. P.,   "The Soundness of Chert as Measured by Bulk Specific Gravity and Absorpbon," *Proceedings, ASTM*,   Vol 40 (1940)

Yamaguchi, G.and Takagi,S.,   "The Analysis of Portland Cement Clinker," *Proceedings of the Fifth International Symposium on the Chemistry of Cement*, Part I, *Chemistry of Cement Clinker*, Tokyo (1969)

Yamaguchi, G. and Takagi,S.,   "Present-Day Methods of Investigation of the Clinker Formation Mechanism and Clinker Phase Composition," A Principal Paper, *VIth International Congress on the Chemistry of Cement*, Moscow (1974)

Yokomichi, H.,   "Influence of High Temperatures of Mixing Water on Setting, Consistency, Strength and Heat of Hydration of Concrete," *Winter Concreting Theory and Pratice*, Session D, Special Report, Danish National Institute of Building Research, Copenhagen (1956)

Young, D. M., and Crowell, A. D., *Physical Adsorption of Gases*, Butterworths, London (1962)

Young, R. B.,   "Some Theoretical Studies on Proportioning Concrete by the Method of Surface Area of Aggregates" *Proceedings* ASTM, Vol. 19, Part II (1919)

Zietsman, C. F., " Mortar and Concrete Making
Properties of Natural Sands Related to Their Physical
Attributes" *ACI Journal*, Proc. Vol. 53 (1957)

Index

A values by Kluge – 401–402
A_h – 410, 433, 436
A_{lin} – 408
A_{log} – 408, 411, 413, 416, 417
Absolute volume – 289–297, 303, 333, 391, 403, 432
Absorption (see Aggregate, absorption of)
Absorptivity – 339
Accelerating admixture (see Accelerator)
Accelerator – 174, 181, 221, 223, 236–239, 241–248, 272
chloride–free – 242–247
Acidity – 217
Activator – 181
Admixture – 19, 27, 55–56, 69, 110, 120–123, 129, 165, 219, 221–273, 532
accelerating – 174, 180, 236–246
air–detraining – 227
air–entraining (AEA) – 27, 92, 169, 176, 178, 190, 221–273
materials for – 224–227
batching of – 224

bonding – 268
calcium chloride (CaCl$_2$) as – 53, 73, 238–242, 245
for concrete – 221–251
expansion–producing – 268
finely divided mineral – 31, 122–123, 158–159, 178–190, 221, 269–271, 349–352, 364–366, 522, 530
fly ash – 29, 159, 181–190, 227, 269, 271, 477, 521, 542
future of – 271
gas forming – 267
grouting – 268
"inspector" – 271
permeability–reducing – 175
polymers as – 258–266
sampling of – 270
selection of – 270
set–controlling – 174, 223, 254, 272
special – 223
storage of – 270
surface–active – 122, 221, 251–257
testing of – 270

Admixture (continued)
  water–reducing – 169, 221–249,
    272
    superplasticizer – 248, 252–
      254
Adsorption – 255
  of air–entraining agent – 190
  with argon – 134
  BET method – 337
  determination of specific surface
    from – 132
  isotherms – 129, 337
  with nitrogen – 132
  theory of – 131–132
  with water vapor – 132
AEA (see Air–entraining
  admixture)
Aeration – 80
Agent (see Admixture)
Aggregate – 105, 218, 274–542
  abrasion of – 311–313, 335,
    340, 359
  absorption of – 287, 297–303,
    323, 332, 336, 518–526
    average – 300
    rate of – 298, 523
  absorptivity of – 339
  for bituminous mixes – 432
  blending of – 476, 483–511
  blending proportions for (see
    Blending proportions for
    aggregate)
  bond to – 379, 522
  bound water in – 518, 527–532
  bulking of – 287, 303–306
  chemical properties of – 349–
    374
  chemical resistance of – 351
  classification of – 274–280
  coarse – 274–277, 280–289,
    296, 301–307, 312, 329,
    340, 347, 359–371, 379–
    382, 390, 398, 420, 423,
    429, 439, 445–455, 458–
    462, 469–474, 475–479,

    493, 514, 524, 530, 531,
      533, 534, 535, 537
  coating on – 354, 366, 368
  crushed – 278, 279, 312, 345,
    378, 381, 382, 415–416,
    429, 455–456, 478–479,
    481, 530, 537
  deformability of – 309–319
  degree of saturation of – 298–
    303, 321–325
  deleterious materials in – 218,
    349–356, 373, 524, 532,
    538
  density of (see also
    Aggregate, specific gravity
    of) – 292, 309, 310
  deterioration of – 372
  durability of – 288, 297, 299,
    321–333, 485, 538
    durability factor – 328
    selection of aggregate for –
      325
  fine – 228–234, 274–286, 303–
    308, 347, 358, 379, 381,
    389, 391, 398, 420, 437,
    442, 447–465, 458–462,
    469–474, 476–480, 493,
    506, 508, 514, 523, 524,
    530, 533–538
  fraction (see Size fraction)
  friction in – 380
  frost resistance of (see
    Aggregate, durability of)
  future of – 542
  in general – 274–286
  geometric properties of – 375–
    410
  grading of – 251, 274–282,
    306–307, 341, 351, 375–
    510, 515, 523, 524, 530,
    532, 533–542
    coarseness of – 251, 376,
      399, 410
    combined – 280, 446–474,
      485–510, 524, 532

Aggregate (continued)
continuous – 389, 444–445,
460–461, 468–474, 476–
483
criteria – 396
evaluation of – 411, 420,
436–438, 442–474
gap – 389, 439, 444–447,
461, 469–474, 482–484,
530
linear – 418, 435
logarithmic – 418, 435
logarithmic normal
distribution – 278–279,
416
maximum density – 443,
468–474
normal distribution – 278
one–size – 400, 420, 434–435
oversanded – 378, 458
parabolic – 418, 435
percentage – 443, 464–468,
472, 474
significance of – 376, 389
specifications for – 399, 442–
474, 488–510, 524
tolerances – 485
undersanded – 389
variance of – 282, 399, 434,
439
variations in (see also
Fineness modulus;
Grading point; Numerical
characterization of
grading; Sieve curve;
Specific surface, of
aggregate) – 282–283,
533–538
gravel as – 380, 381
handling of – 533–538
hardness of – 306–319, 379,
540
heavyweight – 276, 297, 312,
390, 512–513, 527–532
inert – 278, 349–350

internal structure of – 517–522
lightweight – 169, 228, 276–
278, 293–306, 314–320,
339, 349, 390, 461, 512–
527, 541
lightweight particles in – 349,
354
manufactured – 275, 512, 517–
526, 532, 542
maximum size of (see
Maximum particle size)
modulus of elasticity of – 309–
319, 335, 540–591
moisture content in – 287, 297–
303
natural lightweight – 517, 521–
522
normal–weight – 278, 312, 336,
391, 461, 512–517, 523,
526, 530, 532
ore as – 513, 527–529
organic impurities in – 349,
355, 524
organic material as – 517–518,
521–522
packing of – 379, 382–384,
468, 475, 480
particle shape of – 228, 288,
304, 341, 375–386, 396,
415, 423, 424–434, 439,
448, 455, 469, 473, 475–
483, 515–524, 531, 540,
541
angularity factor – 382–383
angularity number – 382–383
particle size of (d) – 281, 375–
410, 411–420, 421–427,
430–435, 439–440, 449–
468, 488–504, 519, 523
average – 376, 386, 389, 399,
404–410, 411, 420, 426–
428, 438, 448
maximum (see Maximum
particle size)
minimum (see Minimum

Aggregate (continued)
    particle size)
        relative – 385, 411, 426–428,
            452–456, 466
        specific surface diameter –
            382–386
        Stokes diameter – 384
        surface diameter – 384, 386
        volume diameter – 384, 396,
            415
    particle size distribution of (see
        Aggregate, grading of)
    particles finer than No. 200
        sieve – 349–359, 540
    percent of voids in (see Percent
        of voids)
    permeability of – 297–303, 540
        coefficient of – 302
    petrography of – 276, 330
    physical properties of – 287–
        347
    polishing characteristic of –
        345–346, 541
    porosity of – 288–291, 297–
        300, 309, 320–325, 333–
        339, 379, 384, 517–522,
        526, 540
        characteristics of – 339
        distribution of – 324–325,
            334, 338
        methods of testing of – 334–
            338
    Pyrex glass as – 371
    reactive – 241, 278, 349–350,
        355–374, 485, 538, 540
    role of – 275, 287–288
    sample size of – 283
    sampling of – 274, 280–286
    segregation in – 281, 533–538
    selection of – 533, 538–541
    service record of – 329, 332,
        369, 533, 538
    skid resistance of – 339–347,
        485
    soft particles in – 349, 354,
        524
    solid volume of – 289–297,
        303, 335, 384, 391, 403,
        432
    soundness of – 288, 299, 321–
        333
        sulfate soundness test for –
            331–333
    specific gravity of – 278, 287–
        297, 303, 310, 335, 375,
        428–435, 439, 451, 485,
        487, 512–518, 523–532,
        541
        apparent – 290
        average – 293–296, 512, 514
        specific gravity factor – 526
    specific surface of (see Specific
        surface, of aggregate)
    steel as – 513, 528–532
    storing of – 533–538
    strength of – 287, 306, 311–
        319, 326, 517, 524–530,
        538, 540
        test method for – 309–315,
            526
    surface texture of – 375–383,
        524, 538
    surface treatment of – 300, 325,
        522
    swelling of – 362
    thermal properties of – 319–
        321, 327, 538–542
    toughness of – 309–319
    uniformity of – 534–538
    unit weight of – 287, 303–
        309, 516, 517, 524, 525
    voids content of (see Percent
        of voids)
    washing of – 214, 219
    waste material as – 276, 521–
        522, 542
Aggregate–alkali reaction – 23,
    29, 159, 189, 241, 271, 356–
    374
Aggregate–cement ratio – 424–

Aggregate–cement ratio (con't)
  431, 440, 451–457, 459,
  523, 530
Air content – 32, 33–34, 226–
  234, 239, 446
  effects on – 228–234, 479,
  514
Air–detraining admixture – 227
Air–entrained concrete – 224–
  236, 332, 477
Air–entraining admixture (AEA)
  (see also Air entrainment) –
  27, 92, 165, 169, 176, 179,
  190, 221–271
Air–entraining natural cement –
  177
Air–entraining portland cement –
  227
Air–entraining slag cement – 180
Air–entrainment (see also
  Admixture, air–entraining) –
  92, 95, 124, 177–178, 189–
  190, 224–236, 325, 523, 531,
  539
  mechanism of – 232–236
Air–permeability method – 41–42
Air voids (see Air entrainment;
  Cement paste, porosity of)
Air–water interface – 234
Alite (see also Tricalcium
  silicate) – 3, 15, 19, 25,
  103, 108–115
Alkali activation – 186
Alkali–carbonate rock reaction –
  350, 367–371
  mechanism of – 367
  selection of aggregate concerning – 368
Alkali chloride – 217
Alkali reactivity – 92
Alkali–silica reaction – 350,
  361, 362–374
  mechanism of – 361, 362
  selection of aggregate concerning – 365, 366

Alkalinity – 217, 370, 372
Alkalis – 6, 14, 23, 29, 108,
  189, 349, 350, 361–272
Alumina (Al_2O_3 or A) – 6, 14,
  15–17, 188
Alumina cement (see High-
  alumina cement)
Alumina–lime ratio – 160
Alumina ratio – 17
American Society of Civil
  Engineers – 32
Andesite – 363
Anhydrite – 23, 111, 167, 174,
  179
Anhydrous calcium sulfoalumin-
  ate – 167
Anionic surface–active agent –
  224, 226
Apparent specific gravity (see
  also Aggregate, specific
  gravity of) – 290
Agrillaceous dolomitic lime-
  stone – 350, 367
Argillaceous limestone – 177
Argon adsorption – 134
Arithmetic average particle size –
  386, 404–408
Attenuation of radiation – 512,
  527–531
Autoclave curing – 123, 130, 242,
  369
Autoclave soundness test – 33, 62,
  70, 74
Average specific gravity (see also
  Aggregate, specific gravity
  of) – 292, 295, 512

Barium aluminous cement – 177
Barium portland cement – 176
Basalt – 277, 339, 340, 363
Batching – 533–538
Bauxite – 160, 167
Belite (see also Dicalcium
  silicate) – 3, 15, 19, 21,
  111, 115, 177

Bentonite – 269
BET method – 132, 337
Blaine specific surface – 36–44
Blast-furnace slag – 29, 159, 167, 178, 184–187, 274–275, 359, 368, 518
foamed – 512–518, 523, 524
Bleeding – 37, 83, 94–95, 228, 441
Blended aggregate – 280, 446–474, 483–510, 524, 532
Blended portland cement – 3, 26–31
Blending proportions for aggregate – 293, 300, 418, 475, 485–511
graphical methods for – 494–511
numerical method for – 485–494
Bogue formulas – 19–21
Bolomey grading – 442, 453–458
Boron – 356, 528, 531
Bound water – 112, 124, 161, 168
British method for blending proportions – 497–503, 511
Bulk volume – 292, 333
Bulking – 287, 304–306

Calcination – 188
Calcium aluminate (CaO·Al$_2$O$_3$) – 158–165, 167
Calcium aluminate hydrate (CaO·Al$_2$O$_3$·6H$_2$O and (CaO·Al$_2$O$_3$·10H$_2$O) – 159, 161, 163
Calcium aluminate sulfate hydrate (C$_3$A·3CaSO$_4$·31H$_2$O, ettringite, or cement bacillus) – 108, 111, 113, 168, 174–175, 179–181
Calcium aluminosilicate – 185
Calcium carbonate (CaCO$_3$) (see also Limestone) – 134, 173, 346

Calcium chloride (CaCl$_2$) (see also Admixture, accelerating) – 52, 73, 236– 242, 225
Calcium haloaluminate (11CaO·7Al$_2$O$_3$·CaX) – 174
Calcium hydroxide [Ca(OH)$_2$] – 108–116, 123, 134, 162, 171–190, 368, 372
Calcium lignosulfonate (see also Admixture, water–reducing) – 221, 256
Calcium oxide (CaO or C) – 6, 15, 20, 62, 134, 176, 349, 368,
Calcium silicate (see also Tricalcium silicate; Dicalcium silicate) – 3, 6, 22, 112–115, 242
Calcium silicate hydrate (C$_3$S$_2$H$_3$, CSH or tobermorite) gel (see also Cement gel) – 103, 111–113, 122, 129–133
Calcium sulfate (CaSO$_4$) (see also Gypsum) – 23, 111, 167–169, 179
Calcium sulfoaluminate hydrate (see Calcium aluminate sulfate hydrate)
Calorimeter – 76
Capillary condensation – 127
Capillary pores – 120–127, 235
Capillary water (see also Free water) – 119–121
Carbonate rocks – 367
Carbonation – 87, 88
Carbonation shrinkage – 87, 88
Celite (see also Tetracalcium aluminoferrite) – 15, 23
Cement (see also Expansive cement; High–alumina cement; Portland cement; Special cements)
future of – 212
Cement–aggregate mixture – 423–428, 450–459

Cement–aggregate ratio – 428,
    450–457
Cement–aggregate reaction –
    361–372, 540
Cement bacillus – 168
Cement gel (see also Calcium
    silicate hydrate gel) –
    122–131, 161
  structure of (see also
      Cement paste, structure
      of) – 132–133
Cement manufacturing – 6–14
Cement models – 133–157
Cement paste – 119–121
  air entrainment in (see Air
      entrainment)
  expansion of: from alkali–
      aggregate reaction – 361–
      366, 370–372
    from expansive cement –
      166–171
    from moisture – 83–90
    from sulfate attack – 92, 93
    from unsoundness – 18, 61–
      63
  fresh – 122
    dispersion in – 251
    electrical resistance of –
      52, 55
    normal consistency of (see
      Normal consistency)
    pH of – 31
    stiffening of (see Time of
      setting)
    viscosity of – 173, 228–229,
      251
  hardened – 122–135, 234
    bonds in – 133–135
    density of – 119
    modulus of elasticity of – 134
    strength of – 133–135
  hardening of (see Hardening)
  porosity of – 117–129, 228–234
    capillary – 117–129
    distribution of – 127–129

effects of – 135, 334–335
  entrained (see Air entrain-
      ment)
  gel – 104, 124, 129, 239
  of high–alumina cement
      158
  test methods for – 127, 132
setting of (see Setting)
structure of – 103, 104, 122–
    135
  models for – 133–135
Cement sampling – 98–100
Cementitious mineral admixtures –
    183–190, 270
Chemical analysis of portland
    cement – 14–17
Chemical bonds – 133–135
Chemical prestressing – 166–171
Chemical resistance of portland
    cement – 23–29, 92, 190,
    270–271
Chemically bound water – 106,
    111–114, 119–127, 161, 168,
    527
Chert – 297, 323, 347, 359, 363
Chloride–free accelerator (see
    Accelerator, chloride–free)
Ciment alumineux – 160
Ciment fondu – 160
Ciment de haut fourneau – 31
Ciment metallurgique sursulfate –
    179
Ciment portland de fer – 30
Clay – 280, 351–354, 359, 367
Clinker (see Expansive cement;
    High–alumina cement; Port-
    land cement, clinker of)
Clinker minerals – 3, 19, 21
Coal – 354–359
Coarse aggregate (see Aggregate,
    coarse)
Coating on aggregate – 354
Coefficient of expansion – 319–
    321
Coefficient of variation – 98

Colored cements – 172
Combined aggregate – 280, 446–474, 485–511, 524, 532
Combined grading – 280, 446–474, 485–511, 524, 532
Compressive strength (see Strength, compressive)
Concrete
  admixture for (see Admixture)
  air-entrained – 224–228, 232–235, 272, 332, 351, 477
    production of – 227
  air voids in (see Cement paste, porosity of)
  black – 173
  for blocks – 518, 521, 524
  cellular – 233
  compaction of – 476–479
  concrete strength–mortar strength ratio – 68, 69
  consistency of (see Consistency)
  consolidation of – 228, 476–479, 530
  cracking in – 349–350, 358
  denseness of – 420, 472
  durability of – 228, 538–542
  expansion in (see also Cement paste, expansion of) – 271, 349–350, 368
  flow of – 420, 422, 441
  freezing in – 234
  frost resistance – 92, 224–236, 321–333, 340
  heavyweight – 527–532
  impermeability of (see Permeability)
  internal structure of – 475–483
  lightweight – 512–527
  mass – 79, 190, 320
  modulus of elasticity of – 541
  for pavement – 339–348
  permeability of (see Permeability)
  prepacked – 462, 530
  segregation of – 523, 530

  self-stressing – 170
  shrinkage of (see also Shrinkage) – 272, 340
  for shielding – 527–532
  skid resistance of – 343–348, 517, 541
  slump of (see Slump)
  soundness of (see Soundness)
  steam-cured (see Curing)
  structural lightweight – 512–527
  unit weight of – 512–514, 527–529, 541
  water requirement of (see Water content)
  wear resistance of – 339–540
  workability of (see Workability)
Concrete aggregate (see Aggregate)
Consistency (see also Slump; Workability) – 44, 227, 246–254, 420, 422, 440
Continuous grading – 389, 444, 454–463, 476–483
Corrosion of reinforcement – 218
Critical size effect – 323–325
Crushed aggregate (see Aggregate, crushed)
Crushing machine – 278–279, 378
Crushing test – 311–319
CSH gel (see also Cement gel) – 109–135
Cumulative percentage passing (see Total percentage passing)
Cumulative percentage retained (see Total percentage retained)
Curing
  at high temperature – 72, 73, 164
  low-pressure steam – 71, 75
  at low temperature – 72, 73

D (see Maximum particle size)
d (see Aggregate, particle size of)
d'_e equivalent mean diameter by

d'_e equivalent (continued)
Hughes – 401–402
$D-m-s$ method – 412, 438–441,
444, 471–474, 483
Darex AEA – 230
Dedolomitization – 367
Deformability of rocks – 311, 318
Degree of hydration – 114–121
Degree of saturation – 291, 300–
302, 322, 323, 340, 523
Deleterious materials – 217, 349–
356, 373
Density (see Specific gravity)
Deval test – 312
Dicalcium silicate ($2CaO \cdot SiO_2$ or
C_2S) (see also Belite; Calcium
silicate) – 16, 18, 20, 28, 105,
114–117
beta – 15, 105
effect of
on heat of hydration – 23
on specific surface – 131
on strength – 23
DIF (see Differential curve)
Differential curve – 394
Differential wear – 347
Diffusion – 116–119
Diffusivity (thermal) – 319, 320
Direct freezing and thawing test –
327
Dolomite – 367
Dormant period – 47, 48, 75, 109,
110
Drying shrinkage (see also
Shrinkage) – 24, 83–90
Durability
of aggregate (see Aggregate,
durability of)
factor – 272, 328
of hardened concrete (see
Concrete, durability of)
Durability factor – 272, 328
Dynamic modulus – 328

e water–requirement value by

Bolomey –401–403
Efflorescence – 95
Eisenportlandzement – 30
Electrical resistance of paste –
52, 55
EMPA grading – 453, 458
Entrained air (see Air entrainment)
Entrapped air – 224, 235
Ettringite – 108, 115, 168, 179
Evaporable water – 107, 124
Expanded clays, shales, and
slates – 514–524
Expanded perlite – 514, 525
Expanding cement (see Expansive
cement)
Expanditure effect – 324, 327
Expansion of cement paste (see
Cement paste, expansion of)
Expansive cement – 159, 166–171
Exponential cement model – 138–
157

$f(d)$ (see Sieve curve, equation of)
f_s surface index by Murdock –
401, 403
False set – 32, 33, 61, 109
Faury grading – 453–456
Ferric oxide (Fe_2O_3 or F) – 14,
15, 19, 171
Ferrite phase (see Tetracalcium
aluminoferrite)
Final set (see also Time of setting) – 47–53, 57–60, 74, 78,
109, 111, 173, 272
Fine aggregate (see Aggregate,
fine)
Fine aggregate–coarse aggregate
ratio – 448, 472, 473
Fine sand (see Sand, fine)
Finely divided mineral admixtures
(see also Pozzolanic material) – 29, 180, 183–210,
269–270, 366
Fineness – 36–44
Fineness modulus (m) – 397, 400,

Fineness modulus *(m)* – (con't)
    411–430, 438–441, 443, 469–
    474, 478, 493, 508, 523, 530
    combined *(M)* – (see also Con-
    crete, total solid volume
    of) – 424–428
    oprimum *(m_o)*, 423–430
Flexural strength (see Strength,
    flexural)
Flow – 64, 65, 94
Flow table – 60, 94
Fly ash (see also Pozzolanic
    material) – 29, 183–190, 227,
    521
Foamed blast–furnace slag – 514–
    518, 525
Fractional rate of hydration – 115
Free lime
    as Ca(OH)$_2$ – 108, 111, 113,
        123, 183
    in clinker – 18
Free moisture – 299–304
Free water – 107–111, 119–124,
    163, 215
Freezing and thawing resistance
    (see Frost resistance)
Freezing and thawing test, direct –
    327
Fresh concrete (see Concrete,
    fresh)
Fresh paste (see Cement paste,
    fresh)
Frost action – 235, 321–325
Frost resistance (see also Aggre-
    gate, durability of; Concrete,
    durability of) – 189, 234,
    321–334
Frost susceptibility test (see also
    Aggregate, durability of) –
    330
Fuller grading – 418, 435, 453–
    458, 464–469
Future of cements – 212

Gamma ray – 527–531

Gap grading – 389, 443–446,
    461, 469–474, 480, 482
Gel (see Cement gel)
Gel pores – 124, 127
Gel–space ratio – 126
Gel water – 120
Geometric average particle size
    (d̄_g) – 405–407, 409, 414,
    425
Gilmore needles – 49, 51
Gradation (see Aggregate,
    grading of)
Grading (see Aggregate, grading
    of)
Grading characterization (see
    Numerical characterization
    of grading)
Grading condition – 475, 476,
    485–487, 494–511
Grading curve (see Sieve curve)
Grading point – 396–399, 506–
    509
Grading requirement (see Grading
    condition)
Granite – 309, 347
Granulated blast–furnace slag –
    159, 179–184
Gravel (see also Aggregate,
    coarse) – 380, 382, 429,
    504–508, 537
Gypsum – (see also Calcium sul-
    fate) – 4, 23, 108–114, 167
    optimum amount of – 24

Hardened cement paste (see
    Cement paste, hardened)
Hardened concrete (see Concrete,
    hardened)
Hardening (see also Strength) –
    47, 110, 112, 116
    rate of – 236
Hardness – 309–319
Harmonic average particle size
    (d̄_h) – 40, 405–409, 433–
    434

specific surface as a
measure of – 40, 433
Heat curing (see Curing)
Heat development (see Heat of
hydration)
Heat evolution (see Heat of
hydration)
Heat of hydration – 22–34, 75–83,
108, 160–161, 180
Heavyweight aggregate – 512–
513, 527–532
High–alumina cement – 160–165,
172, 176
effects of admixtures on – 249
High–pressure steam curing – 130
High–range water–reducing admixture (HRWR) – 251–254
History of portland cement – 4
Hydrated lime – 178
Hydration
development of hydration
products – 108–115
of high–alumina cement – 160–
161
mechanism of – 116–121
of portland cement – 103–135,
242
fractional rate of – 115
kinetics of – 71–74, 106, 121
of pozzolan – 190
retardation of – 255
Hydration products – 108–116
Hydraulic cements – 4
Hydraulic lime – 177–178
Hydraulic radius – 334, 337
Hydrophilic materials – 226, 251
Hydrophobic cement – 175

i index by Faury – 401
Ideal grading (see Sieve curve,
theoretical)
Igneous rocks (see Rocks)
Impact test – 312, 526
Impurities in water – 214–219
Initial set – 47–53, 56–60, 74,

109, 111, 162, 272
Intercepts of mortar layers – 475–
483
Internal porosity (see Aggregate,
porosity of; Cement paste,
porosity of)
Iron modulus – 17
Iron oxide – 19

K (Klein) cement – 167, 170
Kiln – 9, 11
Kinetics of hydration – 71–74
Klein turbidimeter – 40

λ distribution number by
Solvey – 401–402
Latent hydraulic materials – 183–
190
Le Chatelier flask – 90, 91
Lightweight aggregate – 228,
276–278, 514–527
Lignite – 354–357
Lignosulfonic acids – 248
Lime – 15
free – 61
free hydrated – 111–115
Lime–saturation factor – 17
Lime–silica ratio (CaO/SiO_2
or C/S) – 112, 115, 130
Limestone – 160, 277, 322, 323,
345
Limit curves – 321, 458–464,
469–474
Limonite – 528–529
Linear travers method – 336, 479,
483
Logarithmic average particle size
$(\bar{d}_l)$ – 386, 404–407, 411–
415, 423–427
fineness modulus as a measure
of – 414
Logarithmic normal distribution –
278
Los Angeles test – 313–319, 340,
526, 532

Loss of ignition – 24, 100, 178
Lossier expansive cement – 167
Low–alkali cement – 24, 241, 249
Low–pressure steam curing – 70,
   75, 121, 180, 242, 249
Low–silica cement – 163
Lubricating ability – 190, 215,
   479, 480
Lumps in cement – 102, 176

m (see Fineness modulus)
M (Soviet or Mikhailov) cement –
   167
McLeod gauge porosimeter – 336
Magnesia (MgO or M) – 14, 15,
   23, 62, 166, 178, 356
Magnesia based cement – 181–
   183, 519, 522
Magnesium sulfate – 180, 331
Magnesium sulfate test – 331–332
Magnetite – 528–531
Manganese – 171
Manufacture
   of expansive cement – 167
   of high–alumina cement – 160
   of lightweight aggregate – 517–
      522
   of portland cement – 6–14
   process flow in – 12
Manufactured aggregate – 275,
   512, 514–527
Masonry cement – 178
Maximum density – 443, 468–474
Maximum particle size (D) – 283,
   325, 368, 385–390, 395, 404–
   410, 411–414, 420–440, 442–
   461, 469, 473, 539
Meniscus phenomenon – 338
Mercury porosimetry – 127
Metamorphic rocks (see Rocks)
Mica – 356
Microscopic sizing – 38
Mineral admixture – 178–189,
   269, 271, 366
Mineral aggregate (see Aggregate)

Mineral water – 216
Minimum particle size (d_{min}) –
   385–386, 404–410, 419, 433
Mix proportion (1:n) – 424–
   428, 449–451, 459
Mixing – 233
   sequence – 256
   time – 232
   water – 215–219
Model
   for cement gel – 133–135
   for grading – 469
   for pore systems – 333
   for portland cement – 135–157
Modulus of elasticity – 273,
   309–311
Mohs's relative hardness scale –
   347
Moisture content – 297–303
Monocalcium aluminate
   ($CaO·Al_2O_3$) – 10, 16, 161,
   167
Monosulfate hydrate (see also
   Calcium aluminatie sulfate
   hydrate) – 175
Mortar – 477–483, 530
Mortar bar test – 370
Mortar flow – 64, 65, 94
Mortar layer intercept – 475–483

NaOH – 355, 371
Natural cement – 177–178
Natural pozzolan – 29
Neutralized Vinsol resin – 230
Neutron radiation – 527–531
Nitrogen adsorption – 132
Nonevaporable water – 79, 106,
   124–126, 161, 168
Nonionic surface–active agent –
   226
Normal consistency – 43–47, 67
   of high–alumina cement – 162
Normal distribution – 278
Normal–weight aggregate – 278

Numerical characterization of
    grading (see also Fineness
    modulus) – 399–443

Oilwell cement – 61, 173
Oligomer – 135
Opal – 363
Opaline chert – 188, 358
Optimum fineness modulus (m_o) –
    423–430, 437, 469–472, 477
Optimum grading (see also Sieve
    curve, theoretical) – 389,
    423–430, 443–446, 450–474
Ordinary portland cement – 25
Organic impurities
    in aggregate – 355
    in water – 214–220
Orifice flow test – 381, 383
Ottawa sand – 63
Ottawa–sand mortar – 63–64, 188
Oversize – 385, 534
Oxide composition – 14, 16

Particle interference – 467, 472
Particle–by–particle wear – 347
Particle shape (see Aggregate,
    particle shape of)
Particle size (see Aggregate,
    particle size of)
Particle size distribution (see
    Aggregate, grading of)
Particles finer than No. 200
    sieve – 351–359
Paste (see Cement paste)
Pat test – 62
Penetration resistance – 49, 50,
    52, 55
Percent of voids – 303–307,
    382, 468–469
Percentage of grading
    by absolute volume – 391–398,
    448–451, 523
    by number of particles – 391,
    395
Percentage grading – 464–467,

472, 474
Percentage passing–retained
    (PPR) – 391–395, 407–408,
    462–464, 472–474
Periclase (see also Magnesia) –
    23, 61
Perlite – 90, 512–526
Permeability – 124, 189, 228, 338,
    446, 461
Petrographic examination – 276,
    370, 372
pH value
    of water – 217
Pigments – 172
Plastic shrinkage – 86
Plastifier (see Admixture, water–
    reducing)
Point–count method – 336
Poisson's ratio – 319
Polymer – 258–266
Polymer–modified concrete –
    258–262
Popout – 324, 356
Porosity (see Aggregate, porosity
    of; Cement paste, porosity of)
Portland blast–furnace slag
    cement – 26, 30, 58
Portland cement
    air content of (see Air content)
    air–entraining – 26, 227
    ASTM types – 24–29, 85
    chemical requirements for – 15–
    16, 32
    chemical resistance of – 22–23,
    25, 92
    clinker of
        composition of – 9–28
        hydration of (see Hydration)
        major constituents of – 18–23
        minor constituents of – 23
        potential compound composi-
        tion of – 11, 18–22
    density of (see Specific gravity,
    of cement)
    electrical resistance of – 52, 55

false set of – 32, 34, 61, 109
fineness of (see Fineness)
heat of hydration of (see
    Heat of hydration)
history of – 4
hydration of (see Hydration)
of low–alkali content – 365,
    368
lumps in – 100, 176
manufacture of – 6–14
modified – 158
ordinary – 25, 29
particle size distribution of –
    36–39, 43, 122
physical requirements for – 31–
    34
rapid hardening – 29
sampling of – 98–100
soundness of (see Soundness,
    of cement)
specific surface of – 36, 40–44,
    70, 100
specific surface of the gel of –
    129–135, 168, 239
storage of – 100–102, 171
strength of (see Strength)
sulfate expansion of – 34, 92,
    93
time of setting of (see Time of
    setting)
types of – 24–29
uniformity of – 95
white – 171–172
Portland–pozzolan cement – 26,
    29, 62, 178, 227, 249
Potassa (K₂O or K) (see Alkalis)
Powers's freezing test – 329
Pozzolan (see Pozzolanic material)
Pozzolanic material – 29, 123,
    159, 178–190, 227, 241, 271,
    349–350, 366
    fly ash – 29, 159, 181–190,
        227, 271, 477
    test of – 188–189
Prestressing, chemical – 166

Pulse velocity – 328
Pumice – 512–517, 526
Pyrex glass – 371
Pyrite – 349, 356, 368

Quartering – 285, 286
Quartz – 341, 363
Quartzite – 323
Quick set – 61, 103, 109

Radiation absorption – 527–529
Rammler–Rosin diagram – 38
Rapid–hardening cememt (see
    also Hardening) – 29, 163
Rapid–setting portland cement
    (see also Setting) – 173–174
Rate of early hardening – 162,
    165, 236
Rate of heat evolution (see also
    Heat of hydration) – 25, 76,
    239
Rate of hydration – 106, 112–118
Rate of setting – 56, 236
Rate of stiffening – 52
Reactions between cement and
    water (see Hydration)
Reactive aggregate (see Aggre-
    gate, reactive)
Rebound hammer test – 165
Rectangular diagram – 398, 497–
    503, 509
Reef shell as aggregate – 478–
    481
Refractory – 160, 177
Refractory bonding – 162
Regulated–set cement – 174–175
Reinforcement
    corrosion of – 31, 241
Relative dynamic modulus of
    elasticity – 328
Required grading – 442–474,
    490–510, 524
Resistance to penetration – 49,
    50, 55

Resistivity against sulfates –
    25, 92, 159, 189
Resonant–frequency method – 328
Retarder (see Admixture, set–
    controlling)
Retarding admixture (see Admix-
    ture, set–controlling)
Revibration – 175
ρ stiffening coefficient by
    Leviant – 401–403
Rhyolite – 347, 363
Riffle sampler – 284
Rocks – 277, 363, 371, 515, 519
Rothfuchs' method for blending
    proportions – 494–497, 511
Roughness (see also Aggregate,
    surface texture of) – 375–
    380
Roundness (see also Aggregate,
    particle shape of) – 375–383

S (Portland Cement Association)
    cement – 166
Sampler – 284
Sampling – 98–100, 270, 280–286
    error – 281
    sample size – 283
Sand (see also Aggregate, fine)
    beach – 479, 480
    coarse – 449, 462, 473, 504–
        507
    fine – 280, 449, 462, 473, 504–
        507
    for masonry mortar – 462, 523
Sand–equivalent test – 353
Sandstone – 277, 323, 339, 347–
    348
Saturated surface–dry condition
    (see also Aggregate, specific
    gravity of) – 290, 296, 302
Screening – 276
Seawater – 189, 218, 219
Seawater resistivity – 176
Secondary bond – 122, 133–135
Sedimentary rocks (see Rocks)

Sedimentation – 390
Segregation
    of aggregate – 281, 533–538
    of concrete – 228, 249, 389,
        441, 512, 513, 523, 530
Self–desiccation – 119
Self–stressing (see also Expan-
    sive cement) – 159, 166, 170
Semilog system – 394, 406–410,
    413, 421, 433, 438
Set (see Time of setting)
Set–controlling admixture (see
    also Admixture) – 174, 254–
    257, 272
Setting (see also Time of
    setting) 47–63, 94, 121, 165,
    522, 531
    false – 32, 33, 61, 109
    quick – 61, 103, 109
    rapid – 173, 174
    rate of – 236, 242
Setting time (see Time of setting)
Shale – 188, 514–520
Shotcrete – 174
Shrinkage – 24, 83–90, 105, 160–
    169, 189, 239, 275, 351, 367,
    541
    carbonation – 87
    of concrete made with
        admixture – 272
    drying – 24, 86–89, 189, 239
    effect of – 86–89
    plastic – 86
Shrinkage–compensating cement
    (see also Expansive cement) –
    159, 168–171
Sieve analysis (see Sieve test)
Sieve curve (see also Limit
    curves) – 376, 391–399, 408–
    409, 411–417, 421, 438–441,
    442–474, 482, 490–503, 509
    differential curve of – 395, 396,
        404–410
    equation of – 393, 396, 404–
        410, 414, 419, 433, 442,

Sieve curve (continued)
  449–458, 464–467
  graphical form of – 450, 455,
    458–462
  theoretical (see also Bolomey
    grading; Fuller grading) –
    393, 442–444, 449–471
Sieve size (see Aggregate,
  particle size of)
Sieve-size line – 497–503
Sieve test – 275, 376, 381–399,
  439
Sieves – 40, 376, 381–396, 401,
  463–467
Silica (SiO$_2$ or S) – 6, 14–20,
  161, 188, 366, 370–373
  low–crystalline – 350
Silica ratio – 17
Silicate anion polymer – 134
Silt – 280, 351
Silylation – 135
Size fraction – 375, 383–403,
  410, 411, 415–419, 442, 453,
  462–469, 476, 504–511, 523,
  530, 533, 534
Size ratio – 466
Skid resistance
  of aggregate – 339–348, 485
  of concrete – 339–348, 517,
    541
Slag (see Blast–furnace slag)
Slag cement – 159, 179–181
Slump (see also Consistency) –
  44, 52–55, 170–171, 234,
  249, 399, 429, 441, 453,
  459, 477, 513, 531, 534
Soda (Na$_2$O or N) (see Alkalis)
Sodium chloride – 217
Sodium sulfate – 331–332
Sodium sulfate test – 331
Soil stabilization – 176
Solid volume of aggregate – 289–
  297, 303, 333–335, 384, 390,
  403, 432
  specific – 292

Sonic modulus of elasticity – 328
Sorption method (see also
  Adsorption) – 337
Soundness
  of aggregate – 276, 288, 325–
    330, 359, 518, 530
  of cement – 18, 23, 32, 33, 44,
    61–63, 162–165, 177–178,
    218
Spacing factor 230–231, 235
Special cements – 158–213
Special portland cements – 166–
  177
Specific gravity (G) (see also
  Test methods)
  of aggregate (see also Aggre-
    gate, specific gravity of) –
    287–297, 335, 428–435,
    522–532
  specific gravity factor – 526
  of cement – 90, 162, 172
    in kerosene – 90
    in water – 90
Specific heat of aggregate – 319–
  321
Specific surface (s)
  of aggregate – 381, 400, 412,
    419, 430–441, 443, 469–
    474, 483, 493, 504
  of air voids – 228, 231–232,
    334–337
  of cement – 36–44, 70, 71, 100
  of cement gel – 129–135, 239
Specification
  for admixtures – 270, 272
  for aggregates – 289–319, 327–
    333, 356–359, 383, 442–
    474, 524–532
  for high–alumina cement – 161
  for portland cement – 31–34
  for sieves – 387, 388
  for water – 217–219
Sphericity (see also Aggregate,
  particle shape of) – 375–383
Sphericity factor – 382

Splitting – 284, 286
Standard types of portland
    cement – 24–27, 71, 80, 85
Steam curing – 70, 73, 121, 130,
    180, 242, 249
Stiffening (see Time of setting)
Stiffening versus curing time –
    48
    approximation by power
        function – 50–55
Stiffening of fresh paste
    rate of – 52, 56
Stockpiling of aggregate – 282,
    533–535
Stone dust – 351
Storage
    of admixtures – 270
    of aggregate – 533–538
    of cement – 100–102, 171
    effect of – 100
Storage bin – 533–536
Strength (see also Concrete,
    strength of) – 22–26, 32–33,
    43, 63–75, 100, 105, 119–
    124, 133, 161, 174–180, 218,
    271, 311, 314, 325, 375, 379,
    389, 399, 412, 420, 437, 446,
    448, 458–462, 512–513, 524–
    527, 539–542
    coefficient of variation of – 98
    compressive – 64, 68–80, 133,
        164–165, 170, 174–179,
        240, 244–245, 272, 340,
        420, 422, 440, 441, 477,
        520
    final – 142–145
    of concrete versus mortar – 68
    developing ability – 163
    early – 121, 162, 236
    factors affecting
        age – 33, 47–55, 68, 71, 73,
            250, 256
        $CaCl_2$ – 73, 240
        cement content – 240
        cement type – 70, 73

compound composition – 70,
    135–157
fineness – 36, 43, 70, 71
impurities of water – 215–
    219
kinetics of hardening – 121
length of storage – 100, 176
set–retarding admixture –
    248, 255
temperature – 70–74, 158,
    161
water–reducing admixture –
    249
flexural – 67–68, 170, 239,
    272, 354, 375, 379–381,
    477
of high–alumina cement – 160–
    165
of mortars cured in autoclaves –
    70, 74
reduction in – 100, 158, 164–
    165, 215
tensile – 44, 66, 67, 170, 235,
    239, 362, 375, 379–381
test for – 63–66
variation in – 68, 98
Strontium aluminous cement –
    177
Strontium portland cement – 176
Structural lightweight aggregate
    concrete – 228, 512–517,
    520–527
Structure of cement paste – 103–
    104, 122–135
    models for – 133–135
Sugar – 214–217, 255, 372
Sulfate
    in aggregate – 356
    attack – 25, 162, 168, 189,
        241, 372
    expansion – 34, 92, 93
    in mixing water – 217
    resistance – 23, 92, 159, 189
    soundness test – 331–333
Sulfate soundness test – 331–333

Sulfathuttenzement – 179
Sulfoaluminate clinker – 167
Sulfoaluminate hydrate – 159, 166
Sulfuric anhydride (SO_3 or S) –
    15, 108, 217
  optimum amount of – 24
Superplastifier (see also Admix-
    ture, water–reducing) – 248,
    251–254
Supersulfated cement – 31, 159,
    179–180
Surface–active agents (see also
    Admixture) – 122, 225–234
  anionic, cationic and non-
    ionic – 225–226
Surface moisture – 286, 287
Surface texture of aggregate –
    375–383, 524, 538, 540
Surface treatment of aggregate –
    300, 325
Surfactant (see Surface–active
    agents)
Swelling of aggregate – 362
Synthetic detergents – 225, 226

Talc – 178
Tannic acid – 214, 217, 349, 355
Temperature (see Curing Temper-
    ature
Tensile strength – 44, 66, 67,
    170, 239, 235, 362, 375,
    379–381
Test method
  for admixture – 270–271, 272
  for aggregate
    alkali–carbonate rock
      reaction – 368–372
    alkali–silica reaction – 350
    deleterious materials – 352–
      354
    elastic constants – 316–318
    frost resistance – 327–331
    grading – 389–391
    heavyweight – 532
    lightweight – 524–527

  moisture content – 302
  particle shape – 381
  porosity – 335–337
  rock – 311, 314
  soundness – 331–333
  specific – 295
  strength – 311–314
  surface texture – 381
  unit weight – 304
  voids content – 303–306
  for cement
    air entrainment – 32, 92
    alkali reactivity – 92
    bleeding – 94–95
    conversion of high–alumina –
      163–165
    efflorescence – 95, 218
    false set – 34, 61
    fineness – 36–44
    flow – 64, 94
    heat of hydration – 34, 75–76
    hydration – 106
    identification of clinker
      minerals – 19
    normal consistency – 44–47
    oligomers – 135
    porosity of paste – 126–128
    soundness – 32, 61
    specific gravity – 91
    specific surface of gel – 129–
      133
    strength – 32, 63–68, 70–74,
      80, 95, 311
    sulfate resistance – 92
    time of setting – 33, 47–51,
      56
  for concrete
    average mortar intercept –
      475–483
    compressive strength – 68
    Proctor penetration – 50, 52,
      55
    pulse velocity – 52, 55
    slump – 52–55
Testing variance – 68

Tetracalcium aluminoferrite
(see also Celite)
($4CaO \cdot Al_2O_3 \cdot Fe_2O_3$, C_4AF, or
ferrite phase) – 20, 28, 108
Theory of adsorption – 131–133
Thermal conductivity of lightweight concrete – 512–517,
521, 541
Thermal properties of aggregate –
319–321, 327, 538–540
Thermoplastic resin – 273
Thermosetting resin – 273
Thickening time (see also Time
of setting) – 257
Time of setting (see also Setting) – 4, 24, 32–33, 44–63,
100, 101, 103–111, 162, 173–
178, 189, 214–217, 240, 248
of concrete – 50, 240
factors affecting
calcium chloride – 53, 236,
239
consistency – 232
fineness – 60
organic compounds – 257,
522
set–retarding admixture –
255
storage length – 101
temperature – 32, 57–59, 77,
80, 173, 232
water–cement ratio – 53, 58,
60
water–reducing admixture –
54
final setting (see Final set)
initial setting (see Initial set)
Tobermorite gel (see Calcium silicate hydrate gel)
Tonerdeschmelzzement (see also
High–alumina cement) – 160
Tonerdezement (see also High–
alumina cement) – 160
Tortuosity – 334
Total percentage passing (TPP) –

376, 391–396, 408–409, 411–
413, 419, 421, 440, 442–467,
489, 495–503
Total percentage retained (TPR) –
376, 391–396, 411–417, 442
Toughness of aggregate – 309–
319
Traprock – 323, 347
Trass – 188
Tremie – 249
Triangular diagram – 16, 307,
376, 396–399, 418, 504–510
Triangular method for blending
proportions – 476, 494, 504–
510
Tricalcium aluminate ($3CaO \cdot Al_2O_3$
or C_3A) – 15–28, 92, 93,
103–114, 167–175, 242
as catalyst – 23, 138–140
effect of
autoclave expansion – 62
on heat of hydration – 23
on strength – 22, 70–75
on sulfate resistance – 23
Tricalcium silicate ($3CaO \cdot SiO_2$ or
C_3S) (see also Alite; Calcium
silicate) – 15, 19, 22, 23, 28,
103, 105, 113, 114, 117, 242
effect of
on hardening – 112–113
on heat of hydration – 81–82,
83
on setting – 112
on specific surface of gel –
131
on strength – 70–75
Trief cement – 31
Triethanolamine – 238
Tristrontium silicate – 177
Trisulfate hydrate (see also Calcium aluminate sulfate hydrate) – 168
True specific gravity (see also
Specific gravity, of aggregate) – 290, 296

Turbidimeter – 38, 42
Turbidity – 217
Tyler sieve series – 411, 412, 466

Unconfined freezing and thawing test (see Direct freezing and thawing test)
Underliming – 18
Undersize – 385, 534
Uniaxial restrained expansion – 170
Uniformity
of aggregate – 534–538
of cement – 95
of concrete – 536
Unit weight
of aggregate (see Aggregate, unit weight of)
of concrete – 420, 422, 441, 477, 512–514, 528, 541
Unrestrained expansion – 170
Unsoundness (see Soundness)

Van der Waals bond – 122, 133–135
Variance of grading – 283, 399, 435
Variations in grading – 534–538
Vermiculite – 512–516, 519, 524
Vibration – 227, 235, 480, 523
Vicat
apparatus – 41, 44, 49, 50
penetration – 46, 53, 55
Vinsol resin – 230–232
Viscosity – 47, 142–144, 173, 251, 337, 338
Voids content (see Percent of voids)
Volcanic glass – 269, 372
Volcanic tuff – 188
Volume instability (see Cement paste, expansion of; Shrinkage)

Wagner specific surface – 41–42
Wagner turbidimeter – 42
Wall effect – 452–455
Washing of aggregate – 214, 219, 275
Water (H$_2$O or H) – 15, 103–104, 111, 164, 214–220
absorption (see Aggregate, absorption of)
adsorbed – 106, 111, 322
in aggregate – 287, 297–301
capillary (see also Water, free) – 120–125
chemically bound – 79, 106, 124–126, 161, 527
for curing – 214, 219
evaporable – 106–111, 119–121
free – 106–111, 119–124, 164, 215
freezing – 235
gel – 116–120
as ice – 165, 235
impurities in – 214–219
mineral – 215
for mixing – 126, 214–219
nonevaporable (see Water, chemically bound)
repellant material – 175–176
sea – 189, 218, 219
test for – 215–219
vapor adsorption – 132
for washing aggregate – 214, 219, 275
Water–cement ratio (W/C) – 52, 53, 67, 72, 89, 104, 112, 119–129, 162, 165, 169–172
Water content – 44, 64, 246, 231, 272, 420, 524, 527, 539
Water–reducing admixture (see Admixture, water–reducing)
Water requirement (see Water content)
Waterproofed cement – 175
Wear resistance

Wear resistance (continued)
  of aggregate – 309–311, 335,
    339–343
  of concrete – 340, 540
Wetting and drying – 302, 321–
    327, 331, 354, 358–359
White alumina cement – 162
White portland cement – 171–172
Workability (see also Consis-
    tency) – 162, 215, 233, 239,
    351, 352, 379, 389, 420, 423–

428, 437, 446, 448, 458, 472,
  475–479, 512, 513, 523, 531,
  539

X–ray – 176, 513, 527
X–ray diffraction – 167

Yield stress of paste – 228

Zeolites – 372